Tennis Science & Technology

Tennis Science & Technology

Edited by

S.J. Haake
Department of Mechanical Engineering
The University of Sheffield

A. Coe
The International Tennis Federation

b

Blackwell
Science

Editorial Offices:
Osney Mead, Oxford OX2 0EL
25 John Street, London WC1N 2BL
23 Ainslie Place, Edinburgh EH3 6AJ
350 Main Street, Malden
MA 02148 5018, USA
54 University Street, Carlton
Victoria 3053, Australia
10, rue Casimir Delavigne
75006 Paris, France

Other Editorial Offices:

Blackwell Wissenschafts-Verlag GmbH
Kurfürstendamm 57
10707 Berlin, Germany

Blackwell Science KK
MG Kodenmacho Building
7–10 Kodenmacho Nihombashi
Chuo-ku, Tokyo 104, Japan

First published 2000

Set in 10/11.5pt Times
Printed and bound in Great Britain by
the University Press, Cambridge

DISTRIBUTORS

Marston Book Services Ltd
PO Box 269
Abingdon
Oxon OX14 4YN
(*Orders:* Tel: 01235 465500
Fax: 01235 465555)

USA
Blackwell Science, Inc.
Commerce Place
350 Main Street
Malden, MA 02148 5018
(*Orders:* Tel: 800 759 6102
781 388 8250
Fax: 781 388 8255)

Canada
Login Brothers Book Company
324 Saulteaux Crescent
Winnipeg, Manitoba R3J 3T2
(*Orders:* Tel: 204 837-2987
Fax: 204 837-3116)

Australia
Blackwell Science Pty Ltd
54 University Street
Carlton, Victoria 3053
(*Orders:* Tel: 03 9347 0300
Fax: 03 9347 5001)

A catalogue record for this title
is available from the British Library

ISBN 0-632-05638X

Library of Congress
Cataloging-in-Publication Data
is available

For further information on
Blackwell Science, visit our website:
www.blackwell-science.com

Contents

Preface

The 1st International Congress on Tennis Science and Technology (TST) took place in London between 1 and 4 August 2000. Organised by the International Tennis Federation (ITF), the world-wide governing body of tennis, it is hoped that this first Congress will develop into a valuable event and source of information on tennis science and technology, which is freely available to all those with an interest in and love for this great sport.

TST 2000 attracted around 250 delegates including over 70 speakers from more than 15 nations. Such interest, in this first event of its kind, serves to underline the longstanding need for such a congress and the global nature of the sport of tennis.

TST is intended to provide a forum for the presentation and discussion of all aspects of tennis science and technology, and a benchmark for the current status of tennis related research activity from around the world. Hopefully, those who attend this event will be inspired and encouraged by the work presented to take their own ideas and research forward with renewed enthusiasm and vigour.

For many years the level of global research into tennis-related topics has been limited in relation to other sports of comparative size. This is particularly surprising when one considers that the origins of lawn tennis suggest that the game evolved into its current form on the garden lawns of Victorian England during the early 1870s, largely as a result of technological developments to the rubber balls and lawnmowers of that era.

Since the establishment of the ITF Technical Centre in 1997 the profile of tennis science and technology has been raised through the encouragement of research in the areas of tennis equipment, tennis facilities and the way in which the tennis player is affected by developments in these areas. The ITF is fully committed to continuing its support of such research and believes that technology can, and must, play an important role in the future development, growth and health of tennis.

The ITF would like to offer its sincere thanks and appreciation to all those who have supported this first TST Congress and in particular to those who took the time and made the effort to present their work.

Andrew Coe
ITF Head of Product Development and Technical

1 Introduction

The balance between technology and tradition in tennis

A. Coe
ITF Head of Technical and Product Development, ITF, London, UK

ABSTRACT: This paper describes a strategy for establishing a true balance between Technology and Tradition in our sport. By implementing the changes described in this presentation, some of which have recently been initiated, e.g. the bigger, slower ball, we will be adopting a mechanism required to balance Technology and Tradition, within tennis, over future years.

The changes proposed, will raise a number of important philosophical questions that tennis will need to address in the near future. For example:-

Q. Is there a need to address the "Speed of the Game" at the professional level?

Q. The measures proposed have the potential to engineer the sport to be more similar in nature, when played on different court surfaces. Is such a development thought to be desirable?

Q. Is a permanent separation of the rules for the Professional and Recreational game desirable?

Q. Would the introduction of different equipment rules or requirements for Men and Women be desirable, acceptable or practical?

The proposals described in this paper are not designed to bring about major changes to the nature of tennis. On the contrary, they are designed to prevent the nature of our sport changing in a way that is beyond our control.

It is estimated that Tennis is currently played by up to 60 million players in 200 countries worldwide. As guardians of the Rules and specifications of Tennis the ITF holds a unique responsibility over the sport and over an associated industry. That industry has helped build perhaps 750,000 tennis courts, supplies 300 million tennis balls annually, millions of rackets, shoes and a variety of other equipment that results in an annual gross "national" product figure, for Tennis, of several billion U.S. dollars worldwide. The potential force for change that such an industry represents can be beneficial, or harmful, to the health of tennis. Such a responsibility should not be taken lightly.

The strategy outlined in this paper may be an important factor in ensuring the continuing health and growth of this great sport of tennis, at both the professional and recreational levels.

INTRODUCTION

The issue that this paper addresses is one which has done more than any other to define the game of tennis, as we know it today. This issue is the balance between the two forces of TECHNOLOGY and TRADITION and the degree to which these two forces remain in balance as tennis faces the challenges of the next millennium.

A sport that allows this delicate balance to swing too far away from a position of equilibrium risks squandering the very elements which make up it's fundamental appeal. On the one hand, a sport that, through it's Rules, adopts Technology too readily, risks sacrificing the traditional skills, which contribute so much to the enjoyment of the player in learning and playing the game. Conversely, a sport that precludes innovation, relying only on it's so-called Traditions, risks obsolescence in a world which is increasingly dominated by demands for the new and improved.

This discussion paper provides an overview of some of the work that has been carried out to date (and which is continuing) to address what is for many people, the most prevalent example of the Technology versus Tradition debate within tennis, that is, the so-called "Speed of the Game" issue. This paper will then propose and illustrate a vision of how tennis may address this subject, ironically perhaps, through the adoption of new technology designed to help us establish and preserve the precarious balance between Technology and Tradition, for the ultimate benefit of all who play and enjoy the sport.

But it will also address what is, for those of us with an interest in the future of tennis, a more important issue. The proposals discussed will involve new Technology that is designed to benefit the game at all levels by providing a platform for the development of improved tennis products and facilities.

We could be forgiven for believing that the subject of the "Speed of the Game", and in particular the increasing dominance of the Serve in the Men's professional game, is contemporary, arising from the almost simultaneous introduction of graphite rackets and improving physiology of professional tennis players since the 1970's. Since the mid-1980's the subject has become a favourite of the media and appears as regularly as clockwork around the time of Wimbledon and the grass court season.

In truth, this subject is as old as the game itself.

In 1932, Ellsworth Vines of the United States, serving an ace at match point, won the final of the Championships at Wimbledon. The winning serve was reportedly delivered with such ferocity that the receiver, Bunny Austin, later complained that he did not know whether the ball had passed him on the forehand or backhand side.

During an interview in 1920, Bill Tilden speculated that, as a means of addressing the dominance of the serve, tennis would eventually move to eliminate the second serve and allow only one serve per point.

In 1877, the first ever winner of the Wimbledon Championships, one Spencer Gore, in responding to an interviewer who had commented on the strength of his serve relative to his opponents, agreed that the serve perhaps provided too much of an advantage in this new game of Lawn Tennis: -

"...Did you know that Mr. Jones has figured out that 376 games have been won on serves and only 225 games on returns. Does that seem fair to you?"

The nature of tennis, whether it is played by men or by women, has of course evolved and changed dramatically since it's inception, as a result of wide ranging influences which include improved athletic physiology, diet, training and coaching, better facilities, improved equipment and apparel, etc.

Interestingly, this final anecdote and statistic from the 1877 tournament shows that the concern over the dominance of the serve in the very early days was based on a probability of holding one's serve, of 62.5% (376 games out of a total of 601 games played). Quite what Mr. Gore would have thought of the modern game, where the percentage of games won on serve by the top players on the ATP Tour approaches 90% over a season, is anyone's guess!

In addressing the subject of Technology and Tradition, it is necessary to consider any proposals for the future in the context of the following questions: -

- How has the history of this balance between Technology and Tradition in equipment contributed to the game of tennis, as we know it today?
- What are the forces that are altering the landscape of tennis and adjusting the balance of Technology and Tradition?
- Are the perceptions of those who believe that equipment technology alone has brought tennis to the edge of some form of catastrophe, reconcilable with the facts?
- Finally, what measures should be taken to address the issues, what implications will these have and how do we ensure that all interests within tennis (professional and recreational, men and women), are not disadvantaged by the process of change?

THE EXISTING BALANCE BETWEEN TECHNOLOGY AND TRADITION

It is said that the only thing that is constant, is change.

Throughout the history of tennis, the rackets, balls, surfaces, shoes and clothing, which were at one time regarded as traditions, have constantly blended and interfaced with new and innovating technologies. We may conclude therefore that, "the Technologies of yesterday are the Traditions of today".

The question that we must now ask is "Have we let the balance swing too far either way, in favour of Technology or Tradition?"

In order to answer that question in an objective way it is necessary to first consider the contribution that Technology has made to the popularity of tennis in the present day. Let us consider the three basic tennis products, court surfaces, tennis balls and tennis rackets.

TENNIS COURT SURFACES

There are few sports where the surface, upon which the activity takes place, is as influential on the nature of that activity, as the court surface is to tennis. There are no other major sports that are played on as wide a variety of surface types, including at the highest levels, as with the sport of tennis.

The importance of the court surface to the nature of tennis is the very reason for the proliferation of tennis surface types, which have developed from the early days of the garden lawn. The limitations provided by varying climate, the seasons and the maintenance demands required to produce a natural turf surface of acceptable quality soon led to alternatives being sought in the form of clay (originally crushed brick), followed by other granular mineral surfaces, cement, asphalt, macadam, and wood. As polymer and material technology has developed in the twentieth century, the introduction of new man-made, synthetic surface types has accelerated; hardcourts with polymer coatings, cushioned coatings, rubberised shock pads, textiles, artificial turf, plastic modular systems, etc.

The governing authorities of most sports in which the surface plays an important role, e.g. soccer, hockey, athletics (track and field), golf, basketball, bowls, cricket, etc. have introduced, or are in the process of developing, formal qualifications, of one kind or another, for the type of surface which is considered suitable for use at various levels of competition.

Tennis is one of the last major sport's to embark on such a path, and in being so, has attracted the attentions of international legal bodies (the European Commission) who have recently considered legislating on behalf of, but not necessarily in the interests of, our sport.

As we have seen, tennis is perhaps the most liberal sport in this respect, with there being few, if any, official requirements or limitations on the nature of the tennis court surface, contained in the Rules of Tennis.

Such liberalism in the Rules of a sport can only be interpreted, at face value, as the clearest possible endorsement of new technology.

With an estimated three-quarter of a million tennis courts in over 200 countries worldwide, few would dispute that such a policy has contributed to the growth of tennis internationally and greatly increased access to the sport.

As a counter-balance to this liberalism however, tennis is sometimes perceived to be unwilling to allow new and improved surface types to be tried at the professional level as a result of specific Tour and tournament regulations (e.g. Davis Cup, Fed Cup, ATP Tour, etc.). Such regulations act to preclude surface types or brands that have not been used or tried previously.

Such a policy at the highest level may, now or in the future, contribute to a reduced willingness on behalf of the world's sport surface manufacturing industry to invest in the development of new and improved tennis specific surfaces. Research and innovation is crucial in any industry and without the commercial opportunity provided by new product development an industry can stagnate and fade.

At a time when tennis participation in many of the traditionally strong markets is declining, the necessity to innovate has never been more important. Major opportunities exist for the development of surfaces that are safer, more uniform and predictable in performance, more durable, have lower construction and maintenance costs, are suitable for multi-sport use, etc. With particular regard to safety, recent ITF research (ITF, 1999) has shown that tennis related injuries are amongst the most widely quoted reasons why players currently give up playing or play less tennis.

A danger resulting from lack of regulation is that surfaces may be used which are not suitable for tennis. Ironically, this is currently most likely in high level tournaments where the lack of purpose built, permanent tennis facilities, requires the construction of temporary courts which are often "engineered" to provide extreme performance characteristics to suit particular playing styles. This long-standing policy of maximising "home advantage" in events such as the ITF's own Davis Cup and Fed Cup, is now regarded within tennis as a Tradition, that is rarely questioned. This may be an example of a lack of quality control of, and subsequent risk-taking with, with the highest profile media events in our sport. This may be counterproductive at a time when the over-riding aim of tennis should be to set the highest possible standards of professionalism and to establish the best possible image for the sport.

In the case of court surfaces it may be argued, therefore, that there is little evidence of the existence of a true balance between Technology and Tradition. Moreover, the approach that tennis has inherited from it's past is perhaps one where long held Traditions outweigh the true needs of the sport in the new Millennium.

TENNIS BALLS

The tennis ball is an important indicator of the health of the sport, as ball sales statistics are widely accepted as being an accurate reflection of playing activity. It is reasonable to assume that the relative cost of tennis balls and other consumables such as rackets, shoes, clothing, etc. may therefore be related to accessibility to the sport by the consumer and ultimately, it's popularity.

Using historical information it is possible to show how the comparative cost of tennis balls has changed over the years.

In 1906, the UK retail price of one dozen Slazenger tennis balls, as used at Wimbledon that year, was 13 shillings, or, in modern parlance 65 pence. This ball was made with a pressurised rubber core and cemented covers, in fact looking very much as a modern ball would look today, with the exception perhaps of relatively modern pressurised packaging. In the following 44 years the ball changed little and the retail cost increased only marginally to 14 shillings and sixpence (72.5 pence) for a box containing a half dozen Slazenger Wimbledon balls.

The U.K. retail price index for the similar period of 1910 to 1950 shows an inflation rate of just 116%, so tennis ball costs in the U.K. were very much in line with prices in general over that period, which includes the so-called golden period of tennis, the 1920's and 30's.

Since 1950, the cost of the Slazenger Wimbledon ball, now in it's pressurised can, but a substantially similar product, has increased dramatically to it's present day recommended retail cost of around £26 per dozen. However, using the RPI figures to November 1998 it is possible to show that the cost should be £31.95 per dozen. In other words premium tennis balls are some 19% cheaper today, in comparative terms, than they were in 1906 and due, in the main, to increased competition and new manufacturing technology.

Whilst this might be interpreted by most as a positive factor for tennis accessibility, the U.K., and other parts of Europe, are perhaps not an ideal model on which to draw conclusions for tennis ball costs, as anyone who has purchased balls in the U.S. will testify. A similar exercise to the above in the U.S. is revealing.

In 1921, the Wilson "Official" top grade ball was advertised at a cost of $6 per dozen. Around 75 years of U.S. RPI increases should therefore result in a present day cost for tennis balls of $54 per dozen. A dozen Wilson US Open balls can be picked up in most sports outlets in the U.S. for perhaps $12 or less - around 80% less than the cost of balls in the 1920's.

How much of this is due to competition and how much to improved manufacturing technology, only the manufacturers themselves will know. But can there be any question that if balls were at the prices they could be, then tennis participation would be significantly lower than it is today and that we have technology to thank, at least in part, for this.

To what extent the historical difference in tennis ball prices (and other equipment costs) between Europe and the United States has contributed to the relative popularity of tennis is not known (with 2 to 4% of the population of a mature European country playing tennis and 7 to 8% of the U.S. population currently playing the sport). The relationship between equipment costs and accessibility is however, inferred.

But in assuming that equipment costs really are a major force for democratisation within a sport, we must also ask whether our sport has benefited to the same extent as other sports.

The golf ball has probably been the subject of more technological development and evolution than any other sporting product - witnessed by the vast number of patents available in this area. The benefits to the golfer have been numerous, including better durability through the use of no-cut Surlyn cover material, longer drives and improved aerodynamics resulting from fiendishly complex dimple patterns, improved uniformity and lower costs. In 1923, the suggested retail cost of one dozen rubber core golf balls was $12, compared, you will recall from above, to $6 for a dozen tennis balls. In line with the RPI over 75 years, the cost of one dozen golf balls should have increased to $108, but today it is possible to purchase one dozen high performance, long lasting balls for around $8 - around 93% lower in cost than in previous times!

The current price of golf balls as a result of new technology and increasing volume of production, has been gradually introduced under the strict control of the governing authorities of that sport. This must represent value for money for the consumer and must be unique in the world of sport!

This is one aspect of the impact of technology on sport which is not often reported but which must be weighed on the side of Technology over Tradition.

Recently quoted profitability figures for U.S. golf ball manufacturers fall between 50 and 75% (Golf Digest, 1999). Profit levels on tennis balls in the U.S. are known to be significantly lower than these figures, a fact that underlines an apparent lack of initiative by the industry.

The sport of tennis must therefore throw down the gauntlet to manufacturers to make the investment necessary to produce better performing products which enhance the experience of playing our sport, but which are still affordable to a wider audience of potential tennis players of the future.

Equally, the sport of tennis, and the tennis player, must begin to relinquish the sometimes inflexible and rigid thinking which, more than most other sports, forces the tennis industry to adhere to Rules, many of which were drafted in the latter part of the 19th century.

It should be within the capabilities of the industry to find ways of producing tennis balls that perform in a similar way to current balls, that perform more consistently, that do not damage the nature of our sport, that tennis players prefer to play with and which are substantially easier and lower in cost to produce, than the balls of today.

This inflexibility within tennis may be a flaw that has often rendered the balance between Technology and Tradition inoperable throughout the recent history of our sport.

TENNIS RACKETS

Few sports products can have evolved as quickly and as far as tennis rackets have over the last 25 years. Many would cite the tennis racket as the ultimate example of Technology outweighing the interests of Tradition in a sport. A simple comparison of the typical performance characteristics of a racket of the 1960's and one of the 1990's is given later under the section "Racket "Power"" of this paper.

How does the cost of space age technology in 2000 compare with what today we would all see as the ultimate Tradition of tennis - a wooden racket - but which was, in reality, the Technology of yesteryear?

A similar exercise to that described above in the section "Tennis Balls" on tennis racket costs over the years, reveals that the 21 shillings retail cost of a Slazenger & Sons state of the art solid wood bend "Demon" tennis racket in 1910 had increased at roughly twice the UK retail price index (RPI) rate over the intervening 61 years to 1971, when a Slazenger laminated wood Challenge No.1 retailed at £11.50. According to the RPI, a top quality racket of 1998 should cost £47.80. In fact, today's high technology premium rackets in the U.K. cost anything between £170 and £230 - a comparative increase of over 400%!

Even in the U.S., a premium quality Wright and Ditson, Wilson or Spalding wooden racket of the 1920's cost in the region of $12 which, with U.S. RPI increases, equates to a cost in 1998 of $108. With the latest top grade rackets costing up to $300, it is clear that today's hi-tech rackets, with performance benefits that are a boon to the majority of recreational players, come at a cost!

With regard to the tennis racket, where does the balance between Technology and Tradition currently lie? As this paper will attempt to show, the answer to this question is still far from clear.

SUMMARY

The above discussion should have begun to highlight the nature of the proposal that this paper will provide. The inference to be drawn is that the evolution of tennis over the last 120 years has taken place largely as a result of a pragmatic approach to innovation, rather than a conscious knowledge of the implications of new Technology or as a result of any developed strategy for the future advancement of the game.

Pragmatism may no longer be sufficient for a sport entering the 21st century.

Successful sports in the new millennium are likely to be those who recognise that sport is, amongst other things, simply a form of entertainment, and, who succeed in providing that entertainment to existing and potential audiences.

Technology and Tradition are the key elements which define the appeal of a sport and it is vital therefore that Tennis continues to debate these issues and recognises the need to establish a balance between these two elements.

Such a philosophy should be applied to all aspects of the game.

Through Technology it is possible to develop an understanding of the various elements and interactions which contribute to the nature of Tennis. Consider some of these interactions: -

- Ball interaction with racket strings,
- String interaction with racket frame,
- Player interaction with racket grip,
- Ball aerodynamics,
- Ball interaction with surface,
- Player interaction with surface (through the shoe).

These elements are required to be included in the rules, regulations and specifications that provide the control and balance in tennis. The proposals included in this paper will therefore, of necessity, involve the court surface, the ball and the racket and a case will be made for further involvement by the ITF in technical classification of other products in the longer-term future.

THE CHANGING NATURE OF TENNIS

Earlier it was stated that "The only thing that is constant is change". The name of this process is Evolution. Tennis has evolved dramatically over recent years due to a wide range of forces acting on the nature of the sport.

Let us look at two of these forces; tennis equipment and the tennis player.

THE EVOLUTION OF THE TENNIS BALL

As previously discussed, no item of tennis equipment has been required to meet such well-defined regulations over the years as has the ball. Despite these regulations however, and perhaps to the surprise of many associated with the game, the ball has changed significantly over the last 30 years. In particular, the tennis ball has become harder (and therefore faster) since the 1960's.

Until the mid-1960's, the specification for the forward deformation of a tennis ball was: -

0.265" to 0.290"

With the advent of the pressureless ball and at the request of manufacturers, the LTA (who, at the time, were largely responsible for regulating tennis balls on behalf of the sport) and the ITF, agreed to amend the forward deformation specification by extending the range as follows to allow harder balls to be used: -

0.220" to 0.290"

Designed to accommodate the harder pressureless types of ball that were being introduced at the time, it was perhaps not envisaged that manufacturers would eventually be capable of using this extended deformation range to produce harder pressurised balls. Such a move was, and continues to be, a desirable objective for manufacturers as it addressed the biggest cause of customer complaints, returned product and therefore higher costs. That is, tennis balls which have, or are perceived to have, gone "soft".

Making pressurised balls harder from new, increases the shelf life of the product and has all but eliminated complaints about softer balls.

The professional or recreational tennis player of the 1990's has now become accustomed to these balls and would find the experience of playing with the softer balls of past decades, rather unappealing.

ITF research has shown that harder balls are "faster". This effect results from the interaction between such balls and the court surface where there is a reduced amount of deformation of the ball when it impacts with the surface. The reduced contact area of the ball with the surface results in lower frictional forces between ball and court that, in turn results in a lower angle of rebound and a reduced reaction time for the receiver. In the future, it may be possible to quantify the effect that such evolution in the tennis ball has had upon the change to the nature of tennis. However, at this point, it is sufficient to say that this is a further contributory factor to the evolution of the sport.

But changes to the tennis ball continue to take place as manufacturers seek to produce balls that either perform better, or differently, for example, on specific court surfaces such as clay. As tennis sought to maintain the traditional nature of the ball, manufacturers tried new ideas, driven by the need to find some new competitive commercial edge.

ITF research has revealed for the first time, the extent to which the Rules of Tennis have become less relevant to tennis ball manufacturers and tennis players over the past 30 years.

Figure 1 shows, in graphical terms, the results of random testing of over 150 types of premium quality tennis ball obtained from a variety of sources, including the highest level tennis tournaments (Grand Slams, Davis Cup / Fed Cup, ATP Tour Super 9, etc.) and from retail outlets around the world. Almost 1,500 individual balls from nearly 40 countries have been tested with revealing results.

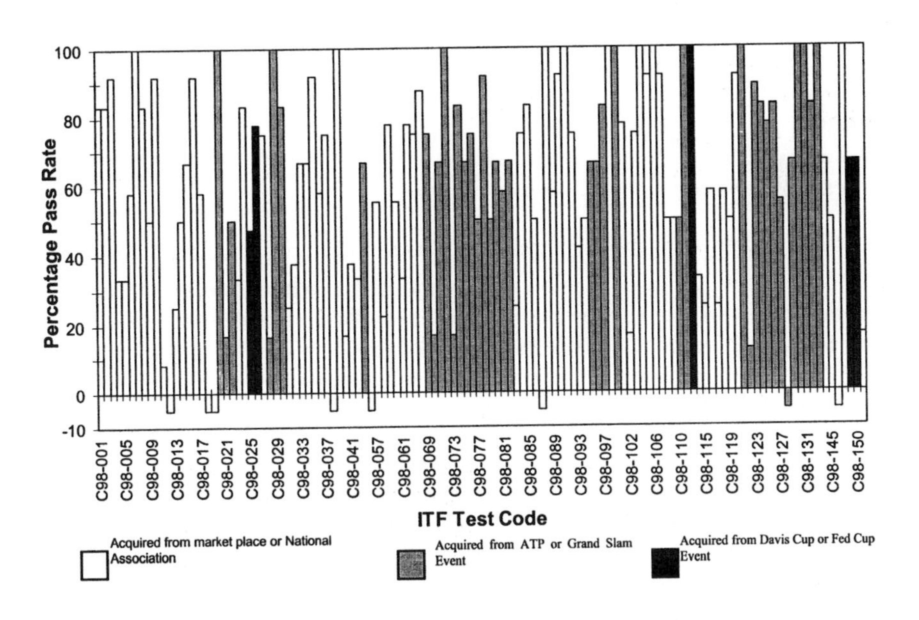

Fig. 1 Percentage of randomly selected balls conforming to Rules of Tennis - 1998

Only *56%* of premium tennis balls have been found to conform to the Rules of Tennis.

If balls from less well recognised manufacturers were to be included in this exercise it is envisaged that the conformance figure would reduce to *50% or less*.

At the highest levels of the sport the situation is occasionally worse than this.

Such an unsatisfactory situation exists in no other major sport.

THE EVOLUTION AND EFFECT OF MODERN RACKET TECHNOLOGY

In relation to the "Speed of the Game" no subject has been discussed more, and understood less, than the effects of the modern racket on the nature of tennis. Many commentators and authorities on tennis continue to assign most of the problems of the appeal of the Men's professional game on the development of the modern graphite racket to the exclusion of many other factors, most notably those relating to the development of the athlete as briefly discussed above.

What is indisputable is the fact that modern tennis rackets are almost unrecognisable from those of just 25 years ago. A few simple comparisons of some easily measurable properties of the wooden rackets, which were typical of 1973, and the composite creations of 1998, will adequately demonstrate this (Table 1).

Table 1 A comparison of tennis racket characteristics between 1973 and 1998.

Property / Specification	**1973**	**1998**
Material	Laminated wood	Carbon fibre composite
Weight (strung)	13 to 15 oz.	7.5 to 10 oz.
Balance from butt end	12.5 to 13 inches	14.5 to 15 inches
Head size	65 to 70 sq. ins.	90 to 135 sq. ins.
Length	26 to 27 inches	27 to 29 inches
Swing Weight	370 RDC	290 to 310 RDC
Vibration frequency	Approx. 90 Hz	Approx. 150 to 220 Hz

In summary, a typical modern day racket is likely to be between 25 and 40% lighter in weight than those of just 25 years ago. But as we can see, this reduction has been achieved at the same time as developments which have seen rackets increase in head size by between 30 and 100%, racket length increased by up to 2 inches and significant increases in frame stiffness (closely related to power). In addition to this, racket designers have managed to incorporate any number of performance enhancing, vibration dampening devices, at the same time as improving the durability of modern rackets, which for most players will now survive several years without breaking or warping, compared to the old wood rackets.

This fact is witnessed in part, by the reduction in the worldwide market for tennis rackets over the last 25 years. In 1980, most estimates put the worldwide market for premium quality tennis rackets at 15 million units per year. In 1999, the latest estimate for worldwide sales is 7.1 million units - a reduction of almost 53% in less than 20 years.

But with all of these developments how do we objectively measure the true effect or benefit of new racket technology on the performance of players and on the nature of the game?

Racket "Power"

For the 99.9% of tennis players who play the game for fun and recreation there is only one characteristic which is important in their judgement of the quality of a racket and that is, how hard it enables them to hit the ball for a given effort. In a word it is *Power.*

The racket manufacturers recognised this fact some years ago and most of the technological innovations to tennis rackets in the past 25 years have been aimed at providing and improving this one performance characteristic - sometimes to the exclusion of all others.

It is possible to carry out a relatively simple comparative measure of the "power" of a tennis racket in the laboratory that can give some indication of how this performance characteristic of rackets has improved over the years. This is achieved by simply projecting a ball at the strings of a suspended racket and measuring the ball velocity both before and after impact. The ratio of: -

$$\textbf{Apparent Coefficient of Restitution} = \frac{\text{ball velocity after impact}}{\text{ball velocity before impact}} \times 100\%$$

The Apparent Coefficient of Restitution (ACOR), as measured in the way described above, is a reasonable method of comparing the "power" of different rackets.

Not surprisingly, the ACOR of all wood rackets made up till 1980, being of the same approximate head size (65-70 square inches) and construction, was very similar. The maximum ACOR of a typical laminated wood racket circa 1980 was 35%.

The maximum ACOR of a typical oversize graphite racket, artificially weighted up to the same weight as the wood racket, is in the region of 45%.

If one considers a ball hitting the racket at 25 ms^{-1} then the ball velocity when leaving the racket would be:-

a) Laminated wood racket (ACOR 35%) $\frac{25\text{m/s} \times 35}{100}$ = 8.75m/s or 19.25 mph

b) Graphite racket (ACOR 45%) $\frac{25\text{m/s} \times 45}{100}$ = 11.25m/s or 24.75 mph

In percentage terms, this simple model shows that the modern racket provides an approximate increase in power of: -

$$\frac{11.25\text{ m/s} - 8.75\text{ m/s}}{8.75\text{ m/s}} \times 100 = 28.5\%$$

Whilst an increase in the maximum power of a racket of this magnitude is advantageous to most players, perhaps an even more important advantage results from the larger area of the racket head over which such increases in power occur. For example, whilst the maximum ACOR of a wood racket may be 35%, such power will only apply to an area on the racket face of perhaps a few square millimetres or centimetres. Outside this area the available power reduces quickly. Such rackets therefore require a high degree of accuracy and skill to ensure maximum performance.

Modern graphite rackets, with increased head sizes, may have an equivalent 35% ACOR area that is many times larger than the wood racket. This will produce a racket that is much more powerful on off-centre hits than the traditional racket and will be of tremendous advantage to the recreational player.

The effect on the serve

The situation with regard to the serve is quite different to the above. For a serve we can regard the ball as stationary whilst the racket head has a velocity of perhaps 40m/s (88 mph).

Using the above two rackets as examples, the resulting ball speed for a serve, again using this simple model, would be as follows: -

a) Laminated wood racket (ACOR 35%)

54 m/s or 118.8 mph

b) Oversize graphite racket (ACOR 45%)

58 m/s or 127.6 mph

In percentage terms, the modern racket therefore provides an approximate increase in power on the serve of 7.4%.

This assumes that the wood racket and the modern graphite racket are the same weight, which they are not. A modern racket is significantly lighter and may therefore be capable of being swung faster than a heavier wood racket (see the above figures on relative swingweight measurements of wood and composite rackets). If we assume that the modern racket can be swung, for example, 10% faster than the wood racket, then the racket head speed in the above calculation is increased from 40 m/s to 44m/s. The resulting ball speed then becomes: 63.8 ms^{-1} or 140.4 mph

This represents an increase in power of 18.2% over the wood racket.

Unfortunately, this is largely conjecture. Little research has been done to show whether players are willing, or able, to swing the new lighter rackets faster and thus maximise their power. Many current professional players still use devices to artificially increase the weight of their racket. The ITF Technical Centre is currently involved in a large-scale research programme which will attempt to determine the effect of racket characteristics on dynamic performance.

Player Reaction Time

What does this apparent increase in serve velocity from modern rackets mean for the player receiving the ball? How much less time does a receiver have to react to a serve in 1998 than in 1978?

Assuming that the receiver stands about 8 feet back from the baseline to receive a serve, then the distance from the server to the receiver is around 86 feet or 26 metres.

Taking a simplistic approach in assuming that the ball decelerates at a constant rate to 60% of the initial speed by the time it reaches the receiver, then the time for the ball to travel from server to receiver would be as follows: -

a) Wood racket - 0.602 seconds
b) Graphite racket - 0.509 seconds

This is a decrease in available reaction time for the receiver of over 15%.

In practice the ball may slow down to less than 60% of it's initial starting velocity but this reduction of 15% in reaction time is actually an understatement of the real effect. This is because there is a limit to the speed at which any human being is able to react. For example, studies have shown that few, if any, athletes can react in less than 100 milliseconds (0.1 seconds). Even the best athletes, who are trained to develop the best possible reaction times e.g. sprinters, are only capable of reacting within 0.1 and 0.2 seconds of hearing the starting gun.

A tennis player has to make a judgement as to the velocity, angle and spin of the ball and then move their body and racket to make the return. If we assume that the minimum time taken by the receiver to react is, for example, 0.35 seconds, then in the case given above the percentage reduction in reaction time is actually 37%.

It is therefore possible to show, with a number of reasonable assumptions, that modern rackets may have reduced the available reaction time of the service receiver, by more than 30%.

It is reasonable to assume that this effect may be more significant on faster surfaces than on slower surfaces. An attempt has been made in subsequent sections of this paper to corroborate this effect, (which is based on the simple approach of the laboratory testing of rackets) by examining any available data relating to an increase in the dominance of the serve. These include reports on the increase in serve velocities, number of tie-break sets, etc.

Dynamic Considerations in the Racket Power Issue

It is now understood that the relatively simple approach to determining racket ACOR (power) which has been adopted by most of the major international racket manufacturers is flawed. Many of the assumptions on which the calculations above are based are now being questioned, with the result that there is currently little agreement on the true effects of new racket technology on the sport. The major issues which require resolving before any credible means of assessing, or regulating, tennis racket power can be proposed are as follows:-

- **Impact position on serve?**

The Serve is largely responsible for determining the "speed" of the point. The traditional method of measuring racket power will always produce a maximum power potential at a position on the racket head that corresponds to the traditional "sweet spot" (actually, it has been shown by Howard Brody (Brody) that there are three separate sweet spots on a tennis racket. For the purposes of this paper we will consider the position of maximum coefficient of restitution to be the sweet spot). The location of this point of maximum power is normally close to the geometric centre of the racket head, but may be moved vertically up or down the face of the racket depending on how the weight of the racket frame is distributed.

The difficulty is that video analysis of the serve shows conclusively that players do not strike the ball with the racket at it's maximum power point (or sweet spot). It has been shown that during the service stroke, players generally strike the ball at a point on the racket head which is significantly higher (towards the tip) than the position of maximum power. This is an attempt by the player to take advantage of the physics of the serve action, which produces higher racket head velocities, and hence higher energy being imparted to the ball, than an impact between ball and racket at the "sweet spot".

For the serve then, the traditional method of measuring maximum racket power may be of limited value.

- **Ultra-Lightweight Rackets**

One of the major implications of the use of composite materials is the potential to reduce weight in products by putting a higher proportion of weight in the head and at the same time retaining the necessary strength and stiffness. This aspect has been fully exploited by tennis racket designers in recent years as highlighted above. It has also been shown that the velocity of the racket head during the racket / ball impact is of paramount importance in determining the energy imparted to the ball i.e. the "power" of the tennis stroke.

The extent to which lighter weight rackets allow an increase in the acceleration and velocity of the racket head during a tennis stroke, such as the serve, is only beginning to be understood. The net effect on the energy imparted to a ball, taking into account both the increased racket head velocity and a reduced momentum from the lower mass of the racket, needs to be examined in greater detail before the true implications of modern racket technology can be assessed.

- **Racket Length**

In recent year's, tennis rackets with an overall length of up to 29 inches have been developed. Tennis rackets now commonly vary between 27 and 29 inches in length. Longer length rackets generally have a higher swing-weight and are therefore more difficult to swing. However, preliminary studies have indicated that an increase in racket length of 1 inch results in up to 10% greater ball speed. To what extent these two mutually dependent factors outweigh each other requires further study and definition.

Comment

It is clear therefore, that a dynamic method of defining racket power, which simulates the energy invested into a tennis stroke, such as the serve, by the player, is required. Such a test method will allow variables such as the impact position of the ball on the racket strings, in addition to the acceleration and velocity profile of the racket head, to be more properly measured, understood and controlled. As inferred above, the ITF Technical Centre is currently involved in a project designed to develop such a test method.

HUMAN PHYSIOLOGY

Nowhere is the improvement in human physiology more apparent than in the elite athletes of modern sport. Tennis is a good example. Improved diet, training, higher rewards and keener competition, amongst many things, have led to significant improvements in the physiology of the elite professional players, evidenced by the following data (source: ATP and WTA yearbooks).

Men

Average height of top 50 men on ATP Tour (1998) - 1.86 metres (6 feet 1¼ inches)
Average body weight of top 50 men on ATP Tour (1998) - 77 kgs (170 lbs)

Compare these facts with available data on champions from the past (Table 2).
Perhaps even more revealing is a comparison of the physiology of today's elite Men with some of the tallest (and finest) players of past generations (Table 2).

Table 2 Heights and weights of male tennis champions from the past

Player	**Height (ins)**	**Weight (lbs)**
Henri Cochet	5'6"	145
Little Bill" Johnston	5'6"	121
Ken Rosewall	5'7"	135
Bobby Riggs	5'8"	?
Rod Laver	5'8½"	145
Jimmy Connors	5'10	155
Norman Brookes	5'11"	150
Frank Sedgman	5'11"	170
John McEnroe	5'11	164
Bjorn Borg	5'11	160
Vitas Gerulaitis	6'0"	155
Arthur Ashe	6'1"	151
Donald Budge	6'1"	160
Fred Perry	6'1"	171
Ellsworth Vines	6'2"	143
Bill Tilde	6'3"	155

The *average* body weight of players ranked in the top 100 of the current ATP Tour with equivalent heights to these past champions is 79 kgs, (or 174.2 lbs). This represents an increase in body weight for the modern male player of between 3 and 31 lbs (or an increase of between 2% and 22%) compared with the leading male players of the 1920's and 1930's.

Women

Height

The average height of the top 20 money earners on the Virginia Slims Circuit in 1975 was 5'6½" (1.69 metres).

The average height of the top 10 ranked players on the Corel WTA Tour in 1998 was 5'9½" (1.76 metres).

This represents an increase in the average height of the leading Women players, of 3 inches (7 centimetres), or 4.5%, in 23 years.

Weight

The average weight of the top 20 money earners on the Virginia Slims Circuit 1975 was 133.8lbs (60.7kgs).

The average weight of the top 10 ranked players on the Corel WTA Tour in 1998 was 143lbs (64.9kgs).

This represents an increase in the average weight of the leading Women players, of 9.2lbs (4.2kgs), or 6.9%, in 23 years.

Two questions are in order: -

- To what extent will this process of evolution in human physiology continue?

- Would such continuing evolution result in a fundamental change to the nature of tennis at the highest level and will this be beneficial or detrimental to the appeal of the sport?

SERVE VELOCITIES

All other things being equal, the taller, heavier and better conditioned players of today, whether they be male or female, are likely to be capable of producing bigger serves and a "faster" game than in the past and, perhaps, be capable of sustaining this higher expenditure of energy and power, for longer.

Evidence to support this postulation is available in the inexorable year on year increase in the recorded maximum serve velocities for both men and women professional players: -

Men

- In 1990, when serve velocities first began to be recorded in the professional game, the number of male players who registered serves over 200 kph (124.3 mph), was just 5.
- By 1995, this number had risen to 38.
- Current figures are not available but it is likely that a majority of players in the ATP Tour top 200 are now capable of serving at over 200 kph.
- Rusedski now holds the world record with a serve of 239.7 kph (149 mph) during Indian Wells, 1998.

Women

- The fastest serve ever recorded for women is now very close to 200 kph (124.3 mph). The record at the time of writing belongs to Brenda Schultz-McCarthy with 197.9 kph (123 mph) during Wimbledon. The Williams sisters are now regularly recording serves over 120 mph.
- More than 20 players on the Corel WTA Tour recorded serves over 170 kph (105.7 mph) during 1998.

It should be noted many critics consider that the modern Women's game is a more attractive and appealing spectacle than in previous times.

TRENDS IN TENNIS PARTICIPATION AT RECREATIONAL LEVEL

Another factor in the changing landscape of tennis is the reported decline in world-wide participation in the sport at the recreational level. Most market research surveys throughout the 1990's, including the ITF's own "Tennis Towards 2000" survey (ITF, 1999) show that there has been a marked reduction in tennis activity, particularly in the traditionally mature tennis playing nations such as Germany, France, the U.S., Japan and Spain. This research is reflective of the reported decline in sales of tennis equipment and apparel that has been emanating from manufacturers since the early 1990's.

The reasons for this trend are numerous and are not speculated upon in this document. However, previous research has sought to highlight some of the major reasons why existing and potential players choose either to play less tennis than before, or, elect to spend their leisure time and money on other sports and activities.

Some of the problems that tennis faces are the same problems faced by all other sports. The number one reason given by tennis players, for playing less, or giving up tennis, is *"a lack of time"*. Such socio-economic changes are beyond the control or influence of any sport, but they must be understood and acknowledged through the measures taken to address those problems that *are* within the scope of influence of a sport.

A sport that fails to understand and react to the accelerating pace of evolution in socio-economic culture may expect a gradual reduction in the activities traditionally associated with that sport. This will be witnessed through reduced participation, particularly among those groups who are most sensitive to changes in social culture, i.e. the under 25's, and reduced commercial activity, e.g. sponsorship, investment, etc.

Two of the most commonly quoted reasons for players becoming disillusioned with tennis, and which are capable of being addressed through the use of Technology, are: -

- Tennis related injuries, and,
- Tennis is a difficult game to master, relative to other sports and activities.

Ideally then, any proposals to establish a better balance between Technology and Tradition in tennis, should acknowledge the difficulty that our sport faces in attracting and keeping recreational tennis players.

RECONCILING FACTS WITH PERCEPTIONS

The histories of most sports are sprinkled with examples of debates being promoted by those who perceive that Technology has outweighed Tradition. Tennis has had it's fair share of these debates, most vociferously over the past 10 years, but these issues tend to remain largely unresolved due to the imponderable nature of such debate and the lack of facts to prove the change in balance either way.

The lack of a fundamental understanding of tennis often leads to beliefs and perceptions, even from leading players, that are difficult to reconcile with the truth.

This section will attempt to reconcile some commonly held perceptions on the Technology question in tennis with the available facts.

COMMON PERCEPTIONS

The Views of Tennis Experts

Some of the views of leading tennis commentators and exponents on the Technology versus Tradition question are given below: -

"You watch him do that four, five times and you go 'that's incredible!' But after the fifteenth or so you kind of start to go into a coma." *A tennis fan on Goran Ivanisevic*, (Wall Street Journal).

"I don't think they should put any constraints on the rackets. You just have to practice more." *(Goran Ivanisevic on the speed of the Men's game - 1991).*

"You can see kids of 12 and 13 playing with wide-bodied rackets, hitting the ball harder than me. They are going to improve those rackets more and more. They have unbelievable power and in 10 years I don't know what's going to happen in tennis. It's going to be tough to see the ball. I don't know if they're going to have linesmen or robots." *(Goran Ivanisevic commenting on widebody rackets after having beaten Stefan Edberg, serving 32 aces in 21 service games - 1992).*

"Everyone knows there's a problem, it's a question of whether anyone's got the guts to do something about it." *(John McEnroe on the speed of the Men's game, 1992).*

"Something has to be done. It's very evident that tennis is losing popularity, not just in America but also in England and Japan." *John McEnroe on the fading appeal of tennis, 1997.*

"I just play with whatever they give me." *Jim Courier, 1992, after defeating Edberg.*

In 1997, Courier was attributed by the London Times (supporting Sampras) with comments criticising the balls used in the Australian Open for being too soft and for causing strain on the arm. (Times, 1997)

"It's become a power game. No-one is thinking out points. No-one is using finesse." *Chris Evert on the modern Women's game (whose views were supported by Martina Navratilova).*

Ivan Lendl on the prospects of watching a match between Kraijcek and Ivanisevic, "I wouldn't watch it. I'd just read the papers the next day to see what speed their serves were and how many aces they hit."

"I think you have to look at what it will be like in ten years down the road. I have seen some statistics that say that in one big final last year 54% of the points finished after the first two strokes. I don't think there's anything pretty to look at there." *Ivan Lendl*

Boris Becker, aged 23, commenting on Lendl's views, "It's the guy's who are not that fast who say it's too fast."

"I started talking about the softer balls two years ago…I hit the ball harder and harder today….For power players like me it's a killer. I didn't feel I was getting any power from the ball and when I tried to force it I lost my timing." *Boris Becker, aged 28, commenting on the balls used in the Australian Open, after having served 22 aces in a first round match, which he lost to Carlos Moya.*

"I don't know why they are complaining - Becker, I don't know, but maybe it was because of his loss, no?" *Carlos Moya, 1997.*

"I think high-powered rackets are destroying tennis and making it really boring. When you saw the matches that McEnroe and Connors played at Wimbledon, that was real tennis." *Michael Stich, following his defeat of Boris Becker in the 1991 Wimbledon final.*

"There could be two sets of rules, one for the majority and another for the Tour players where physique, diet and equipment have gone beyond what we know now. …I watched Philippoussis - Rusedski and was staggered." *David Lloyd, ex-player and ex-GB Davis Cup captain.*

"The big guys have an edge…..We should not be frightened of changing the rules if it developed so that 90 per cent of the top 50 were more than 6'6" and that finesse was being lost." *John Newcombe*

"I just wish they had been available years ago. It would have made it a lot easier on me physically and I might have added some additional titles to my record." *Jimmy Connors, on modern rackets, 1995.*

Jim Baugh, President of Wilson Sporting Goods on the ITF decision to ban longer rackets for the Professional game, 1996, "The actions the ITF is taking for the professional game is too late. The pro's that are playing today are playing with rackets from ten years ago. The goals of the Wilson's, Prince's and Dunlop's are to bring up new kids and have them start out with the latest technology frames. That would mean in five to ten years we are going to have young pro players with very large, stiff, head heavy rackets. Then that power level would reach the pro game in the years to come….. So my fear is that in five to ten years the professional game may be too quick."

Commenting on the 1994 Sampras - Ivanisevic Wimbledon final, in which only three of 206 points played lasted more than four shots, Fred Perry called it "..one of the most boring finals in history."

"I would have been a totally different player if I had started off with a modern graphite racket. ... Today's racket lets you put a lot more speed on the ball... Tennis now is all speed and power - and that's killing the game." *Rod Laver, 1996.*

"Maybe the modern game is far more advanced than the inventor's imagined possible. I'm sure we will be far more concerned next century than we are today, and we're a lot more concerned today than we were a decade ago. ...The rules and the court remain constant and I guess now there is an imbalance as the player's take every advantage they can." *Neale Fraser, 1996.*

" I don't know the answer. ...perhaps the size of the service area could be reduced, or players forced to serve from further back. ...perhaps a rule change is the only answer." *Frank Sedgman, 1996*

"Wimbledon has become a shooting gallery where only the strong survive. Grass is such a rarefied surface and almost nobody on the tour really knows how to play it. You get all power and no finesse." *Bud Collins, tennis author and journalist.*

"Today, power is all. On every surface, from grass to clay, the modern game is a one-dimensional slugfest - exciting at times, occasionally brilliant, but tediously one-dimensional. ...Power has killed the artist." *John Barrett, TV commentator and author.*

"What they're playing now isn't tennis. They should call it by a different name so that audiences don't get confused." *Gianni Clerici, tennis author and journalist.*

The Views of Tennis Fans

The ITF has sought the views and perceptions of ordinary tennis players on the speed of the Men's game as part of the large-scale international survey entitled "Tennis Towards 2000" (ITF, 1999). This research is designed to provide data on the participation and attitude trends of tennis players on a variety of issues and to date, has included detailed qualitative interviews with over 3,500 tennis playing households in 13 countries.

Respondents were asked to say to what extent they agreed with the following statement: -

"The Men's professional game is too fast"

Responses were divided into one of three categories, i.e. agree, disagree or no opinion.

Fig. 2 shows, in graphical form, a summary of the results by nation.

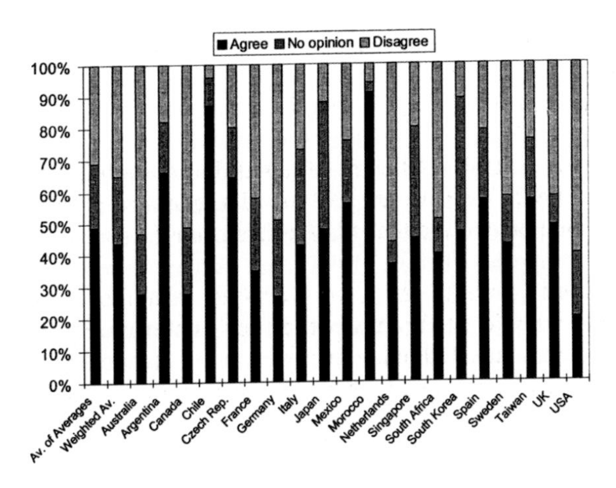

Fig. 2 "The Men's Pro Game is Too Fast" (Percentage of respondents who agree/disagree by nation)

The following observations may be made on this research: -

- The level of agreement with the statement varies widely from nation to nation.
- It appears that the more mature tennis nations (e.g. Germany, France, and Australia) are more likely to disagree with the view that the men's game is too fast. However, there may be a strong correlation between the predominant tennis court surface type and the views held by tennis players and spectators, in a specific nation. For example, both France and Germany are predominantly clay court nations, for which the lack of rallies in the game is not an issue.
- The less mature, but developing, tennis nations are more likely to agree with the view that the Men's game is too fast.
- Overall, almost half, (49%) of players believe that the Men's professional game is too fast. Less than 32% of those questioned disagree with this view.
- When the views of tennis players in the traditional European clay court nations of France, Germany and the Netherlands are removed then the majority who agreed that the Men's game is too fast increases to 54%, whilst 26% disagree - a majority of more than 2:1.

TIE-BREAK DATA

Most sports have convenient methods of assessing the gradual changes which occur as a result of the influences and forces which have been discussed i.e., improving human physiology, conditioning, new equipment technology, rule changes, etc.

Change can be measured through improvements in performance, which are most apparent in improving scores, faster times, etc. Unlike most other sports tennis does not have a convenient method of comparing the performance of players in the 1990's with those of earlier times. The scores of a tennis match do not directly reflect better performance.

Other data that may be capable of giving a strong indication of the trends within tennis, for example, number of aces served, average point length, % of serve games held or broken, etc. has only recently (1990-1 onwards) begun to be collected by the Grand Slams or professional Tours. Using such available data in a historical context is therefore of very limited value and is not included in this paper.

However, a method of monitoring the composite effect of several influences, which may have combined to increase the dominance of the serve in the professional game, has been proposed (Brody, 1998) and further investigated by the ITF Technical Centre.

It has long been known that the pace of a tennis court surface has a significant influence on the probability of a player winning and holding a game on serve. On a fast surface, such as grass, the probability of holding serve is known to be higher than that on a slower pace surface. If both players in a singles match can continue to hold serve, then the probability of a set reaching 6-6, at which stage a tie-break would now occur, or prior to the introduction of the tie-break a set would continue until a player obtained a two game lead, is increased.

Assuming that this conjecture is correct, it should be possible to show that the percentage of sets in a tournament, which go to a score of 6-6 in games, or above, will be higher on faster surfaces than on slower surfaces. Given the major advantage of this technique, which is the availability of basic score data for matches and tournaments going back through the history of tennis, this may also help to show trends which exist to confirm the view that the serve is more dominant today than in previous times.

Using this technique it may also be possible to show, through the rate and period of change, whether changes in Rules, equipment technology, etc. have specifically contributed to a change in the nature of professional tennis, whether it is played by Men or by Women.

An analysis has been carried out of the four Grand Slam tournaments over a thirty-one year period. This involved calculating the number of sets which go to 6-6 or above, as a percentage of the total number of sets played, in both the Men's and Women's singles draw for each Grand Slam event. This timescale is approximately commensurate with the beginning of the Open period of Tennis in 1968, through to the present day.

For the Men, this has involved analysing over 15,000 matches and 55,000 sets of professional tennis, on a variety of surfaces. For Women, the analysis has included approximately 12,500 matches and almost 30,000 sets of professional tennis on the same variety of surfaces.

Verification of the Method

Figures 3 and 4 attempt to verify the basic assumption that the percentage of tie-break sets, and therefore the dominance of the serve, is related to the pace of the surface.

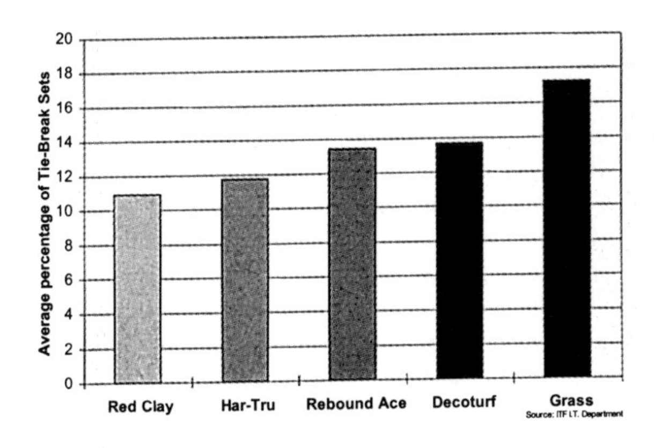

Fig. 3 Average percentage of tie-break sets in Grand Slam men's singles from 1968-1998 - by surface type

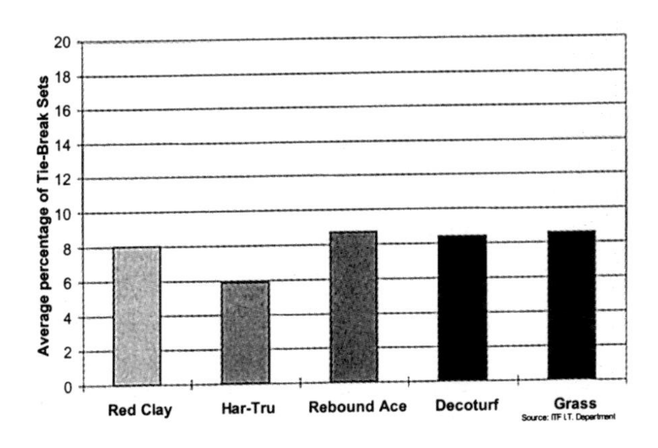

Fig. 4 Average % of tie-break sets in Grand Slam women's singles from 1968-1998 - by surface type

The results from the various surface types on which the four Grand Slam events have been played since 1968, have been combined to give results for five different surfaces. Figure 3 shows the combined results for Men and Fig. 4 the combined results for Women.

RESULTS

As can be seen, the results given in Figure 4 for Men show a very clear relationship between the percentage of tie-break sets and the perceived (and recently measurable and quantifiable) pace of a tennis court surface. The faster surfaces give a significantly higher percentage of tie-break sets for Men.

The red clay of Roland Garros and similar performing unbound mineral Har-Tru (U.S. Open) surfaces are significantly slower than the medium paced hardcourt surfaces of today's Australian and U.S. Open tournaments. Both, in turn, are significantly slower than the natural grass of Wimbledon and the historic venues of Forest Hills and Kooyong.

Figures 5, 6, 7 and 8 show the individual trends in the number of tie-break sets played from 1968 to 1998 for all four Grand Slam Men's Singles events (including 1999 for the Australian Open).

A mathematical "best-fit" linear trendline has been calculated, using Microsoft Excel, for each Grand Slam data series.

A line has been drawn to highlight the approximate date at which the transition from the use of predominantly wood rackets to composite rackets took place (1981 to 1982). In practice, this transition took place gradually through the period of the mid-1970's to the early 1980's.

Over the years, it is clear that a consistent and significant difference exists in the percentage of tie-break sets which occur in the Men's Singles game when played on different surfaces.

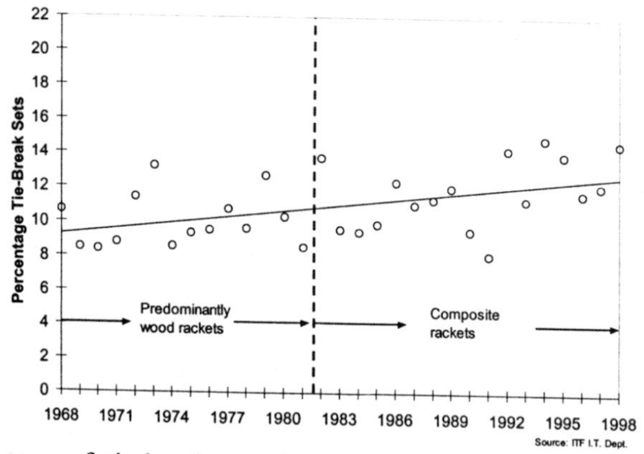

Fig 5. Percentage of tie-break sets in Roland Garros men's singles 1968-1998 (played on red clay)

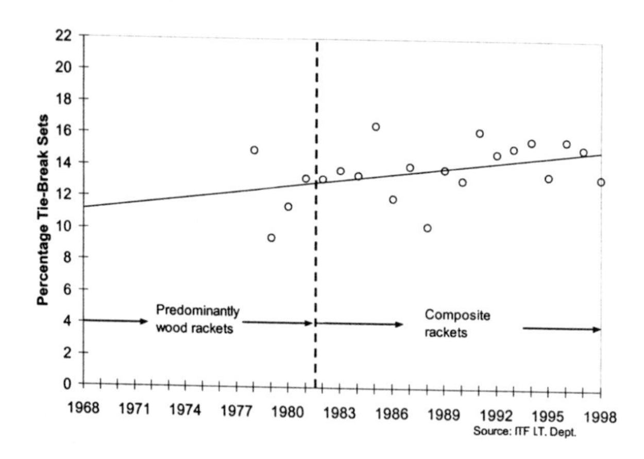

Fig. 6 Percentage of tie-break sets in US Open men's singles 1978-1998 (played on hardcourt)

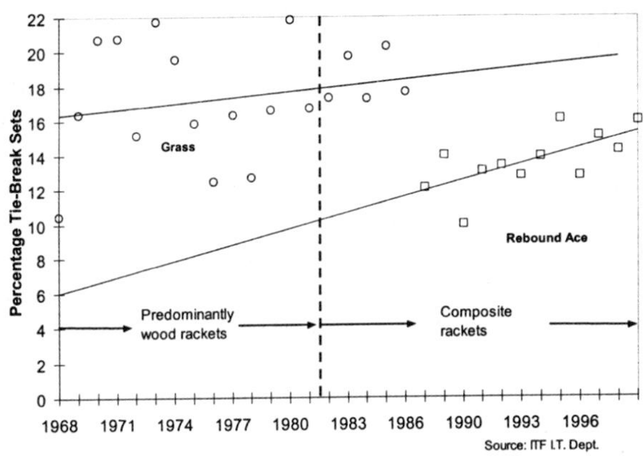

Fig. 7 Percentage of tie-break sets in Australian Open men's singles from 1968-1998 (grass and Rebound Ace)

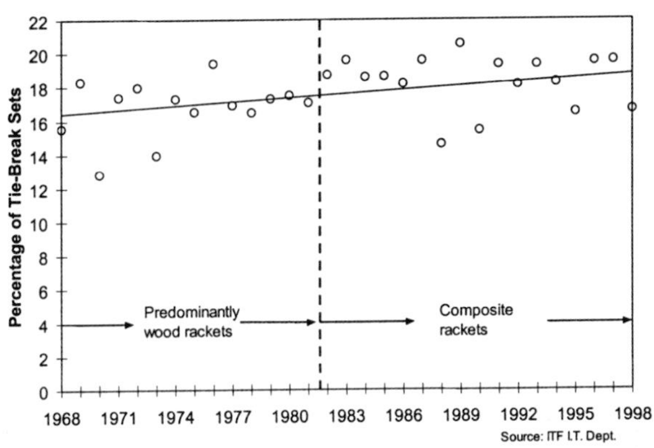

Fig. 8 Percentage of tie-break sets in Wimbledon men's singles 1968-1998 (grass)

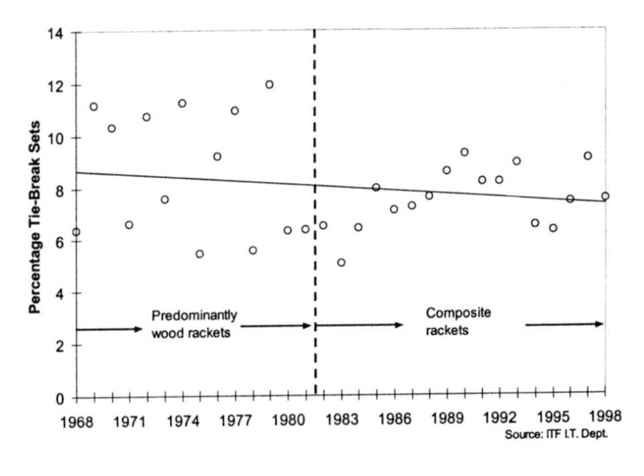

Fig. 9 Percentage of tie-break sets in Roland Garros women's singles 1968-1998

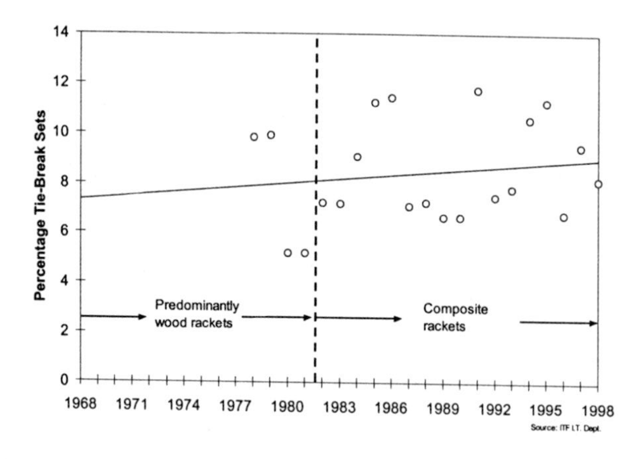

Fig. 10 Percentage of tie-break sets in US Open women's singles 1978-1998

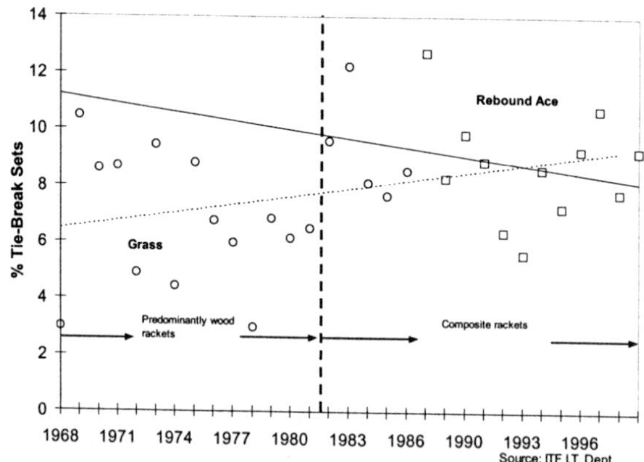

Fig. 11 Percentage of tie-break sets in Australian Open women's singles from 1968-1998

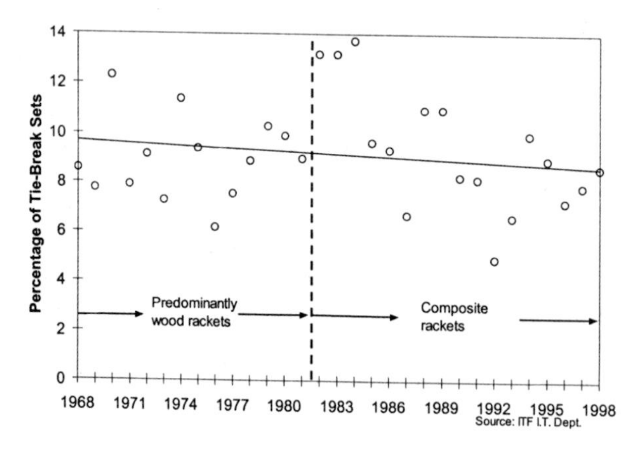

Fig. 12 Percentage of tie-break sets in Wimbledon women's singles 1968-1998

Women

Figure 4 is interesting in that it highlights the lower overall dominance of the serve, as determined by the lower percentage of tie-break sets on all surfaces, in the Women's professional game vis-à-vis the Men's professional game. The number of tie-break sets, and by inference the effectiveness of the serve in the Women's game, is remarkably similar on all four current Grand Slam surfaces.

Figures 9, 10, 11 and 12 show the trends for each individual Grand Slam Women's Single's event over this timescale.

DISCUSSION

The range in time that has been analysed (1968-1998) includes the entire Open period of tennis, with the onset of the professional game, in addition to the period of the introduction of the modern composite racket. A dramatic increase in the dominance of the Serve on any surface is not observable.

For the professional Men the individual trendline for each Grand Slam clearly shows however, that there has been a gradual and steady increase in the number of tie-break sets, and by definition the dominance of the Serve, on all surfaces, throughout the past 31 years.

One may speculate as to why there does not appear to have been as dramatic an increase in the dominance of the serve, during or following the introduction of graphite rackets, as many would imagine. We have seen in previous sections that it is possible to show that modern rackets can be more powerful than traditional wooden rackets and that serve velocities, for both professional Men and Women, are increasing. It is possible therefore to speculate that, as serve velocities have increased, the ability of players to improve the quality of their returns has improved accordingly. Due to the relatively uniform rate at which the number of tie-break sets has increased over the past 31 years, one might speculate that the increasing dominance of the serve may be more as a result of progressive improvements in human physiology than racket technology. In this context however, it should again be pointed out that the professional Men have been, and continue to be, slow in adopting the very latest racket technology.

For the Women the trends are surprisingly different. At three of the four Grand Slams there would appear to be a negative trend, which implies that the serve has become less dominant over the time period in question. The US Open is the only Women's event that shows an increase in the percentage of tie-break sets but the gradient of this line is relatively low.

Speculation as to why the Men's and Women's results are so different must inevitably centre on the serve velocities currently employed and player reaction times. It would appear that, in general, professional Women are currently serving the ball at velocities which are still well within the range in which a player is able to react and make a reasonable quality return. This may be as a result of improving athletic ability and perhaps new racket technology. This is also evidenced by the lack of discernible effect of surface pace on serve potency. The opposite may be true for the Men, as discussed in the section "Performance of Individual Male Players".

Taking the results for the Men only and measuring the gradient of the trendline of each Grand Slam will provide a good indication of the comparative rate of change that has taken place on the various surfaces. These are shown in Table 3.

Table 3 The increase in percentage tie-breaks at the four Grand Slams for 1968 to the present day

Event	Surface	Change in percentage of tie-breaks
Australian Open	Rebound Ace	+ 0.30% / year
US Open	Decoturf	+ 0.13% / year
Roland Garros	Red Clay	+ 0.11% / year
Australian Open	Grass	+ 0.10% / year
Wimbledon	Grass	+ 0.075% / year

Taking Wimbledon as a baseline, it is possible to calculate how long it will take for the percentage of tie-breaks at the other Grand Slams to equal that at Wimbledon in 1998. The periods are as follows:-

Australian Open (Rebound Ace) - 3.7 years
US Open (Decoturf) - 10.8 years
Roland Garros - 34 years

Thus, if we accept that Wimbledon is fast, then two of the other three Grand Slams will be at least as fast within the next decade.

PERFORMANCE OF INDIVIDUAL MALE PLAYERS

As a further means of verifying the tie-break method of measuring trends within tennis an analysis of the records of a number of individual male tennis players was carried out.

Table 4 shows the number of tie-break sets played by eminent individuals as a percentage of the total number of sets each player played at Wimbledon between 1968 and 1998.

DISCUSSION

A number of conclusions may be drawn from the data relating to individual male tie-break set records. The probability of a player holding serve on a fast surface such as the grass used at Wimbledon is greatly enhanced by the player's ability to deliver a very fast serve. Those players with greatly enhanced probabilities of holding serve are exclusively of the modern period. These players are still relatively few in numbers. The greatly improved probability of holding serve, evidenced by these few players, is the result of an ability to deliver their first serve, in a consistent manner, with a velocity approximately 10 to 20 mph (8 to 16%) faster than the majority of other players.

This raises two further points. A relatively small increase in serve speed (8 to 16%) may enable any player to take advantage of a greatly increased likelihood of holding serve. Further minor evolution, e.g. through the adoption of advanced racket technology by the professionals of the future, may result in major changes to the nature of tennis. This point may bear out the theory proposed in the section "Player Reaction Time" of this paper, in relation to player reaction time and the possibility that current ball velocities may be close to reaching some limit of human capability. The ITF is currently embarked upon a programme of research into this subject.

Table 4 Percentage of Tie-Breaks for individual players at Wimbledon 1968 to 1998

Player	Height & Weight		% of Tie-break Sets
G. Rusedski	6'4"	190lbs	36.0%
G. Ivanisevic	6'4"	181lbs	33.1%
M. Philippoussis	6'4"	203lbs	28.6%
T. Henman	6'1"	154lbs	25.5%
T. Martin	6'6"	190lbs	25.5%
P. Sampras	6'1"	170lbs	22.9%
M. Stich	6'4"	175lbs	22.0%
R. Kraijcek	6'5"	190lbs	21.7%
R. Tanner	6'0"	170lbs	21.6%
B. Becker	6'3"	187lbs	20.6%
A. Agassi			18.8%
S. Edberg			18.0%
A. Ashe			17.4%
J. McEnroe			17.3%
K. Rosewall			15.7%
J. Connors			15.3%
J. Newcombe			14.3%
B. Borg			14.1%
M. Chang			13.4%
R. Laver			8.3%

CONCLUSIONS ON TIE-BREAKS

This method is likely to be a useful and relatively simple tool that can be used to monitor trends within the game relating to an increase in the dominance of the Serve.

It may also be used, in conjunction with other more direct measurements (such as serve speeds, point length, number of aces, % of service games held or broken, etc.) to measure the effect of rule amendments designed to change and improve the nature of the game, as proposed in the following section.

THE MEASURES THAT TENNIS NEEDS TO TAKE

The following section of this paper is a proposal designed to address the issues that have been highlighted above and which contains measures designed to establish a mechanism for the balancing of Technology with Tradition.

Over the course of the period 2000 to 2003, it is proposed that substantial amendments be made to the Rules and specifications which define the ball and racket. It is also proposed that specifications and guidelines be introduced for court surfaces, which will define a range of performance and constructional characteristics for which, currently, no such international specifications exist.

It will also be proposed that the ITF assume a higher degree of control over the balance between Technology and Tradition through the introduction of comprehensive approval programmes for rackets and surfaces, in addition to an extended procedure for the control of ball properties. Consideration may also be given to this principle being expanded to include other equipment and apparel (shoes), within a pre-determined timescale.

AMENDMENT TO THE TENNIS BALL REGULATIONS

In July 1999 the ITF Annual General Meeting voted to approve a two-year experiment in which two new types of tennis ball will be permitted for use in tournaments played according to the Rules of Tennis, for the purpose of detailed evaluation and development.

The two new types of ball are designed to have specifications that will result in different performance characteristics derived from their differing dynamic and aerodynamic properties.

These important amendments are closely linked to work that has been carried out by the ITF and some National Associations (and which is continuing) designed to assess the interaction between tennis ball and court surface. This work has resulted in a credible technical method of measuring the pace of a court surface (ITF test method ITF CS 01/01) and a system has been developed (known as the ITF Surface Pace Rating) which enables types of court surface to be classified into one of three types, i.e. slow pace, medium pace or fast pace. It is intended that the two new ball types, in addition to the existing ball type, be introduced and developed to improve the appeal and enjoyment of tennis at all levels for players and spectators alike.

The three types of tennis ball that would result from this amendment are to be known as Type 1, Type 2 and Type 3.

Type 1 balls

The specification of Type 1 balls is largely identical to the specification of balls that has existed for many years up until 2000. The one important exception to this being the hardness, or deformation, of the ball. Type 1 balls are to be produced with a harder specification than has existed previously.

Type 1 (fast) tennis balls are to be produced with forward deformation readings in the range: -

0.195 to 0.235 inches (4.953cm. to 5.969cm.)

Research has shown that, during impact with the court surface, a harder ball behaves in a way that produces a lower angle of bounce than a ball with a more traditional "softer" specification. This characteristic counteracts the tendency of a normal ball to bounce at a relatively high angle off a surface with high frictional characteristics, such as a court with a clay, or similar type of unbound mineral surface.

For ease of understanding a graphical interpretation of this concept is presented in Fig. 13.

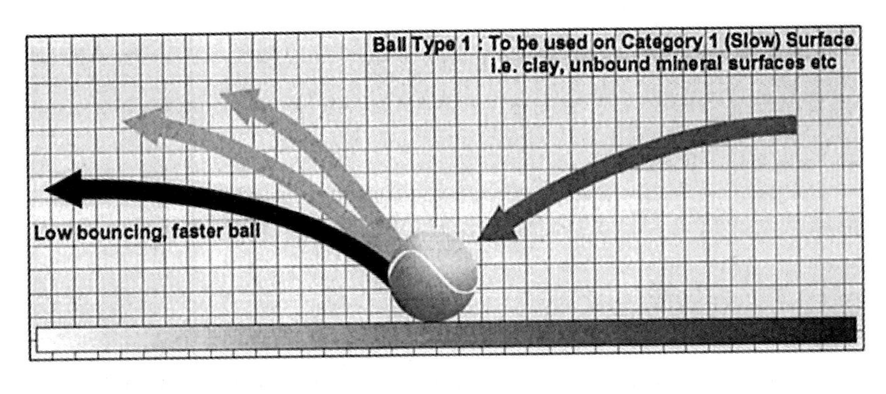

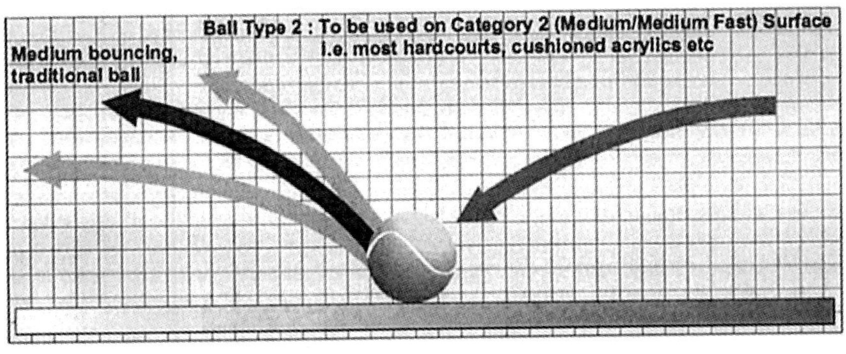

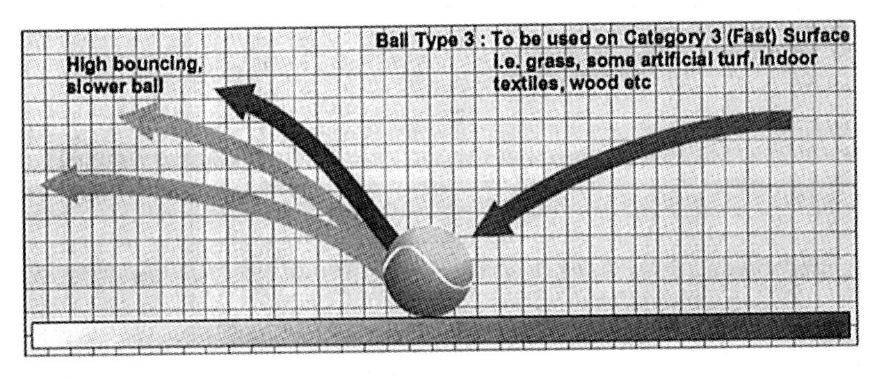

Fig.13 Graphic (ITF, 2000) to explain the introduction of three new tennis balls categories.

Such a ball type will address two issues: -

- Ball type 1 will formalise, within the Rules of Tennis, a number of brands of tennis ball which are currently available in the market but which are too hard to conform to the Rules.
- Ball type 1 will provide a new type of ball for many recreational clay court players, which may improve the enjoyment and appeal of tennis for some.

Type 2 balls

For the proposed period of experimentation (Jan. 2000 to Dec. 2001) the specification of Type 2 balls would be identical to the existing specification.

However, assuming that the experiment with three balls types is successful and is approved by the ITF for official implementation in 2002, it is proposed that the specification of Type 2 balls also be amended. The amendment would be designed to provide a well-defined difference in properties between Type 1 and Type 2 balls. This would be achieved by amending the deformation specification to ensure that Type 2 balls are produced with a significantly higher (softer) deformation than Type 1 balls. Under the terms of the experiment to be carried out between 2000 and 2001 it will be possible for new Type 1 balls to be produced with deformation characteristics which are also possible under the existing ball specification. This is as a result of the existing deformation specification being relatively wide-ranging (0.220 to 0.290 inches).

The forward deformation of Type 1 balls has been set as: -

0.195 to 0.235 inches (4.953cm. to 5.969cm.)

The forward deformation of Type 2 balls (as from 2002) is proposed to be: -

0.240 to 0.280 inches (6.096cm. to 7.112 cm.)

The rationale behind this, and the concept of creating a definite gap of 0.005 inches between the specifications of Type 1 and Type 2 balls, is to ensure that there will eventually be a tangible and measurable difference in the characteristics of both ball types. If this measure were not introduced, then it is considered highly likely that some manufacturers would simply aim to produce balls which straddled both deformation ranges, resulting in a distribution of ball properties which were too similar.

It is intended therefore, that this proposal will result in two ball types, largely identical in appearance, but which have different rebound angle characteristics, resulting from the manufacturer aiming at producing balls whose deformation figures fall in the middle of the above ranges.

Figure 13 shows the comparative rebound characteristics of Type 2 balls with Type 1 balls.

Type 3 balls

The specification of Type 3 balls is different to Type 1 and Type 2 balls in one important way. Type 3 balls are larger in diameter than that which is currently allowed under the Rules of Tennis.

The existing ball diameter specification is: -

2.575 to 2.700 inches (6.541cm. to 6.858cm.)

The proposed size specification for the Type 3 ball is: -

2.750 to 2.875 inches (6.985cm. to 7.303cm.)

Such a ball will be between 6-8% larger in diameter than a conventional ball. The range provided by this proposed specification will allow sufficient latitude for experimentation on ball sizes which may be considered practical for the playing of tennis.

Research has indicated that these larger balls possess performance characteristics that differ from existing balls in two ways: -

- A larger diameter ball possesses different aerodynamic drag characteristics to those of a conventional size ball. Such a ball will therefore slow down more quickly during it's flight and give the Receiver more time to react.
- A second characteristic of the larger ball is one that is related to the rebound angle of the ball after impact with the court surface. This is, in part due to the steeper incoming trajectory of the larger ball and in part due to the different compression of the larger ball on the surface during The rebound angle of this ball has been found to be higher than that of an existing ball, again giving the Receiver marginally more time to react and return the ball. This concept is also shown graphically in Fig. 13.

Both of the above characteristics combine to produce a ball type that is ideally suited to a court surface type that is considered to be "fast". Surfaces that fall into this category would be grass, some artificial turf, some indoor carpets and textiles, wood, etc.

PROPOSAL TO INTRODUCE A CLASSIFICATION SYSTEM FOR COURT SURFACES

It has been shown in the previous section that the proposal for the development of three specific tennis ball types is dependent upon the interaction between ball and surface. The court surface is therefore considered to be of equal importance in this respect.

In 1997 the ITF proposed a test method (ITF CS 01/01), (ITF, 1997) that is designed to classify the "pace" of a tennis court surface through the measurement of the ball velocity and angle both before and after impact with a surface. Sufficient international experience has now been gained with this equipment to allow credible standard's bodies such as CEN (the Standardisation body of the European Commission), to accept this method as an officially recognised test method.

The ITF introduced an official certification scheme for the pace of tennis court surfaces, based on test method ITF CS 01/01 in March 2000.

The system of classification of the pace of a court surface involves the division of all known court surface types into three distinct categories determined by their ITF Surface Pace Rating: -

CATEGORY 1

Category 1 surfaces are those which are measured to have an ITF Surface Pace Rating of between 0 and 35. Court surfaces that typically fall into this category are clay courts and other types of unbound mineral surface.

CATEGORY 2

Category 2 surfaces are those which are measured to have an ITF Surface Pace Rating of between 30 and 45. Court surfaces that typically fall into this category are hardcourts, hardcourts with various forms of coating and cushioning, courts with bound mineral surfaces and some forms of textile and artificial turf surfaces.

CATEGORY 3

Category 3 surfaces are those which are measured to have an ITF Surface Pace Rating of 40 and above. Court surfaces that typically fall into this category are grass, some artificial turf, some indoor textile surfaces and wood.

WHICH BALL TYPE FOR WHICH SURFACE CATEGORY?

- Ball Type 1 is intended for use on Category 1 surfaces.
- Ball Type 2 is intended for use on Category 2 surfaces.
- Ball Type 3 is intended for use on Category 3 surfaces.

GUIDELINES FOR THE DESIGN, CONSTRUCTION AND MAINTENANCE OF TENNIS FACILITIES

An ITF Working Group has been established to produce internationally credible guidelines for the design, construction and maintenance of tennis facilities. This should prove to be of great help to National Associations and other bodies, in establishing a higher quality tennis infrastructure worldwide.

PROPOSAL TO DEFINE AND REGULATE RACKET POWER

The ITF is currently working on a project designed to address the issues raised in section Dynamic Considerations in the Racket Power Issue.

A test method, capable of measuring the dynamic characteristics of a tennis racket, is under development and should be completed in early 2001.

This device should enable us to determine, in a way that is not currently available, the real effects of modern racket technology on changes to the nature of tennis.

If, with the aid of this equipment, it is shown that improved racket performance is a significant factor in the changing landscape of tennis, then this same test method will be available to be used as an advanced technical means of imposing maximum limits on the power of tennis rackets.

The proposed test method will be designed with features that are intended to address a number of issues:-

- Research has been carried out to measure typical racket head accelerations and velocities for a variety of male and female players with rackets of widely varying properties. These studies will continue and be extended to include top professional players. It is proposed that an apparatus be constructed which simulates the energy expended by a player in a serve and which will therefore swing a racket with a determined velocity and acceleration.
- The apparatus will include a tennis ball release mechanism designed to allow the ball and racket face to impact at a pre-determined point. This facility will allow the effect of players using the upper region of the racket head (which travels at a higher velocity) to deliver the serve, to be considered in the racket power issue.
- The facility to swing a racket with a pre-determined energy will enable the effects of varying racket weight, weight distribution and length to be considered.
- The apparatus will be designed to have the facility of varying the energy expended during the simulated serve swing. This facility will allow for studies (possibly leading to different racket power specifications) for different types of player, e.g. professional men, professional women, recreational players, etc.

ITF APPROVAL PROGRAMMES

The proposals detailed in sections "Amendment to the Tennis Ball Regulations", "Proposal to Introduce a Classification System for Court Surfaces", "Guidelines for the Design, Construction and Maintenance of Tennis Facilities" and "Proposal to Define and Regulate Racket Power" are the basis for a mechanism for the establishment of a new balance between Technology and Tradition in tennis. This mechanism can be updated and amended in future years to engineer the sport from the professional to the recreational level.

In order for such a mechanism to function and for the balance to be maintained in the longer term, it will also be necessary to encourage manufacturers and constructors to continue to work within the regulations by providing a credible means of enforcement and incentive. This can be achieved through the implementation of comprehensive testing and approval, or certification programmes for balls, rackets and surfaces, etc.

ITF Approved Balls

The ITF Approval procedure for tennis balls is becoming well established. Over the last three years the number of balls that have been submitted for ITF Approval has increased from 50 to 140 and it is estimated that a significant majority of ball brands available throughout the world are now tested and approved by the ITF.

This fact is of great value in helping to monitor product conformance on an international level.

The next stage in the development of the control mechanism for balls is the introduction of a formal market and tournament testing procedure as part of the criteria that ensures a ball keeps ITF Approved status. Earlier data in this paper showed the importance of such a mechanism.

On January 1st 1999 a draft procedure was introduced which proposes statistical criteria and action limits for the control of the quality of ITF Approved balls selected randomly from markets and tournaments. Manufacturers have been informed of the requirements of this procedure and the plan to fully implement the procedure as from either 2000 or 2001.

ITF Certified Court Surfaces

This recently introduced scheme will fulfil a number of functions: -

- Providing an official list of tennis court surfaces that have been tested and classified according to their ITF Surface Pace Rating and placed into one of three categories.
- Such a classification will eventually provide a simple and official guide for the type of ball that should be used on that surface.
- The procedure is envisaged as the initial stage in an approval scheme for court surfaces with a gradually increasing scope. Other classification criteria of performance, safety and constructional issues for surfaces will be added as and when acceptable test methods and criteria can be developed and agreed.
- An official and credible list of Approved court surfaces will provide a valuable service to National Tennis Associations who often request assistance with the selection of new facilities. Currently no official facility guidelines or information exists for smaller and less experienced National Associations - a factor that may be restricting the development of high quality tennis facilities - and limiting the growth of the sport.
- ITF Approval may provide an incentive for tennis court surface manufacturers to adopt higher standards and develop new and improved products for the benefit of tennis players.

ITF "Approved" Rackets

Assuming that the ITF eventually agrees with the principle of imposing limitations on racket performance (power and spin), the necessity for which will become apparent from work being carried out throughout 1999 and 2000, an ITF testing and Approval procedure for rackets used in competition could be initiated at some future date.

Other Areas of Involvement

For reasons that have been alluded to in earlier sections of this paper, it is considered to be important that the ITF begin to assume responsibility for other important Technology issues that affect the health of tennis and tennis players.

Tennis Shoes

In proposing classification standards for tennis court surfaces it is unrealistic to ignore the relationship between player and surface which is characterised by the tennis shoe. Improved knowledge of tennis shoe properties and the mechanism by which they interact with a wide range of court surfaces, may result in reduced risks of tennis related injuries to players and a greater enjoyment of the game.

Programmes of work are being considered to improve our knowledge of such issues and may lead to international industry standards for products which are found to be more suitable for court surfaces categorised by the ITF.

SOME IMPORTANT PHILOSOPHICAL QUESTIONS

The changes proposed in this document have a number of important implications for the nature of tennis, as we know it today. These philosophical questions will need to be addressed in the near future. For example:-

Q. Is there a need to address the "Speed of the Game" at the professional level?

Q. The measures proposed in this paper have the potential to engineer the sport to be more similar in nature, when played on different court surfaces. Is such a development thought to be desirable?

Q. Is a permanent separation of the rules for the Professional and Recreational game desirable?

Q. Would the introduction of different equipment rules or requirements for Men and Women be desirable, acceptable or practical?

CONCLUDING REMARKS

The proposals described in this paper are not designed to bring about major changes to the nature of tennis. On the contrary, they are designed to prevent the nature of our sport changing in a way that is beyond our control.

It is estimated that Tennis is currently played by up to 60 million players in 200 countries worldwide. As guardians of the Rules and specifications of Tennis the ITF holds a unique responsibility over the sport and over an associated industry. That industry has helped build perhaps 750,000 tennis courts, supplies 300 million tennis balls annually, millions of rackets, shoes and a variety of other equipment that results in an annual gross "national" product figure, for Tennis, of several billion U.S. dollars worldwide.

Such a responsibility should not be taken lightly. The potential force for change that such an industry represents can be beneficial, or harmful, to the health of tennis.

This paper describes a strategy for establishing a true balance between Technology and Tradition in our sport. By implementing the change described in this document, we will be adopting a mechanism required to balance Technology and Tradition, within tennis, over future years. This balance is an important factor in ensuring the continuing health and growth of this great sport of tennis.

ACKNOWLEDGEMENT

Many thanks to John Santa Maria of the ITF's I.T. Department for preparation of the data used to produce the tie-break information used in this paper.

REFERENCES

Brody, H. Tennis Science for Tennis Players
Golf Digest, March (1999)
ITF (1997, 1998 and 1999) Tennis Towards 2000
ITF (1997) An Initial ITF Study on Performance Standards for Tennis Court Surfaces p11-14.
ITF (2000) ITF Approved Tennis Balls, p5.

2 Equipment

An overview of racket technology

H. Brody
Physics Department, University of Pennsylvania, Philadelphia, PA 19104, USA

ABSTRACT: The modern tennis racket has evolved due to the availability of new materials and a better understanding of the basic ball-racket interaction. Today's rackets are bigger, lighter, more comfortable, more powerful, more stable, and more forgiving with respect to small errors made by the player. Many of the properties of a new racket can be determined by laboratory testing and computer simulation, where previously play-testing was the only way to evaluate a racket. This article aims to give a general overview of tennis racket technology and a list a number of the important references on this subject that produced or analyzed the present day racket.

INTRODUCTION

During the course of this century there has been a marked change in the size, shape, weight and weight distribution of tennis rackets. This is due both to the availability of new materials and a better understanding of the basic physics of the tennis racket. One hundred years ago, wood was the material of choice for the racket. Rackets are made of graphite composites as the century ends. The strength to weight ratio of the modern graphite composite has allowed the racket engineer to design frames by considering the physics of the swing and the ball racket interaction without being inhibited by the structural limitations of wood.

SWEET SPOTS OF A RACKET

The sweet spot of a racket is the ball impact location on the string where the players claim the hit "feels good". Brody had defined three distinct sweet spots (Brody 1987a) and the reasons why they feel sweet or good. These are the centre of percussion (COP), the node of the lowest mode of vibration (the node), and the maximum of the coefficient of restitution (COR). One of the aims of a racket designer is to try to place all three sweet spots at the same location, and have that location close to the centre of the head of the frame.

CENTRE OF PERCUSSION

The centre of percussion (COP) is defined for a free, rigid body as a pair of conjugate points such that the translational motion at one of them cancels out when the impact of the ball is at the other point. This concept was well established over a hundred years ago and Nikanow (1925) showed how to move the location of the COP by adding mass to the racket at specific points. Both Durbin (1980) and Frolow (1979) have shown that removing mass from the middle of the frame moves the COP toward the centre of the head. Head (1976) increased the length of the head of the racket, which instead of moving the location of the COP, moved the centre of the head to the COP location. If the translational motion of the racket handle cancels out at the location of the hand, the player feels no shock or jar due to the ball impact.

However, the tennis racket is not a free body, and Hatze (1998) and Cross (1998) have shown what happens to the COP (or the concept of COP) when the hand and the forces produced by the hand are included in the analysis.

THE NODE

The node of the lowest mode of vibration has been investigated by Brody (1987a, 1987b) and recently in great detail in a series of papers by Kawazoe (1998). Ball impacts at the location of the node do not excite vibrations of this lowest mode of oscillation, hence the hit feels sweet. The farther the impact location is from the node, the larger is the amplitude of the vibrations of that mode. The stiffer the frame of the racket, the smaller the amplitude of the vibrations, but the higher their frequency. There are higher modes of vibration in a tennis racket, but they generally are of higher frequency and of smaller amplitude.

Hatze (1994) has shown that player prefer to hit ball at the location of the node as opposed to other locations on the head of a racket, even if the node does not coincide with the OCP or the maximum of the COR.

Various methods for damping out vibrations of the frame of a racket have been proposed for those cases where the ball impact is not close to the node (Lacoste 1976, Kuebler 1980). However, most of these solutions are not as effective as the human hand in damping out vibrations. It has been shown that the tighter the racket is gripped, the quicker the vibrations are damped out (Brody 1989). The small gadgets that are placed in the string pattern of a racket near the throat are very effective in reducing string vibrations, but do essentially nothing to reduce the frame vibrations (Brody 1989).

THE MAXIMUM OF THE COEFFICIENT OF RESTITUTION

The location of the maximum of the coefficient of restitution (COR) is that point on the racket head where the ball rebound speed is a maximum. While the sweet spot does not have the same physical good feeling as the previous two sweet spots, it is a psychologically good feeling to hit the ball hard.

Finding the location of the maximum of the coefficient of restitution (COR) and finding methods to increase the COR have a long history. Donisthorpe (1926) experimented with rackets of all sizes, shapes, and weights, meeting only with moderate success. Head carefully mapped out the COR as a function of position of the ball impact on the strings for both his original oversized racket and a standard racket (Head 1976). He discovered that the enlarged strung area contained a region of greater COR that was not available on a standard size racket.

When a racket is tested in the laboratory for its ball rebound response, it is usually tested while free of all restraints, as opposed to having its handle or head clamped.

The resulting ratio of ball rebound speed to incident ball speed for a free racket is a measure of the "power" of the racket. It is called the apparent coefficient of restitution or ACOR (Hatze 1994) because it neglects the recoil velocity of the racket.

There are good arguments (based on a rigid body model of a racket frame) that show the ACOR maximum should be located near the racket balance point (the centre of mass or CM). Work on this has been recently done by Hatze (1994) and by Cross (1997). Brody (1997) has shown that when the actual swing of the racket is taken into account, the location of the maximum ball rebound speed moves up closer to the centre of the head.

Brody (1995) has derived a formula to take the ACOR data obtained with a stationary, free racket into the tennis court frame of reference, and allow a calculation of how the racket will actually perform in play. The formula

$$V(\text{hit ball}) = V(\text{incident ball}) * ACOR + (1 + ACOR)*V(\text{raq}) \quad (1)$$

assumes that the gripping force due to the hand can be neglected during the 4 milliseconds that the ball is in contact with the strings.

RACKET WEIGHT

With wood as the principal material for a racket frame, it was difficult to construct a racket of sufficient stiffness and durability, yet weighing under 14 ounces (400 grams). With present materials, rackets as light as 7 ounces (200 grams) that have acceptable playing properties are being constructed. For a heavy racket (weights of 12 ounce or greater), it can be shown that the speed of the ball off of the strings (racket power) is a weak function of racket weight. If by making the racket lighter, the racket can be swung faster, there can be a net gain in resultant ball speed or power. For very light rackets (weights in the 7 to 8 ounce range), the dependence of racket power on weight becomes stronger, and unless there is an appreciably faster swing, there will be a net loss of ball speed. The lighter rackets are particularly advantageous for serving, as opposed to ground strokes.

MOMENTS OF INTERTIA

A racket has a moment of inertia about each of its principal axes. The moment about the long axis of the racket is often called the polar moment or the roll moment. The larger this moment, the more stable the racket is against off axis impacts. A larger polar moment also produces less reduction of COR for off axis impacts. Both of these effects tend to give the player more margin of error with respect to impact location of the ball on the strings.

To first order, the magnitude of the polar moment scales with the racket weight multiplied by the square of the racket head diameter (Brody 1985). If a player wants to increase the polar moment of a racket, it is most effective to add additional weight to the head at 3 and 9 o'clock rather than at some other locations.

The moment of inertia for an axis through the handle at the butt end in the plane of the racket head is called the swing weight. The larger the swing weight, the more difficult it is to swing the racket (it has more rotational inertia). However, there is usually an increase in racket power to compensate.

The manoeuvrability of a racket is some function of both its polar and swing moments of inertia. There is no data in the literature (other than anecdotal) correlating these moments with the player's perception of manoeuvrability. Since the tennis racket is close to a two dimensional object (its thickness is small compared to its other dimensions, it can be shown that the moment of inertia about an axis through the CM and perpendicular to the plane of the racket is equal to the sum of the other two moments about axes through the CM (Brody 1985).

RACKET DIMENSIONS AND PERFORMANCE

It is possible to make rackets longer, yet not increase the swing weight of the frame, by either making the entire racket lighter or by removing weight from the head. This is made possible by using the new composite materials and modern manufacturing techniques. While longer, lighter rackets may be of some advantages for ground strokes, elite players will gain a very large advantage for the serve using a longer frame (Brody 1998). If, on the other hand, all that is done is to add two inches to the length of standard weight racket, the resulting frame will have a swing weight that may make it quite hard to swing.

The size of the head, which determines the length of the main and cross strings, influences the power of the racket. As a general rule, longer strings lead to more ball speed (power). This is one of the reasons why the ITF, in Rule 4 of the The Rules of Tennis, limits the maximum length of the strings in both directions (ITF 1999). To get around these limitations, manufactures have designed string suspension systems which interact with the strings on the outside of the head, rather than the inside.

By making the frame of the racket thicker, the stiffness of the frame increases. This means that less energy goes into deformation of the frame and vibrations. This concept was exploited in a series of rackets that had almost twice the thickness of a standard racket (Kuebler 1987) and was very powerful.

The length of the racket has an interesting effect on whether grip tightness has any bearing on the rebound ball speed. There are several arguments and a number of somewhat contradictory experiments published in the literature about the difference in ball rebound from a racket with a clamped handle and a free racket. The most interesting result (Cross 1999) investigates the propagation of the impulse down the racket due to the ball impact and the time it takes for the wave signal to return to the ball location from the handle. If the signal returns to the impact region after the ball has departed the strings, the conditions at the butt end of the racket can have no influence on the rebounding ball. In this context, Cross found that the hand forces can usually be neglected, if one treats the racket not as a rigid body, but as a flexible beam having the measured stiffness of a modern racket frame. Various authors have tried to model a racket mathematically (Brannigan and Adali 1980) and a finite element analysis of the tennis racket has been published (Bitz-Widing and Moeinzadeh 1990), but some of its conclusions are suspect.

CONCLUSIONS

Modern, space age, composite materials have allowed racket manufacturers to design and fabricate tennis frames that are based on the physical principles of mechanics and not the inherent structural limitations of the frame material. This has produced a racket that is lighter, stiffer, more forgiving, more stable, more durable, and capable of imparting higher speeds to the ball with less effort by the player. It has also Resulted in rackets that makes it easier for a beginner to learn the game of tennis and for the recreational player to enjoy the game more.

REFERENCES

Bitz-Widing, M.A. and Moeinzadeh, M.H. (1990) Finite element modelling of a tennis racket. *International Journal of Sports Biomechanics*, 6, 78-91.

Brannigan, M. and Adali, S. (1980) Mathematical modelling and simulation of a tennis racket. *Medicine and Science in Sports and Exercise*, 12.

Brody, H. (1985) The moment of inertia of a tennis racket. *The Physics Teacher*, 23, 213.

Brody, H. (1987a) Tennis Science for Tennis Players, *Univ of Pennsylvania Press*, Philadelphia, PA.

Brody, H. (1987b) Models of tennis racket impacts. *International Journal of Sports Biomechanics*, 3, 293-296.

Brody, H. (1989) Vibration damping in tennis rackets. *International Journal of Sports Biomechanics*, 5, 451.

Brody, H. (1995) How would a physicist design a tennis racket? *Physics Today*, 48, 26-31.

Brody, H. (1997) The physics of tennis:III, The ball racket interaction. *American Journal of Physics*, 65,981-987.

Brody, J. (1998) Improving your serve. In: *The Engineering of Sport*, S. Haake, ed., pp 311-316, Blackwell Science Ltd, Oxford.

Cross, R. (1997) The dead spot of a tennis racket. *American Journal of Physics* 65, 754-764.

Cross, R. (1998) Sweet spots of a tennis racket. *Journal of Sports Engineering*, 1, 63-78.

Cross, R., (1999) Impact of a ball with a bat or racket. *American Journal of Physics*, 67, 692-702.

Donisthorpe, F.W. (1926) The Mechanics of Tennis. *Amateur Sports Publishing*, London.

Durbin, E. (1980) US Patent no. 4,196,901

Frolow, R. (1979) US Patent no. 4,165,071

Groppel, J. (1986) The biomechanics of tennis: an overview. *International Journal of Sports Biomechanics*, 2, 141-155.

Groppel, J. and Roetert, P. (1992) Applied physiology of tennis. *Sports Medicine*, 14, 260-268.

Hatze, H. (1993) The relationship between the coefficient of restitution and energy loss in tennis rackets. *Journal of Applied Biomechanics*, 9, 124-142.

Hatze, H. (1994) Impact probability distribution, sweet spots, and the concept of an effective power region in tennis rackets. *Journal of Applied Biomechanics*, 10, 43-50.

Hatze, H. (1998) The centre of percussion of a tennis racket: a concept of limited applicability. *Journal of Sports Engineering*, 1, 17-25.
Head, H. (1976) US Patent no. 3,999,756
Kawazoe, U. (1998) Performance prediction of tennis rackets. In: *The Engineering of Sport*, S. Haake, ed., pp 325-332 Blackwell Science Ltd, Oxford. There are 9 other references to work by Kawazoe in this article.
Lacoste, F. (1976) US Patent no. 3,941,380
Lair, J. (1974) US Patent no. 3,801,099
Nikanow, J.P. (1925) US Patent no. 1,539,019

Modelling the impact between a tennis ball and racket using rigid body dynamics

S.R. Goodwill, S.J. Haake
Department of Mechanical Engineering, University of Sheffield, UK

ABSTRACT: A model of an impact between a tennis ball and freely suspended racket, normal to the string plane, was derived using rigid body dynamics. The inputs to this model were the ball and racket physical parameters (e.g. racket mass), and the experimentally determined coefficient of restitution for an impact on a head-clamped racket. The predicted values of ball and racket post-impact velocities were compared with experimental results and typically correlated within ~5%.

INTRODUCTION

The aim of this paper is to describe a model to predict the dynamic characteristics of an impact between a tennis ball and a freely suspended racket. Liu (1983) developed a mathematical model based on the laws of the conservation of linear and angular momentum. This model related the theoretical equations for a freely suspended and head-clamped racket for an impact normal to the string plane. A number of assumptions were made to derive the final formulae and very little experimental data was used to verify this model. It was concluded that the post-impact ball velocity was simply a function of the coefficient of restitution and not dependent on the method of supporting the racket.

Brody (1997) also used the conservation of linear and angular momentum laws to derive a theoretical model. This work generated formulae to predict the ball rebound velocity for an impact normal to the string plane of a freely suspended racket. The inputs to this model were the ball and racket physical parameters (e.g. racket mass), and the coefficient of restitution determined for a normal impact on a head-clamped racket. The aim of Brody's work was to generate formulae to predict how a racket will perform on the court using data collected in the laboratory. This could be used to advise tennis players of methods to reduce the errors in their game.

The current paper is an advancement of Brody (1997) since the recoil velocity of the racket, as well as the ball, is predicted for a normal impact on a freely suspended racket. The model uses the same input parameters as Brody.

This work is the initial stage in modelling the impact between a ball and racket for a groundstroke or serve. The model can be used to predict the effect of changing one parameter of the racket or ball on the game of tennis.

Freely suspending a racket may not truly represent a player's grip. However, Brody (1987) and Cross (1997) found that this was the most valid method of supporting a racket in a laboratory. They found that the ball had left the strings before the transverse force wave had travelled from the impact point to the handle and back, so the method of gripping would have no effect on the ball rebound velocity. They also measured the frequency of vibration found when a hand-held, handle-clamped and free racket were excited. It was found that the frequency of the free racket was very close to that of the hand-held racket so it was concluded that this was the best method of simulating a player's grip.

EXPERIMENTAL PROCEDURE

A Fischer Pro Classic 90 tennis racket was strung to 290N (65lbs.) tension. Tennis balls were propelled normal to the geometric string centre of the racket, at velocities between 25 and 60m/s. Two different types of balls are listed below:

(1) Type 1 - Permanent Pressure Balls.
(2) Type 2 - Pressurised Balls.

For experiment 1, the racket was rigidly supported around the head. The velocity of the impacting and rebounding ball was determined using a high speed video operating at 400 frames per second. The coefficient of restitution, *e*, was calculated using,

$$e = -\frac{V_b'}{V_b} \qquad (1)$$

where V_b and V_b' are the pre- and post-impact ball velocities respectively.

For experiment 2, the racket was freely suspended using small Velcro adhesive pads attached to the tip of the racket, with its longitudinal axis vertical. After impact, the racket recoiled with no external forces acting on it. The ball and racket post-impact velocities were determined using high speed video equipment. Figure 1 shows a combined image (side-view) of the stationary and recoiling racket, and three markers which were placed on the frame of the racket. The central marker identified the geometrical string centre of the racket. The top and bottom markers were sampled to determine the racket linear and angular post-impact velocity. The racket rotated around its longitudinal axis so two points on the far edge of the frame were also sampled.

A repeatability study was conducted to identify the accuracy of the manual image analysis method. A randomly selected impact from each experiment was analysed fifteen times and the standard deviation values for the ball and racket velocities was found.

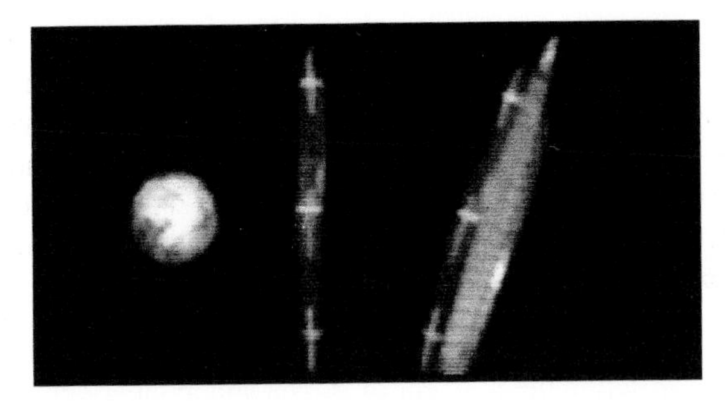

Fig. 1 Combined high speed video image of stationary and recoiling racket.

THEORETICAL PROCEDURE

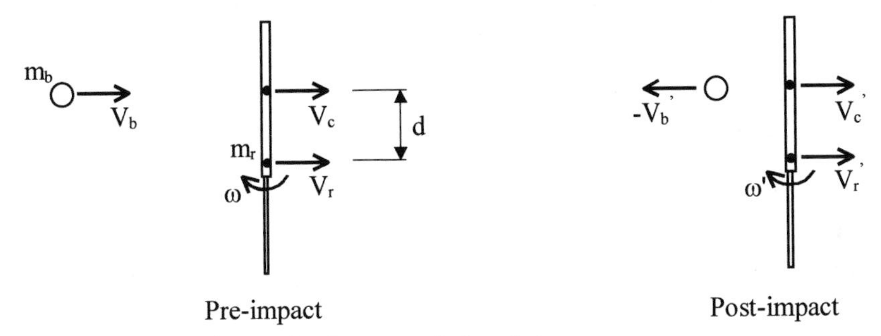

Fig. 2 Ball impacting on freely suspended racket.

Figure 2 shows a ball impacting on a freely suspended racket (as in experiment 2) and defines the notation used in this analysis. V_c is the pre-impact velocity of the contact point on the racket at a length, *d*, from the racket centre of mass. For experiment 2, the pre-impact linear and angular velocity of the racket (V_r and ω respectively) were zero. However, they were included in the analysis to obtain a complete solution for any initial conditions. The separation coefficient of restitution, e_s, is defined as,

$$e_s = \frac{V_c' - V_b'}{V_b - V_c} \tag{2}$$

The velocity of the rebounding ball and racket are given by the following equations,

$$V_b' = \frac{V_b\left(m_b d^2 - I_r\left(e_s - \frac{m_b}{m_r}\right)\right) + (V_r + \omega d)(I_r(1+e_s))}{\left(m_b d^2 + \left(I_r\left(\frac{m_b}{m_r} + 1\right)\right)\right)} \tag{3}$$

$$V_c' = e_s(V_b - V_c) + V_b' \tag{4}$$

$$V_r' = \left(\frac{m_b}{m_r}\left(V_b - V_b'\right)\right) + V_r \tag{5}$$

$$\omega' = \frac{V_c' - V_r'}{d} \tag{6}$$

Equations (3) to (6) are the same as those derived by Brody (1997), using the laws of conservation of linear and angular momentum, assuming that the racket frame was a rigid body. I_r and ω define the moment of inertia and angular velocity of the racket around its centre of mass respectively.

In experiment 1 the coefficient of restitution, e, was determined for impacts on a head-clamped racket. The definition of e is given in eq. (1) which is simply the reduced form of eq. (2), with $V_c = V_c' = 0$.

To predict the values of V_b', V_c', V_r' and ω for experiment 2, the value of e (from experiment 1) was substituted for e_s into eqs. (3)-(6). The measured physical properties of the racket and ball were also required. For this analysis it was assumed that $e = e_s$, and therefore e (or e_s) is not dependant on the method of gripping the racket and is only a function of the relative ball-racket impact velocity.

RESULTS

EXPERIMENT 1

The coefficient of restitution (e) is shown in Fig. 3, for normal impacts on a head-clamped racket. A fourth-order polynomial trendline was plotted to determine a mathematical relationship between e and the ball impact velocity, V_b. The standard deviation of the value of e was 0.004, as determined from the repeatability study.

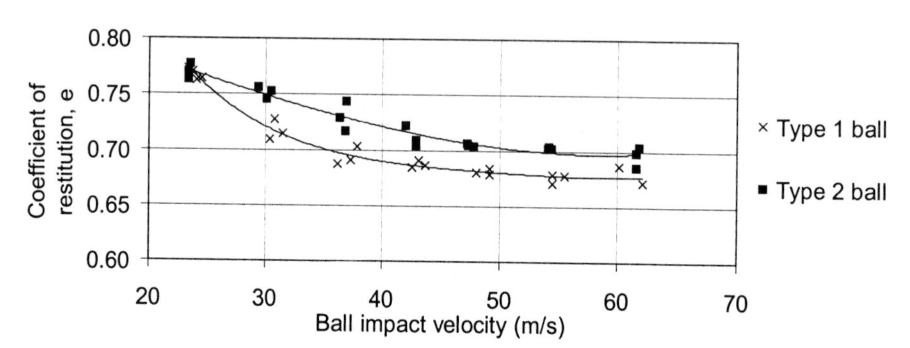

Fig. 3 Coefficient of restitution vs. ball velocity for normal impacts on a head-clamped racket.

EXPERIMENT 2

The experimental and theoretical ball rebound velocity, V_b', is shown in Fig. 4 for normal impacts on a freely suspended racket. The theoretical value of V_b' was determined by substituting e (from the trendline in Fig. 3) for e_s into eq. (3) for all values of V_b. The measured physical parameters were also required (mass of ball, m_b = 0.057kg, mass of racket, m_r = 0.343kg and moment of inertia of racket, I_r = 0.0155 kgm^2). A nominal value of d = 0.2m was used in eq. (3), corresponding to the length between the centre of mass and geometric string centre. All the balls impacted within 10mm of this point. The standard deviation of the value of V_b' was 0.2m/s, as determined from the repeatability study.

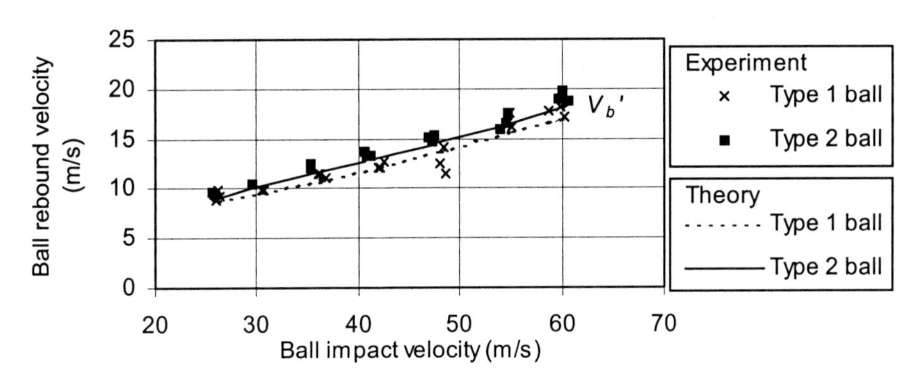

Fig. 4 Comparison of ball pre- and post-impact velocities for impacts on a freely suspended racket.

Figure 5 shows the linear post-impact velocity of the centre of mass (V_r') and contact point of the racket (V_c'). The standard deviation of the values of V_r' and V_c' were 0.3m/s and 0.2m/s respectively. Figure 6 shows the angular post-impact velocity of the racket (ω'). The standard deviation of the value of ω' was 1.5 rad/s.

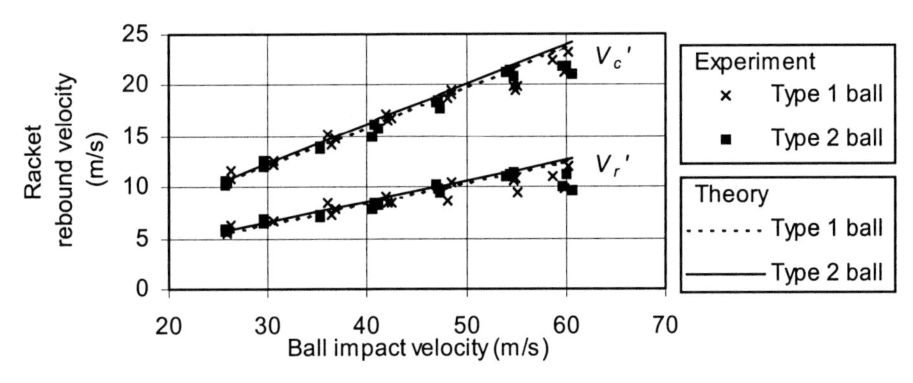

Fig. 5 Post-impact velocity of racket centre of mass and contact point (V_r' and V_c' respectively) for normal impacts on a freely suspended racket.

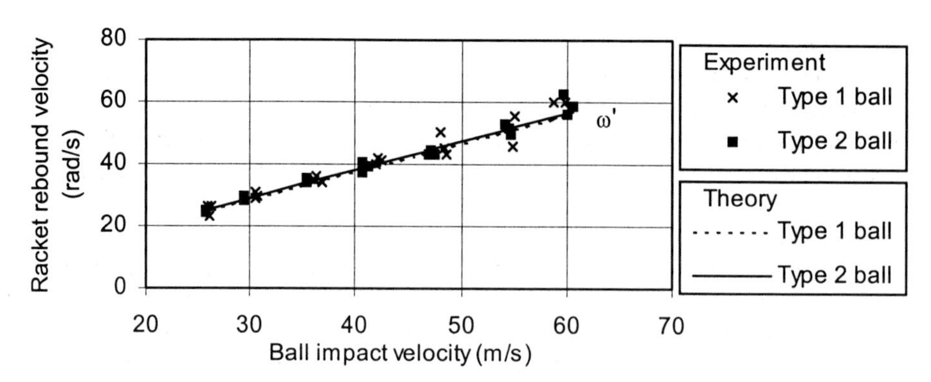

Fig. 6 Angular post-impact velocity of racket, ω', for normal impacts on a freely suspended racket.

DISCUSSION

Figure 3 shows the value of the coefficient of restitution, e, at ball velocities between 25 and 60m/s, for normal impacts on a head-clamped racket. These results, along with the measured physical parameters, were used to predict the post-impact velocities of the ball (V_b') and racket (V_r', V_c' *and* ω') for impacts on a freely suspended racket. These theoretical results were compared with the experimental data in Figs. 4 to 6, and a summary of the differences is given below (the standard deviation (SD) values are shown in brackets):

1. Ball rebound velocity, V_b' - experimental values were approximately 5% higher than the predicted theoretical values. (SD=1.4%).
2. Racket centre of mass and impact point velocity, V_r' and V_c' - experimental and theoretical values were similar at low ball impact velocities, (V_b<50m/s). However, at high ball impact velocities the experimental values were typically 10% lower than the theoretical values. (SD= 3%).
3. Racket angular velocity, ω' - experimental and theoretical values were very similar for all ball impact velocities. (SD=3.8%)

The experimental and theoretical values of ball and racket rebound velocities are generally very similar (within 5%), with maximum differences of approximately 10%. The inaccuracies of the analysis method for determining experimental results, as quantified by the standard deviation values, may partly account for these differences.

The theoretical data in Figs. 4-6 was determined assuming that the ball impacted at the geometric string centre, giving a nominal value of d=0.2m. However, the impact position generally varied by ±10mm. Using eq. (3) it was determined that an increase in d of 10mm, resulted in a decrease in V_b' of 1.5m/s (≈10%). This clearly shows that the ball post-impact velocity is highly dependent on the impact position, so this point should be defined precisely. It also shows, therefore, that it was invalid to assume a nominal value of d. To correct for this, the theoretical values were recalculated using the actual values of d for each impact. These are not shown as the results merely illustrated the same pattern as in Figs. 4-6.

The theoretical values in experiment 2 are dependent on the experimental values of e from experiment 1. If the ball or racket properties changed between experiment 1 and 2 then an incorrect value of e would be put into eq. (3). It was determined that an increase in e of 0.01 resulted in an increase in V_b' of 2.5%.

One difference between the two experiments was that, after impact, the racket did not vibrate in experiment 1 (providing that the racket frame clamp was infinitely stiff), whereas in experiment 2 it was likely to do so. The ball impacted nominally at the geometric string centre of the racket. Cross (1997) and Brody (1995), amongst others, determined that this was generally very close to the nodal point of the fundamental transverse vibration mode for a freely suspended racket. Therefore, impacts on a free racket at this point resulted in minimal energy loss due to vibration and impacts on a head-clamped racket result in no frame vibrations. This may explain why the model predicts the desired results within a reasonable accuracy. Also, any vibration effect is likely to be minimal anyway as Cross (1999) stated that the energy losses due to vibration were only a small fraction of the total energy losses.

Hatze (1994) measured the two-dimensional impact probability density distribution, for a selection of rackets. It was found that the centre of this distribution corresponded to the vibration node for a freely suspended racket. Therefore, although the impact point chosen in these experiments was a special case (as it gave minimal frame vibrations), it closely corresponded to the most common impact point. The model should be further developed so that it could be used for the entire range of impact positions.

Another difference between the two experiments was that, during impact, the free racket could recoil but the head-clamped racket could not, so the maximum deflection of the ball may be different in the two cases. Energy losses in the ball are proportional to the magnitude of deformation, therefore different values of coefficient of restitution (e or e_s) may be obtained for the two cases. Baker and Putnam (1979) concluded that the movement of the contact point on the racket was the same for both handle clamped and freely suspended rackets. In future work, therefore, it may be necessary to perform impact tests on a handle clamped instead of a head clamped racket.

This model is the first stage in modelling the impact between a ball and racket for a groundstroke or serve, using simple empirical data as the input parameters to the model. Brody (1997) used a simple change of reference method to relate laboratory results (racket is initially stationery and the ball is moving) with actual tennis strokes (ball and racket are both moving). The same method can be used for this model as the pre-impact racket velocity (V_r) in eq. (3) could be non-zero. This model, however, is an advancement of Brody(1997) as it allows the post-impact velocity of the racket as well as that of the ball to be found. It assumes that the coefficient of restitution (e or e_s) is only a function of the relative ball-racket pre-impact velocity, i.e. the value of e for an impact where V_c=20m/s and V_b=15m/s, is the same as that for V_c=0m/s and V_b=35m/s. This change of reference has not yet been proven but has been used by many researchers and illustrates the potential use of this model.

CONCLUSIONS

The theoretical values of ball and racket post-impact velocities were compared with experimental results. The theoretical ball velocity was generally higher than the experimental value, but the theoretical value was higher for the racket velocity. The majority of experimental and theoretical results correlated to within 5%.

It was concluded that the ball post-impact velocity was highly dependent on the impact position. Equation (3) showed that a difference of 10mm in position gave a difference of 10% in post-impact ball velocity. Therefore it was concluded that the precise impact position should be determined for the experimental impacts.

The impacts were conducted on the geometric string centre which, from previous work, has been shown to be close to the most common impact position. This point also closely corresponded to a node of vibration for a freely suspended racket. Therefore the energy losses due to vibration were minimal at this point for both the free and head-clamped racket. This may explain the high correlation of the theoretical and model results.

It is shown that the model could be used to predict the ball/racket velocities after an impact between a ball and racket (e.g. a groundstroke) by putting simple empirical data into the formulae. This change of reference technique has been used by many other researchers.

ACKNOWLEDGEMENTS

The authors would like to thank the International Tennis Federation and University of Sheffield for the funding of this work.

REFERENCES

Baker J. & Putnam C. (1979), Tennis racket and ball responses during impact under clamped and free standing conditions. *Research Quarterly*, **50**, 164-170.

Brody H. (1987), Models of Tennis Racket Impacts. *Int. J. of Sport Biomechanics,* **3**, 293-296.

Brody H. (1995), How Would a Physicist Design a Tennis Racket. *Physics Today*, 26-31.

Brody H. (1997), The physics of tennis III. The ball-racket interaction. *Amer. J. Physics*, **65(10)**, 981-986.

Cross R. (1997), The dead spot of a tennis racket. *Amer. J. of Physics*, **65(8),** 754-764.

Cross R. (2000), Optimising the Performance of a Tennis Racket. *In press.*

Hatze H. (1994), Impact Probability Distribution, Sweet Spot, and the Concept of an Effective Power Region in Tennis Rackets. *J. of App. Biomechanics*, **10**, 43-50.

Liu Y. (1983), Mechanical analysis of racquet and ball during impact. *Medicine and Science in Sports and Exercise*, **15(5)**, 388-392.

The influence of racket moment of inertia during the tennis serve: 3-dimensional analysis

S.R. Mitchell, R. Jones, J. Kotze
School of Mechanical & Manufacturing Engineering
Loughborough University

ABSTRACT: This paper presents the initial findings of research funded by the International Tennis Federation to investigate the degree to which racket characteristics magnify a players ability to generate ball speed in the tennis serve. Racket motion during impact with the ball was studied using an active marker tracking system for 6 players. Several trial rackets were employed, including each player's own racket, with mass and moment of inertia (MOI) characteristics adjusted to represent the range of 'playable' characteristics from the lightest commercial racket to an earlier wooden design. As predicted the results indicate that racket head speed increases with reductions in mass and MOI. Further, when normalised with respect to the individual player's performance with their own racket, the results suggest a consistent relationship between the head speed generated and the racket's MOI for all the players tested. The results also indicate that the instantaneous centre of rotation for the racket at impact is surprisingly consistent for most of the subjects despite changes in the racket characteristics. The study is based on a limited population and a wider study is required to confirm these findings.

INTRODUCTION

In 1990 service speeds in the professional game of tennis were recorded for the first time. Only 5 players produced ball speeds in excess of 200 kph but by 1995 this number had risen to 38. As recently as 1998 Greg Rusedski broke the world record with a 240kph serve and it is widely believed that the top 200 players on the ATP tour are now capable of reaching or exceeding 200 kph [Coe (1999)].

This trend is arguably responsible for the results of a survey of over 3,500 tennis playing households in over 13 countries reported by Coe (1999). Of the tennis players interviewed 49% believed that the men's professional game was too fast and only 32% disagreed. The imbalance is even more pronounced in countries where faster court surfaces predominate.

Although the player's skill and athleticism and the ball and court properties affect the speed at which the game is played several researchers have noted the racket's importance, including its mass properties, in generating ball speed [e.g. Brody (1979,

1987, 1997), Elliot (1982), Liu (1983), Hatze (1993), Cross (1997)]. A heavier racket with a higher moment of inertia (MOI) is harder to swing and should result in lower head speeds or greater effort to achieve the same speed. Conversely a lighter racket should result in higher speeds for less effort. Some would argue that modern rackets enable the game to be played at such elevated speeds that even the fastest player reaction times are too slow to produce a good return of serve, particularly on the faster surfaces [Groppel (1986), Arthur (1992), Brody (1995)].

One indicator of a trend towards service dominance would be an increase in service games held causing an increase in sets played to a tie-break. Coe (1999) noted just such an increase in tie-break sets in a study of recent men's Wimbledon championships and a correlation between those sets and players executing faster serves. Service dominance reduces spectator enjoyment, so there is little wonder then that the 1994 Sampras-Ivanesevic Wimbledon final, where only 3 of the 206 points lasted more than 4 shots, was described as "... one of the most boring finals in history" by Fred Perry.

In response to this trend the International Tennis Federation has instigated research into the racket's contribution to the power serve. Several researchers have reported two and three-dimensional studies of tennis serve biomechanics based on cinematographic data [Johnson (1957), Plagenhoef (1970), Johnson (1976), Elliott (1983), Elliott & Wood (1983), Van Gheluwe & Hebbelinck (1985), Elliott *et al* (1986), Van Gheluwe *et al* (1987), Bahamonde (1989), Sprigings *et al* (1994), Elliott *et al* (1995)]. Some also recorded electromyography (EMG) data [Anderson (1979), Miyashita *et al* (1980), Van Gheluwe & Hebbelinck (1986), Buckley & Kerwin (1988)] and in 1994 Cohen et al (1994) also published their study of the correlation between upper extremity strength and serve velocity.

Groppel (1986) and Elliott (1988) reviewed the tennis serve related literature in the late 80's. Even a review of more recently published work reveals that a study of the correlation between head speed and racket mass properties has not been published, although this information is fundamental to an understanding of the racket's effect on player performance. This paper presents the initial findings of an investigation into how racket inertia properties affect head speed generation. One of the uses to which this information might be put is to develop a fair and repeatable comparative test of racket performance under service conditions using a realistic simulation mechanism.

METHOD

Six county standard (and above) players from the Loughborough University tennis team acted as subjects for trials using several rackets with different mass properties, including their own. After a warm-up and familiarisation period with each racket, each player was required to hit 6 'good' first service shots. Only data from shots that the players considered close to their peak performance was recorded.

Three test rackets were used:

(1) A Head Ti.S6 racket chosen because it was the lightest racket available at the time.

(2) A modified Dunlop 200G weighted to a notional acceptable maximum for a modern racket.
(3) A modified Ti.S6 racket weighted to simulate the mass and inertia properties of a wooden racket, but having similar frame stiffness and stringing to racket (1).

Rackets (1) and (3) were chosen to represent the two extremities of the mass/inertia spectrum. The player's own racket was also used to provide a baseline measure of the player's likely performance when they were more familiar with a racket's weight and balance. Table 1 lists the properties of all the rackets used with instrumentation attached and measured using a Babolat Racket Diagnostics Centre.

Table 1 Test racket properties

Racket	Length [m]	Mass [kg]	Balance point [m from heel]	Moment of Inertia [$kg.m^2$ 100mm from heel]
Head Ti.S6	0.705	0.291	0.374	0.0331
Dunlop 200G	0.684	0.392	0.302	0.0340
Head Ti.S6 (weighted)	0.705	0.426	0.338	0.0371
Subject A's	0.677	0.398	0.315	0.0351
Subject B's	0.683	0.409	0.310	0.0343
Subject C's	0.688	0.423	0.298	0.0343
Subject D's	0.685	0.407	0.310	0.0339
Subject E's	0.683	0.388	0.315	0.0337
Subject F's	0.682	0.401	0.313	0.0344

A cartesian optoelectronic dynamic anthropometer (CODA) real time 3D-motion capture system, manufactured by Charnwood Dynamics Ltd., was used to measure the position of 4 infrared LED active markers attached to the racket during the shot. The CODA system was used to burst sample the markers, with a 172 μs interval between markers, at a sampling rate between bursts of 400 Hz. The position of each marker was recorded at a resolution of ±0.1 mm in the horizontal and vertical directions perpendicular to the sensor array viewing direction, and a depth measurement resolution of ±0.6 mm. The technique compares favourably with previous cinematographic studies typically employing 200 fps frame rates and ±4.0 mm resolution [e.g. Sprigings *et al* (1994), Elliott *et al* (1995)].

Figure 1 shows the instrumentation attached to the racket. Two receiver/power modules were attached midway between the head and the grip each wired to 2 markers as shown. Typically the instrumentation increased racket mass by 0.047 kg, MOI by 0.0024 $kg.m^2$ and moved the centre of gravity 12 mm towards the throat. So that all 4 markers remained in view at impact two CODA mpx30, scanner units were positioned facing each other at either end of the service baseline. The players were required to execute their serves aiming just to the left of the centreline from a normal service position.

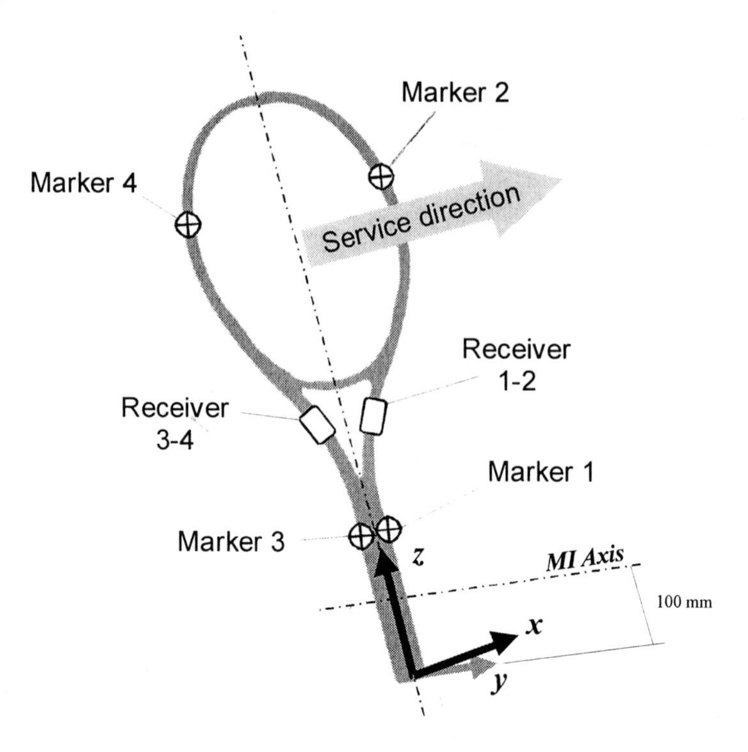

Fig. 1 Racket instrumentation layout and local coordinate system

RESULTS

The CODA system's global coordinate frame was aligned with the x axis down the centreline, the y axis along the baseline and the z axis vertical. A coordinate system local to the racket, shown in figure 1, was also used to analyse the data. Figure 2 shows a sample graph of horizontal (global xy component) velocity versus time. Results for several points on the racket are shown: markers 2 and 4, the racket butt and a notional impact point 0.5 m from the butt chosen to allow a fair comparison of different rackets and each player's achievements with them. The angular velocity of the racket is also shown.

Figure 3 shows the average position of the instantaneous centre of rotation (ICR), in the local xz plane, achieved by each player for each of the three test rackets.

Finally figures 4(a), 5(a) and 6(a) show absolute results at impact for the three test rackets and all six players, averaged from the 6 shots for each racket, plotted against racket MOI. Figures 4(b), 5(b) and 6(b) show similar graphs normalised with respect to the player's performance with their own racket. Figures 4(a) and (b) show the horizontal (global xy component) velocity of the impact point. Figures 5(a) and (b) show the local racket face normal (local x component) impact point velocity. Figures 6(a) and (b) show the rackets angular velocity in the local xz plane.

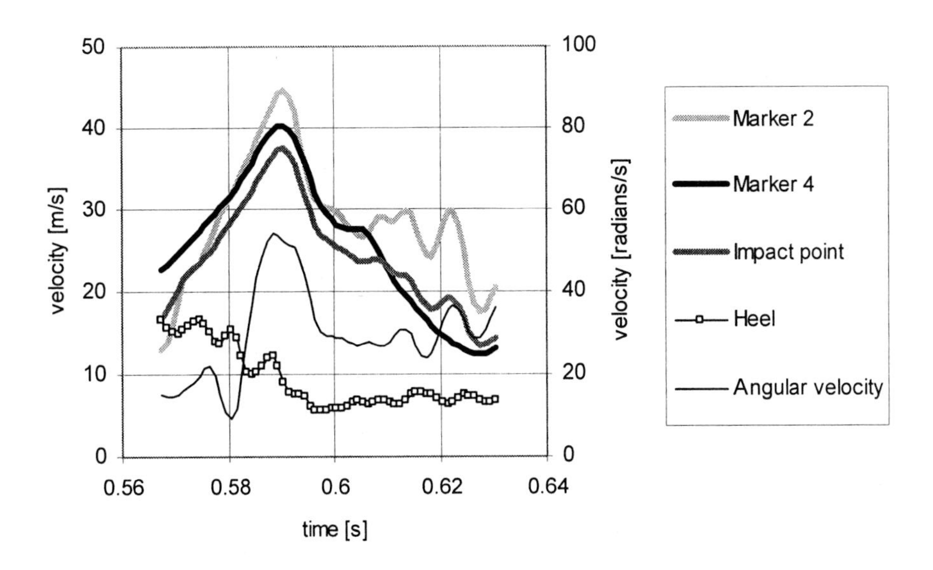

Fig. 2 Example racket point horizontal velocities

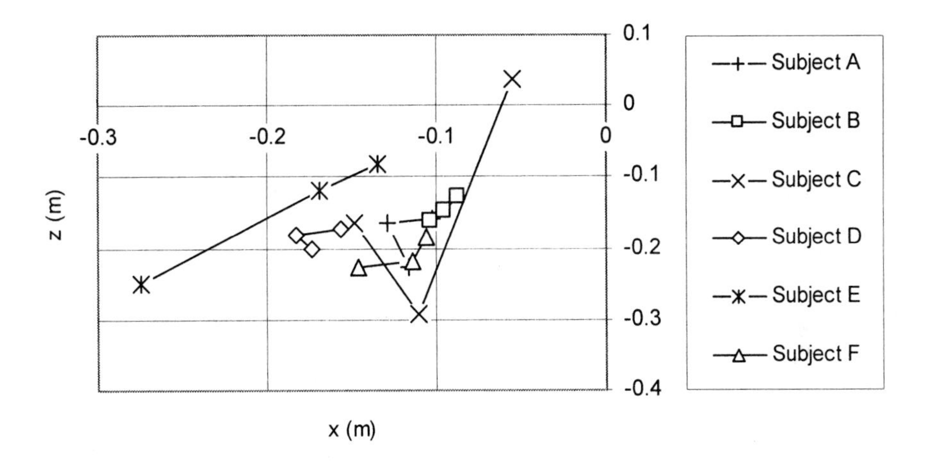

Fig. 3 Local *xz* plane instantaneous centre of rotation

The standard deviation for the impact point velocities with each racket was typically ~3% but could be as large as ~10% for the racket's angular velocity. Given that the player's hand is moving relatively slowly at impact so that the greater racket head speed is due to the racket's angular velocity, the impact point velocity achieved might have been expected to be more variable than the angular velocity causing it. The actual result can be attributed to the players executing different service actions during the trials, so that a high racket rotational speed was coupled with a lower hand speed and vice versa.

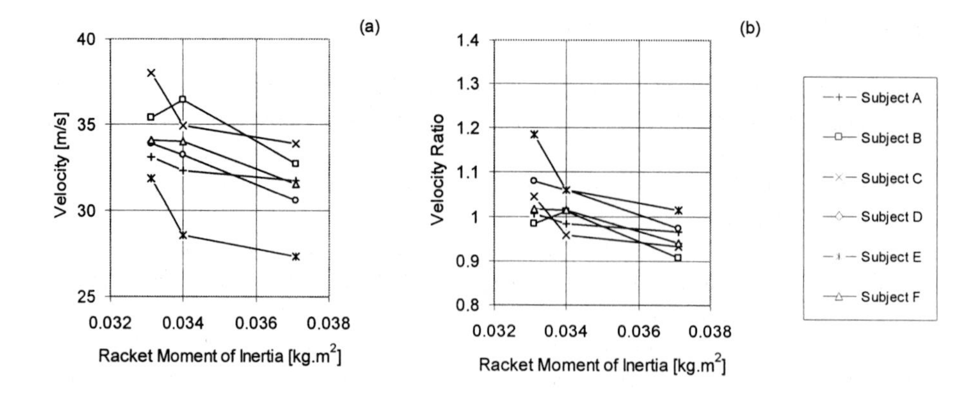

Fig. 4 Impact point horizontal velocity vs. moment of inertia

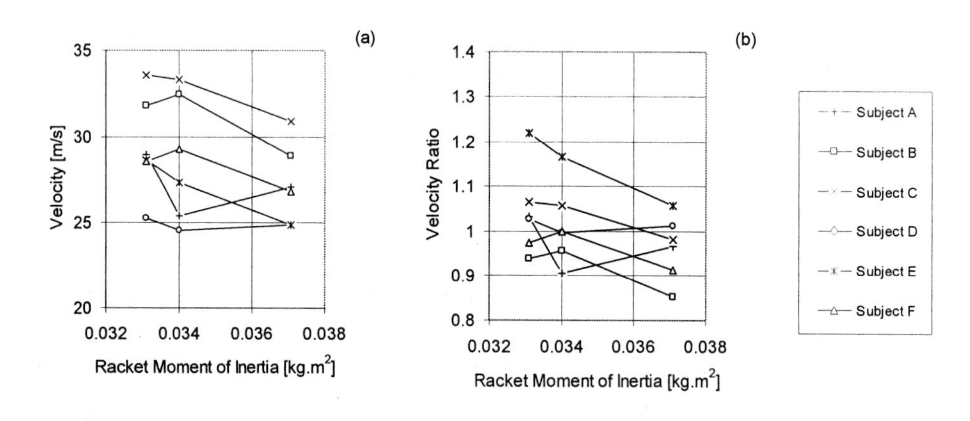

Fig. 5 Impact point local face normal velocity vs. moment of inertia

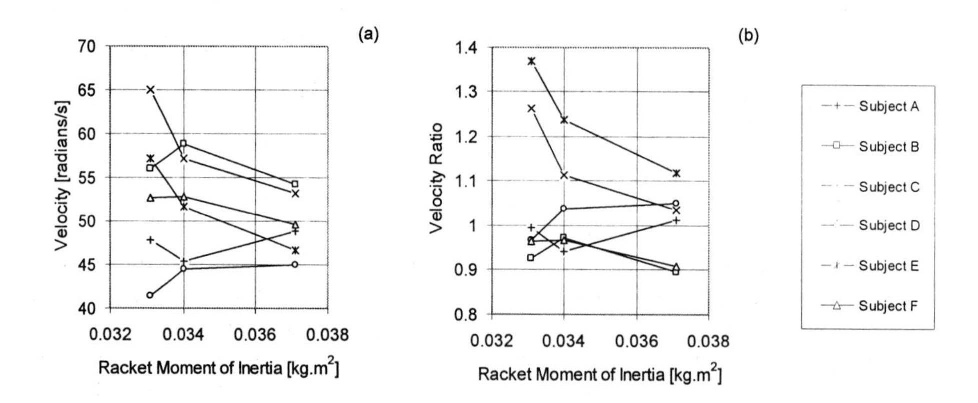

Fig. 6 Racket local xz plane angular velocity vs. moment of inertia

DISCUSSION & CONCLUSIONS

This initial study indicates that a decrease in racket MOI can significantly increase head speed during the tennis serve. How large and controllable this increase may be needs to be investigated further, but it is likely that an assessment of racket performance made under constant impact velocity conditions would be biased in favour of lighter rackets.

Most of the players tested achieved faster head speeds with lighter rackets and exhibited ~10±5% drop in impact velocity for a 12% increase in racket MOI over the range tested. Figure 4(b) shows remarkable consistency between the normalised results for the 6 players but the sample size is, as yet, too small to draw any firm conclusions. Interestingly, the players with the slowest and fastest service actions, subjects E and C, were the most erratic. It may be that the other players were more conservative with the test rackets or that these players had the least skill/control or played with the most abandon. Excluding the data for subjects E and C reveals an impact point velocity change of ~8±2.5% across the group for a 12% MOI increase. The variation in angular velocity change is similarly decreased from ~16±12% to ~9±1%.

The MOI effect on the racket's angular velocity is noticeably more erratic than for the linear impact point speed, indicating that the relationship between MOI, racket angular velocity and head speed is complex. This greater variation may be due to the player's lack of confidence with the test rackets, or the changes in inertia inhibiting the player's timing.

Instead of achieving a maximum head speed with the lightest racket used, Subjects B and F achieved it with the racket most closely matched to their own. It is tempting to conclude that this is an optimum for the player but it may just as easily be due to greater familiarity. Since the conflicting performance of other subjects shows that this is not always so a larger, more carefully designed study will be necessary to evaluate this phenomena further. A more extensive range of rackets and players will be needed to investigate and confirm (or deny) the racket inertia/service speed relationships suggested by this initial study, and the inclusion of elite tournament players would help to establish the upper limit of player performance.

Despite the changes in mass properties and head speeds attained, figure 3 indicates that 4 of the 6 players were able to achieve remarkably consistent ICR positions at impact with the trial rackets. Subjects E and C again exhibited the least consistency. As a whole the results suggest that a service simulation mechanism for racket testing and comparison might only need modest adjustment of the ICR positions at impact, to simulate different players, for a realistic study. However, the mechanism would have to incorporate a means to vary impact in relation to each racket's inertia properties.

ACKNOWLEDGEMENTS

The authors would like to thank the International Tennis Federation for funding this work and encouraging its publication. Our thanks also goes to the players that took part in this study and the staff at the Dan Maskell Tennis Centre for their cooperation.

REFERENCES

Anderson, M.B. (1979) Comparison of Muscle Patterning in the Overarm Throw and Tennis Serve. *Research Quarterly*, **50**(4), 541-553.

Arthur, C. (1992) Anyone for slower tennis. *New Scientist*, 2 May: pp. 24-28.

Bahamonde, R.E. (1989) Kinetic Analysis of the Serving Arm During the Performance of the Tennis Serve. Abstracts of the International Society of Biomechanics XII Congress 1989, in: *Journal of Biomechanics*, **22**(10), pp. 983.

Brody, H. (1979) *Physics of the tennis racquet.* American Journal of Physics, **47**(6), 482-487.

Brody, H. (1987) *Tennis science for tennis players.* University of Pennsylvania Press, Pennsylvania.

Brody, H. (1997) The physics of tennis III: The ball-racket interaction. *American Journal of Physics*, **65**(10), 981-987.

Brody, H. (1995) How Would a Physicist Design a Tennis Racket. *Physics Today*, March: pp 26-31.

Buckley, J.P. & Kerwin, D.G. (1988) The role of the Biceps and Triceps Brachii During Tennis Serving. *Ergonomics*, **31**(11), 1621-1629.

Coe, A. (1999) *The balance Between Technology and Tradition in Tennis*, Discussion paper for ITF Board of Directors, ITF: London. pp. 1-42.

Cohen, D.B., Mont, M.A., Campbell, K.R., Vogelstein, B.N. & Loewy, J.W. (1994) Upper Extremity Physical Factors Affecting Tennis Serve Velocity. *American Journal of Sports Medicine*, **22**(6), 746-750.

Cross, R. (1997) The dead spot of a tennis racket. *American Journal of Physics*, **65**(8), 754-764.

Elliott, B.C. & Wood, G.A. (1983) The Biomechanics of the Foot-up and Foot-back Tennis Serve Techniques. *Australian Journal of Sport Sciences*, **3**(2), 3-6.

Elliott, B.C. (1982) Tennis: The Influence of Grip Tightness on Reaction Impulse and Rebound Velocity. *Medicine and Science in Sports and Exercise,* **14**(5), 348-352.

Elliott, B.C. (1983) Spin and the Power Serve in Tennis. *Journal of Human Movement Studies*, **9**, 97-104.

Elliott, B.C. (1988) Biomechanics of the Serve in Tennis: A Biomedical Perspective. *Sports Medicine*, **6**, 285-294.

Elliott, B.C., Marsh, T. & Blanksby, B. (1986) A Three-Dimensional Cinematographic Analysis of the Tennis Serve. *International Journal of Sports Biomechanics*, **2**, 260-271.

Elliott, B.C., Marshall, R.N. & Noffal, G.J. (1995) Contributions of Upper Limb Segment Rotations During the Power Serve in Tennis. *Journal of Applied Biomechanics*, **11**, 433-442.

Groppel, J.L. (1986) The Biomechanics of Tennis: An Overview. *International Journal of Sports Biomechanics*, **2**, 141-155.

Hatze, H. (1993) The Relationship Between the Coefficient of Restitution and Energy Losses in Tennis Racquets. *Journal of Applied Biomechanics*, **9**, 124-142.

Johnson, J. (1957) Tennis Serve of Advanced Women Players. *Research Quarterly*, **28**(2) 123-131.

Johnson, M.L. (1976) High Velocity Serve. *Athletic Journal*, **56**, 44-46.

Liu, Y. (1983) Mechanical analysis of racquet and ball during impact. *Medicine and Science in Sports and Exercise*, **15**(5), 388-392.

Miyashita, M., Tsunoda, T., Sakurai, S., Nishizono, H. & Mizuno, T. (1980) Muscle Activities in the Tennis Serve and Overhand Throwing. *Scandinavian Journal of Sports Science*, **2**(2), 52-58.

Plagenhoef, S. (1970) *Fundamentals of Tennis*. Prentice-Hall, Englewood Cliffs, New Jersey.

Sprigings, E., Marshall, R, Elliott, B. & Jennings, L. (1994) A Three-Dimensional Kinematic Method for Determining the Effectiveness of Arm Segment Rotations in Producing Racquet-Head Speed. *Journal of Biomechanics*, **27**(3), 245-254.

Van Gheluwe, B. & Hebbelinck, M. (1985) The Kinematics of the Service Movement in Tennis: A Three-Dimensional Cinematographical Approach. In: *Biomechanics IX-B* (Eds. D. Winter, R. Norman, R. Wells, K. Hayes & A. Patla), pp. 521-526. Human Kinetics Publications, Champaign, Illinois.

Van Gheluwe, B. & Hebbelinck, M. (1986) Muscle Actions and Ground Reaction Forces in Tennis. *International Journal of Sports Biomechanics*, **2**, 88-99.

Van Gheluwe, B., De Ruysscher, I. & Craenhals, J. (1987) Pronation and Endorotation of the Racket Arm in a Tennis Serve. In: *International Series on Biomechanics* (Ed. B. Jonsson), **6B** (Biomechanics X-B) pp. 667-672, Human Kinetics Publications, Champaign, Illinois.

Dynamics of the collision between a tennis ball and a tennis racket

R. Cross
Physics Department, University of Sydney, Sydney, Australia

ABSTRACT: The collision of a ball with a tennis racket is usually modelled in terms of rigid body dynamics, assuming that the hand exerts no impulsive reaction force on the handle during the collision. An improved model is described where the racket is regarded as a flexible beam. It is shown that the rebound speed of the ball is essentially independent of the hand force and the string tension but increases with frame stiffness.

INTRODUCTION

The collision between a tennis racket and a ball can be modelled by assuming that the racket is perfectly rigid and that the handle is not subject to any impulsive force during the collision (Brody, 1985; 1997). The racket and ball speeds after the collision can then be calculated from the conservation equations, in terms of the corresponding speeds before the collision, and an assumed or measured coefficient of restitution. However, the model provides no information on the dynamics during the collision, nor does it provide information on the energy coupled to racket vibrations. In this paper, these shortcoming are avoided by modelling the racket as a one-dimensional, flexible beam. The model was tested experimentally using aluminium beams so that the beam length could easily be varied. It is shown that the hand force can be neglected, since the ball rebounds from the racket before the reflected pulse from the hand arrives back at the impact point. It is also shown that racket power increases when the string tension decreases, but the magnitude of this effect is essentially negligible.

BEAM THEORY

The equation of motion for a beam subject to an external force, F_o per unit length, has the form (Graff, 1975; Cross, 1999a)

$$\rho A \frac{\partial^2 y}{\partial t^2} = F_o - \frac{\partial^2}{\partial x^2}\left(EI \frac{\partial^2 y}{\partial x^2} \right) \qquad (1)$$

where ρ is the density of the beam, A is its cross-sectional area, E is Young's modulus, I is the area moment of inertia, and y is the transverse displacement of the beam at coordinate x along the beam. For a uniform beam of mass M and length L, numerical solutions of Eq. (1) can be obtained by dividing the beam into N equal segments each of mass $m = M/N$ and separated in the x direction by a distance $s = L/N$, as shown in Fig. 1. An impacting ball may exert a force acting over several adjacent segments, depending on the ball diameter and the assumed number of segments. For simplicity it was assumed that the ball impacts on only one of the segments, exerting a time-dependent force, F. The equation of motion for that segment (the n'th segment) is obtained by multiplying all terms in Eq. (1) by s, in which case

$$m\frac{\partial^2 y_n}{\partial t^2} = F - (EIs)\frac{\partial^4 y_n}{\partial x^4} \tag{2}$$

assuming that the beam is uniform so that E and I are independent of x. The equation of motion for the other segments is given by Eq. (2) with $F = 0$. The boundary conditions at a freely supported end are given by $\partial^2 y/\partial x^2 = 0$ and $\partial^3 y/\partial x^3 = 0$. The boundary conditions at a rigidly clamped end are $y = 0$ and $\partial y/\partial x = 0$.

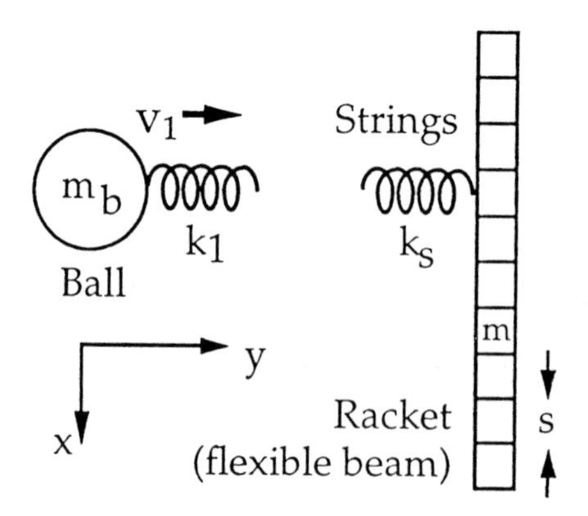

Fig. 1 Model used to describe the ball-racket collision.

The equation of motion of the ball is given by

$$d^2 y_b / dt^2 = -F / m_b \tag{3}$$

where m_b is the ball mass and y_b is the displacement of the centre of mass of the ball. The elastic properties of the ball can be modelled by assuming that $F = k_1 Y_b$ during the compression phase and $F = k_2 Y_b^p$ during the expansion phase, where Y_b is the compression of the ball and p is a parameter describing the effect of hysteresis in the ball. If k_1 and k_2 are constants and if Y_o is the maximum compression of the ball during any given impact, then $k_1 Y_o = k_2 Y_o^p$ so $k_2 = k_1 / Y_o^{p-1}$. The hysteresis loss in the ball is equal to the area enclosed by the F vs Y_b curve for a complete compression and expansion cycle. The parameter p can be chosen to give a total ball loss equal to the experimentally determined loss.

If a soft ball impacts on a hard beam, so that the compression of the beam is negligible, then $Y_b = y_b - y_n$. Alternatively, if the ball impacts on the strings of a racket, and if the string plane is displaced by a distance Y_s relative to the frame during the impact then F is given by $F = k_s Y_s$ where k_s is the spring constant of the strings. In this case, the compression of the ball plus the compression of the strings is given by $Y_T = y_b - y_n$. By equating the forces, is easy to show that $Y_b = k_s Y_T / (k_1 + k_s)$ during the compression phase, and $Y_b + k_2 Y_b^p / k_s = Y_T$ during the expansion phase.

It is assumed that the ball is incident at right angles to the string plane. It is also assumed that at $\tau = 0$, $y_b = 0$, $y = 0$ for all beam segments, the beam is initially at rest and that $dy_b / dt = v_1$. The subsequent motion of the ball and the beam was evaluated by numerical solution of Eqs. (2) and (3). These results were used to determine the rebound speed of the ball, v_2, and the apparent coefficient of restitution (ACOR), $e_A = v_2/v_1$. In normal play, a racket is swung towards the ball and is not normally at rest at the moment of impact. The resulting outgoing speed of the ball is easily determined by a simple coordinate transformation, as described below.

BEAM EXPERIMENTS

The experimental arrangement is shown in Fig. 2. A rectangular cross-section aluminum beam of width 32 mm and thickness 6 mm was supported horizontally, either by a 1.2 m vertical length of string attached to each end or by clamping one end to a rigid support. A 36 mm diameter, 42 gm superball was mounted, as a pendulum bob, at the apex of a V-shaped string support, so that it could impact the beam horizontally and at right angles to the beam. A small (5 mm x 15 mm) rectangular card was glued to the top of the ball. A He-Ne laser beam was directed parallel to the beam so that it could pass sequentially through two small holes in the card, 10 mm apart. The laser beam was detected using a photodiode. From this data, measurements were obtained of the ACOR as a function of impact location along the beam. The ball was incident at low speed, about 1 ms^{-1}.

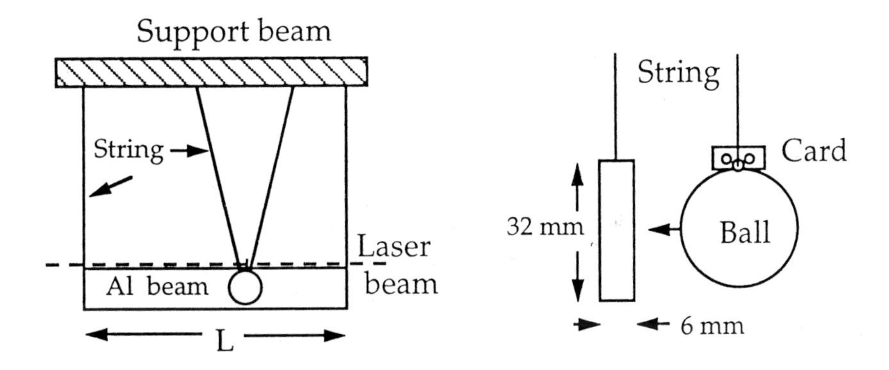

Fig. 2 Experimental arrangement.

Measurements of e_A, and the corresponding theoretical estimates of e_A, are presented in Fig. 3 for aluminum beams of length L = 60 cm or 110 cm. The spring constant of the ball was taken as $k_b = 2\times10^4$ Nm^{-1} to be consistent with the observed impact duration, about 4.2 ms. Apart from a few minor discrepancies, agreement between the theoretical and experimental values of e_A is remarkably good. The results show very clearly that, for a sufficiently long beam, (a) the impact of a ball near one end of a beam is not affected by the length of the beam or the method of support at the other end and (b) e_A for an impact anywhere along the central section of a beam is independent of the impact location and is not affected by the length of the beam or the method of support. e_A remains constant at $e_A = 0.45 \pm 0.02$ along the beam up to a point about 15 cm from each end. This result implies that the rebound speed is affected only if the impulse reflected off one end arrives back at the impact point within the 4.2 ms period of the impact.

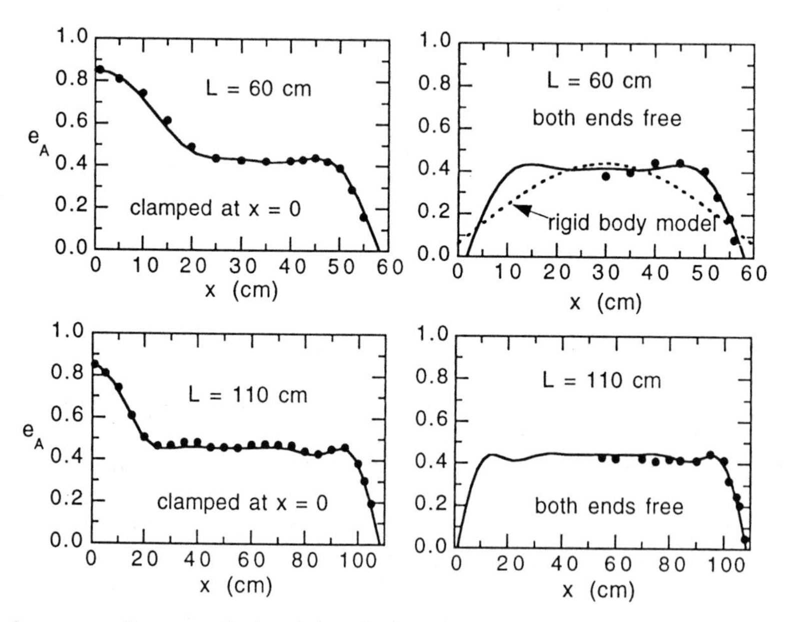

Fig. 3. e_A vs x for several aluminium beams. Experimental data is shown as black dots, and the solid curves are solutions of Eqs. (2) and (3).

The dynamics of the situation are illustrated in Fig. 4 which shows the theoretical beam deflection, for a freely supported beam, at equal time increments during and shortly after the impact. An impulse propagating towards a free end is reflected without phase reversal, so the beam moves further away from the ball, thereby reducing the rebound speed. A pulse propagating towards a clamped end is reflected with a phase reversal, sending the beam back towards the ball, thereby increasing the rebound speed. The reflected pulse has no effect on the ball if the ball rebounds before the reflected pulse reaches the ball. Fig. 4 shows that the beam deflection at the impact point, during the impact, is essentially the same for impacts at x = 55 cm or x = 90 cm, indicating that the pulse reflected at x = 110 cm does not have a significant influence on the impact. However, for an impact at x = 103 cm, the reflected pulse acts to deflect the beam away from the ball during the impact, thereby reducing the ACOR significantly.

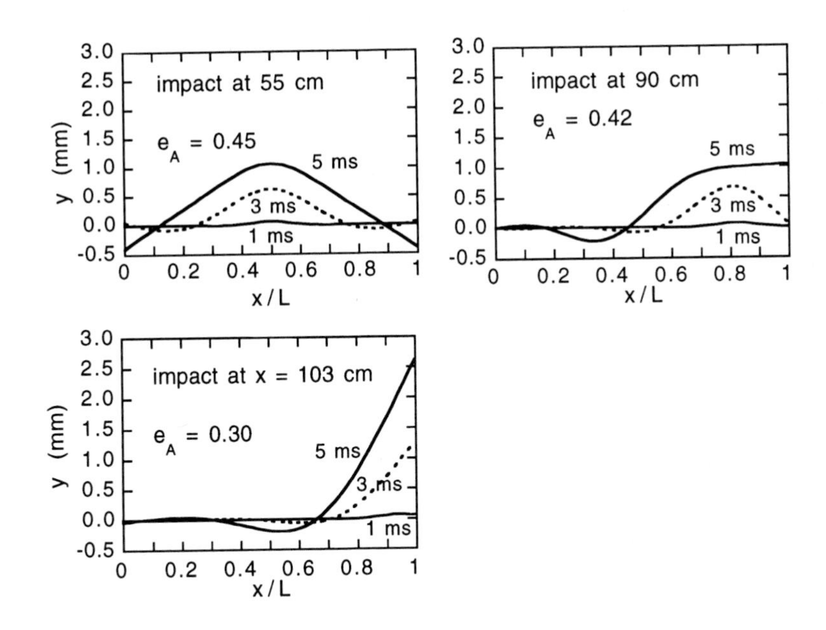

Fig. 4 y displacement of a freely suspended beam.

An attempt to fit the L = 60 cm free beam data using a rigid body approximation is shown in Fig. 3. The dashed curve represents solutions of the conservation equations assuming that the beam mass is 311 gm (ie its actual mass) and that 60% of the incident ball energy is dissipated in the ball and in beam vibrations during the collision. Similar poor fits can be obtained if one assumes that the beam mass is less than the actual mass. A rigid body solution can be forced to fit the data if the energy loss is allowed to vary with x, but the flexible beam solution is obviously superior. Nevertheless, a forced fit provides a valid measure of the fractional energy loss, consistent with the conservation equations. The fractional energy loss is about 0.6 in the middle of the 60 cm free beam and also at x = 3.8 cm and x = 56.2 cm, but it is different at other points along the beam.

CALCULATIONS FOR A TENNIS RACKET

Calculations are presented below for a graphite/epoxy composite tennis racket of length L = 71 cm and mass M = 340 gm, modelled as a uniform beam with EI=150 Nm, giving a fundamental vibration frequency of 125 Hz. The ball was modelled with m_b = 57 gm, $k_1 = 3 \times 10^4$ Nm^{-1} and p = 2.55, corresponding to an impact duration of 4.64 ms on concrete and a COR = 0.751 (Cross, 2000).

Figure 5 shows the variation of e_A along the long axis of the racket (passing through the handle) when the handle is (a) rigidly clamped along a 10 cm length at the end of the handle, or (b) pivoted about an axis through the end of the handle or (c) freely suspended. The results show that e_A is zero at a "dead" spot near the tip of the racket and increases to a broad maximum near the throat of the racket. This behaviour is easily demonstrated experimentally, at least in a qualitative sense, simply by dropping a ball at various spots on the strings and observing the bounce. e_A is independent of the handle support at distances up to 20 cm from the tip of the racket. e_A is affected near the throat of the racket as a result of reflections from a clamped handle, but this is of no consequence in that the hand does not act as a rigid clamp. In practice, the racket tends to pivot about an axis through the wrist (Cross, 1999b).

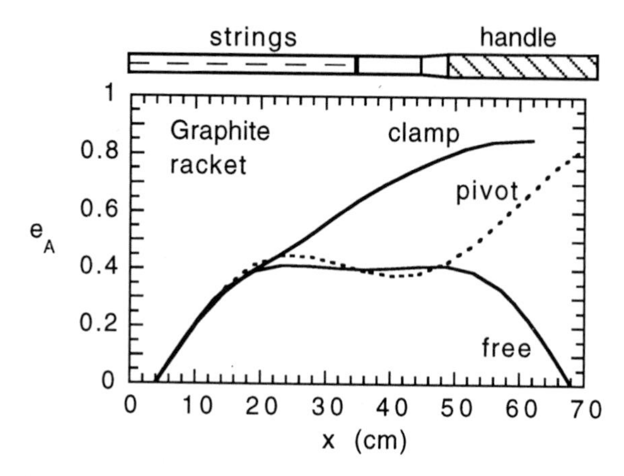

Fig. 5 e_A vs x for a tennis racket.

Figure 6a shows the variation of e_A, as well as the fractional energy loss in the ball (f_{ball}) and the fractional energy loss due to vibration of the racket (f_{vibr}), as a function of string plane stiffness for an impact 6 cm from the tip of a freely suspended racket. The energy fractions are fractions of the incident kinetic energy of the ball. The string plane stiffness is varied over three orders of magnitude in this diagram. An interesting result is that the vibration amplitude can be reduced essentially to zero if the impact duration exceeds about 10 ms. The amplitudes of the various vibration modes of a beam depend not only on the impact location but also on the impact duration. A short impact of duration τ contains a continuous spectrum of frequency components up to a frequency $f \sim 1/\tau$. The amplitude of the spectrum peaks at zero frequency and drops to zero near $f = 1/\tau$. For example, if the impact duration is 5 ms, the frequency spectrum extends to about 200 Hz, which is above the 125 Hz vibration frequency of the fundamental mode but well below the second mode at 345 Hz. Consequently, the second mode is not usually excited in a tennis racket. If the impact duration is increased by reducing the string tension, then the spectrum extends to a lower limit, and the amplitude of the induced 125 Hz vibration is reduced as shown in Fig. 6a.

The energy loss due to vibration is essentially zero for an impact at the fundamental vibration node near the middle of the strings, at $x = 16$ cm. The vibration loss and the ball loss are similar in magnitude at impact points removed from the node, and are both reduced for impacts on softer strings. As a result, e_A is larger at all impact points for soft strings than for stiff strings. The impact duration, τ, is also increased for an impact on the softer strings. The impact duration is smaller near the tip of the racket than near the throat since the tip accelerates away from the ball more rapidly than the throat.

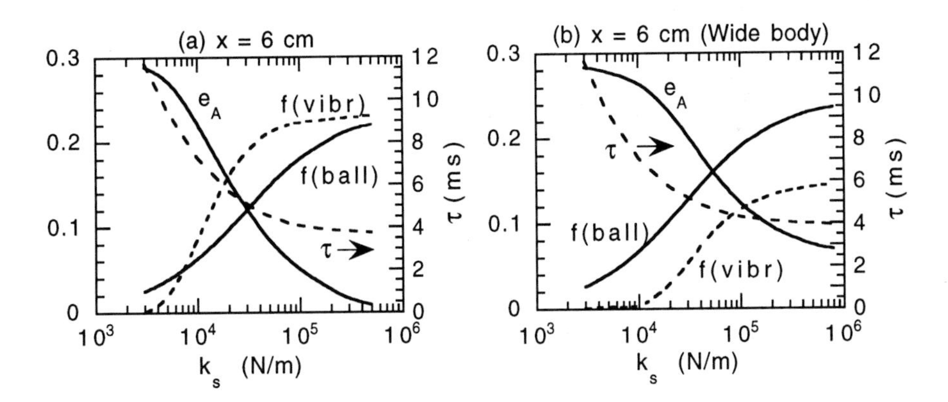

Fig. 6 e_A vs k_s for (a) normal and (b) wide-body rackets.

The change in e_A with k_s is relatively small considering the range in k_s normally used in a tennis racket. For example, for an impact at the centre of the strings, $e_A = 0.44$ when $k_s = 2 \times 10^4$ Nm^{-1} while $e_A = 0.41$ when $k_s = 4 \times 10^4$ Nm^{-1}. This corresponds to a 7% change in the ball speed when it bounces off a racket that is initially at rest. Since the ball rebounds at speed $v_2 = e_A v_1$, then in a reference frame where the ball is initially at rest, the racket will be incident at speed v_1 and the ball will come off the strings at speed $v = v_1 + v_2 = (1 + e_A)v_1$. In this case, corresponding to a serve or overhead smash, the above increase in e_A results in only a 2% increase in the ball speed. Given that k_s is proportional to the string tension, then a reduction in string tension from say 60 to 50 lb will increase the ball speed by only 0.7%. Similarly, in a reference frame where the racket speed is equal to the incident ball speed, the ball will come off the strings at speed $v = (1 + 2\,e_A)v_1$. This is typical of a forehand or backhand, in which case a factor of two decrease in k_s leads to a 3.3% increase in the rebound speed of the ball. A decrease in string tension from 60 to 50 lb would therefore lead to an increase in ball speed of only 1.1%.

Calculations for a wide-body racket are shown in Fig. 6b, assuming that the fundamental vibration frequency is 200 Hz (EI = 384 Nm) and that the mass and length of the racket are the same as that in Fig. 6a. Compared with the vibration loss shown in Fig. 6a, the vibration loss for a wide body racket is much smaller for impacts near the tip or throat of the racket and e_A is therefore significantly larger. For example, at $k_s = 4 \times 10^4$ Nm^{-1}, e_A is increased from 0.10 to 0.18, which translates (because of the $1 + e_A$ factor), to a 7% increase in serve speed from a point near the tip of the racket. Many elite players serve from a point near the tip of the racket, presumably because of the added height advantage and because the racket is moving fastest at the tip. The ability to serve at high speed from a point near the tip is possibly the main reason that modern rackets appear to be more powerful than the old wooden rackets of twenty years ago. There is no significant difference in racket power between stiff and flexible rackets for an impact at the vibration node since the vibration amplitude is essentially zero in both cases.

CONCLUSIONS

In this paper, aluminium beams of different length were used to simulate the behaviour of a ball colliding with a tennis racket. It was found that the apparent coefficient of restitution, for an impact at any point well removed from either end of the beam, is independent of the impact location or the length of the beam or the method of support at the ends. These results support the often quoted assumption that the impulsive reaction force on the handle can be neglected. A rigid body model of the collision yields results that are consistent with the conservation equations but the model provides no information regarding beam vibrations. The flexible beam model is superior and yields results that are in remarkably good agreement with experimental data, at least for a uniform beam. Calculations for a tennis racket show that racket power can be increased by reducing string tension, but the magnitude of this effect is essentially neglible. A more significant increase in racket power can be achieved by increasing the stiffness of the frame, at least for impacts near the tip of the racket. An increase in frame stiffness has no effect on racket power for an impact at the vibration node near the centre of the strings.

REFERENCES

Brody H. (1986) The sweet spot of a baseball bat. *Am. J. Phys.*, **54**, 640-643.

Brody H. (1997) The physics of tennis. III. The ball-racket interaction. *Am. J. Phys.*, **65**, 981-987.

Cross R. (1999a) Impact of a ball with a bat or racket. *Am. J. Phys.*, **67**, 692-702.

Cross R. (1999b) The sweet spots of a tennis racket. *Sports Engineering*, **1**, 63-78.

Cross R. (2000) Effects of string tension and frame stiffness on racket performance. *Sports Engineering* in press.

Graff K.F. (1975) *Wave motion in elastic solids*, Oxford University Press, Oxford pp. 140-210.

Sweet area prediction of tennis racket estimated by power: comparison between two super large sized rackets with different frame mass distribution

Y. Kawazoe
Department of Mechanical Engineering, Saitama Institute of Technology, Saitama, Japan
R. Tomosue
Yasuda Women's College, Hiroshima, Japan

ABSTRACT: This paper investigates the prediction of the tennis racket performance in terms of the sweet area where the post-impact ball velocity is higher when a player strikes a ball. The prediction is based on impact analysis by using the experimental simulation of the racket and ball with a simple swing model. The result of the comparison between the two super-large sized rackets with different mass and mass distribution shows that the sweet area of a super-light racket is wider than that of a conventionally mass distributed racket. However, it also shows that the post-impact ball velocity of the former is lower than that of the latter when a player hits the ball at the near side and off the longitudinal axis of the racket head.

INTRODUCTION

The implementation of material composites has led to increased flexibility in the design and production of sporting goods. The increased freedom has enabled manufacturers to tailor goods to match the different physical characteristics and techniques of users. However, ball and racket impact in tennis is an instantaneous non-linear phenomenon creating frame vibrations and large deformations in the ball/string system in the racket. The problem is further complicated by the involvement of humans in the actual strokes. Therefore, there are many unknown factors involved in the mechanisms explaining how the racket frame influences the racket capabilities. This paper investigates tennis racket performance in terms of the sweet area where the post-impact ball velocity is higher when a player strikes a ball. It predicts the difference in the sweet area in terms of the velocity of the hit ball or power between the two super-large sized rackets with different mass and mass distribution. The prediction is based on the impact analysis by using the experimental identification of a racket and a ball with a simple swing model (Kawazoe et al 1989, 1992, 1993, 1994, 1997).

RACKET PHYSICAL PROPERTIES AND PREDICTION OF THE RESTITUTION COEFFICIENT BETEEEN A BALL AND A RACKET

The main specifications and physical properties of the test rackets made of carbon graphite with a head size of 120 square inches are shown in Table 1. The racket, called EOS120A, is a super-light racket (292 g including the weight of strings), while the racket called EOS120H is a conventional weight and weight balanced racket (349 g). In Table 1, I_{GY} denotes the moment of inertia about the center of mass, I_{GR} the moment of inertia about the grip portion 70 mm from the grip end, I_{GX} the moment of inertia about the longitudinal axis of racket head.

Table 1 Specifications and physical properties of rackets.

Racket	EOS120A	EOS120H
	Super light	Conventional
Total length	690 mm	685 mm
Face area	760 cm^2	760 cm^2
Mass (+ strings)	292 g	349 g
Tension (1 Ib=4.45 N)	79 Ib	79 Ib
Center of Gravity	363mm	323 mm
IGY	14.0 gm^2	16.0 gm^2
IGR	39.0 gm^2	38.0 gm^2
1st freq.	137 Hz	142 Hz
IGX	1.78 gm^2	2.21 gm^2

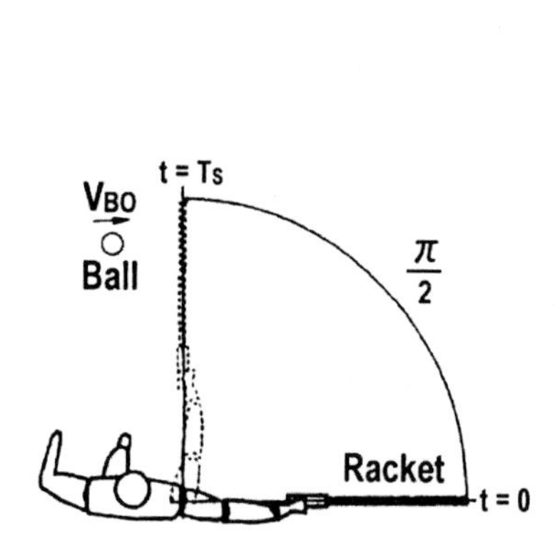

Fig. 1 Simple forehand ground stroke swing model.

1st 142Hz 137Hz
2nd 356Hz 322Hz
3rd 419Hz 391Hz
4th 586Hz 605Hz
m/s 3 0 -3
(a) EOS 120H (b) EOS120A

Fig. 2 Initial velocity amplitude distributions of the four vibration modes of different type of rackets.

The racket vibration characteristics were identified using experimental modal analysis for a racket placed horizontally on a soft sponge (corresponding to mid-air hanging, freely supported racket) and a racket with the handle held firmly by a hand. Since the experimental modal analysis (Kawazoe, 1989,1997) showed that the fundamental vibration mode of a hand-held racket is similar to the mode of a freely supported racket, it is assumed here in this study that the racket is freely supported.

Figure 1 shows a simple forehand ground stroke model used in this study. A player hits a ball arriving with velocity V_{Bo} where the racket head velocity V_{Ro} is given by $L_X(\pi N_s / I_s)^{1/2}$, where L_X denotes the distance between the player's shoulder joint and the impact location on the racket face, N_s the constant torque around the shoulder joint, and I_s the moment of inertia of arm/racket system around the shoulder joint.

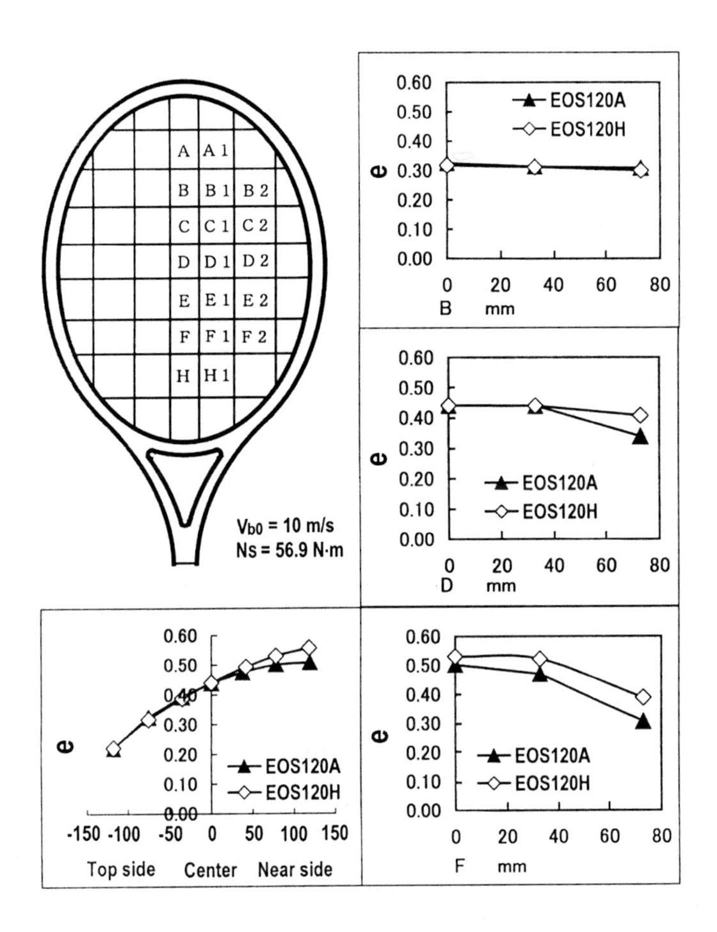

Fig. 3 Predicted rebound power coefficient e when a player hits the ball at the longitudinal axis and off the longitudinal axis ($V_{Bo} = 10$ m/s, $N_s = 56.9$ Nm).

The impulse can be approximately derived using a model assuming that a ball with a concentrated mass and a nonlinear stiffness collides with the nonlinear spring of strings supported by a rigid frame, where the measured restitution coefficient e_{BG} inherent to the materials of ball/strings is employed as one of the sources of energy loss. The contact time T_c can be derived, if it is assumed that the contact time T_c, which is not affected much by the frame stiffness according to the experiment, is determined by the natural period of a whole system composed of the mass m_B of a ball, equivalent compound stiffness K_{GB} of a ball and strings, and the reduced mass M_r of a racket.

On the basis of the approximation of the force-time curve of impact as a half-sine pulse and the application of its Fourier transform to the experimentally identified racket vibration model, the initial amplitude of racket vibration due to impact can be derived. The energy loss due to the racket frame vibration can be derived from the amplitude distribution of the velocity and the mass distribution along a racket frame. Figure 2 shows the initial velocity amplitude distributions of the four vibration modes of the two different types of rackets.

The coefficient of restitution (COR) e_r between a ball and a racket can be estimated by considering the energy loss E_1 due to frame vibration as well as the energy loss E_2 due to large instantaneous deformation of the ball and strings. The coefficient of restitution e_r corresponds to the total energy loss E ($=E_1+E_2$) could be obtained as

$$e_r = \left\{1 - \frac{2E(m_B + M_r)}{m_B M_r V_{BO}^2}\right\}^{\frac{1}{2}} \qquad (1)$$

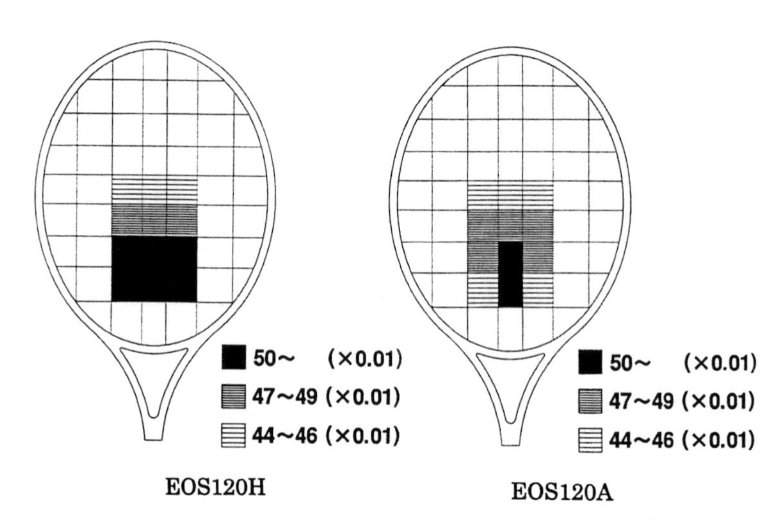

Fig. 4 Sweet area with respect to the rebound power coefficient e (V_{Bo} = 10 m/s, N_s = 56.9 Nm).

The predicted restitution coefficient e_r of a super-light weight racket has been lower than that of a conventional weight and weight-balanced racket, particularly at the near side off-center of the racket, where a player hits the ball (V_{Bo} = 10 m/s, N_s = 56.9 Nm).

PREDICTED POST-IMPACT BALL VELOCITY AND THE SWEET AREA IN TERMS OF POWER WITH TWO TYPES OF SUPER-LARGE SIZED RACKETS

Here we introduce the rebound power coefficient e defined by the ratio of rebound velocity V_B against the incident velocity V_{BO} of a ball when a ball strikes the freely supported racket at rest ($V_{Ro} = 0$) , written as Eq.(3) . The rebound power coefficient e can particularly estimate the rebound power of a racket for a volley,

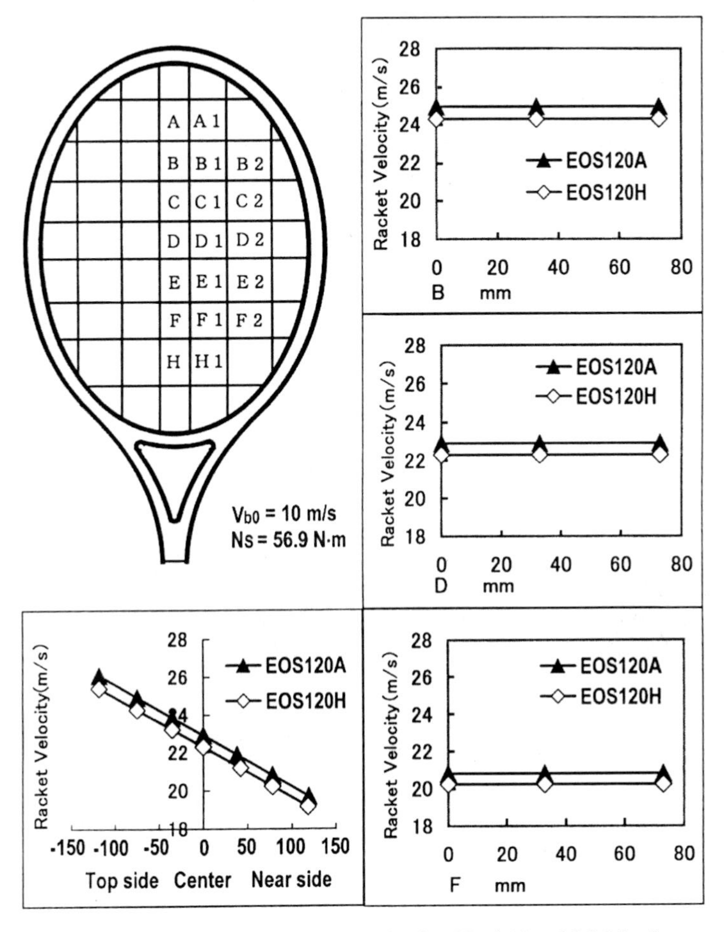

Fig. 5 Predicted pre-impact racket head velocity V_{Ro} ($N_s = 56.9$ Nm).

$$e = -\frac{V_B}{V_{BO}} = \frac{(e_r - m_B / M_r)}{(1 + m_B / M_r)} \tag{2}$$

then a player hits the ball with pre-impact racket head velocity of V_{Ro}, the coefficient e can be expressed as

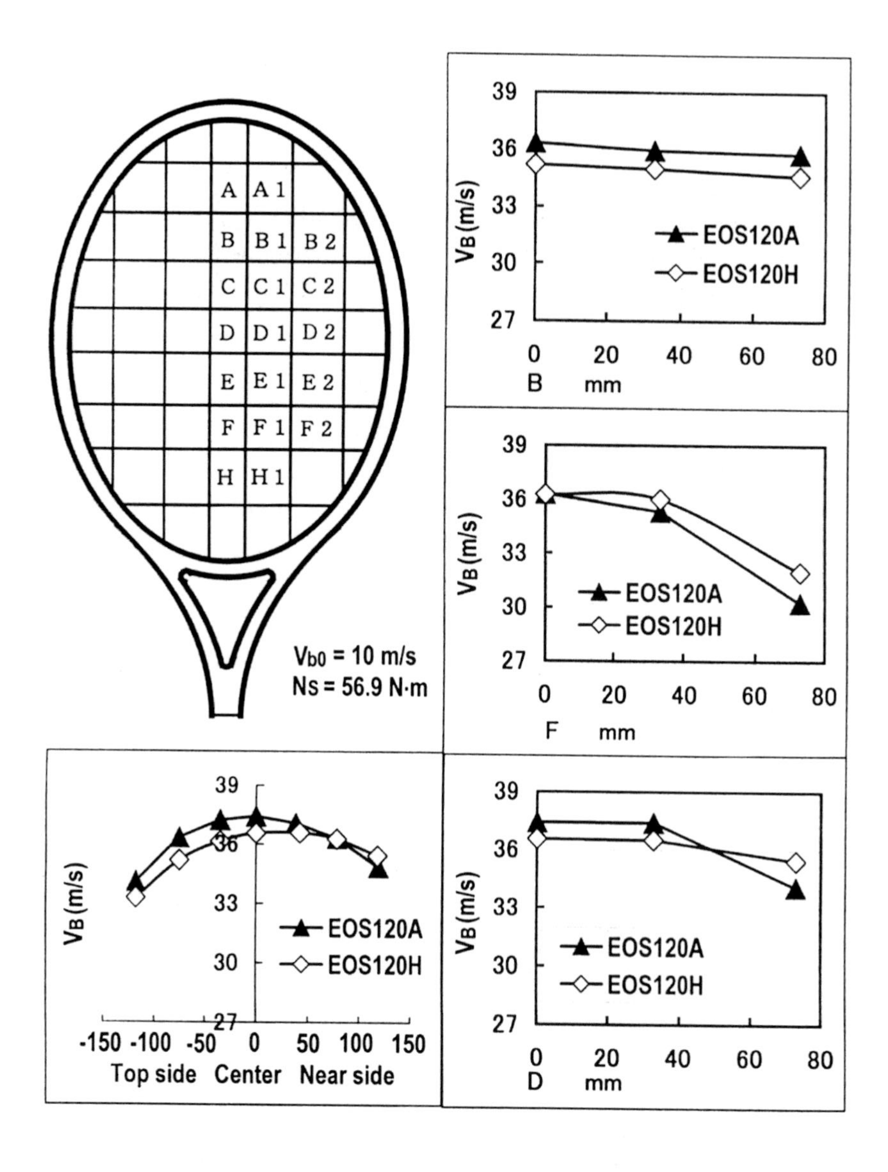

Fig. 6 Predicted post-impact ball velocity V_B when a player hits the ball at the longitudinal axis and off the longitudinal axis ($V_{Bo} = 10$ m/s, $N_s = 56.9$ Nm).

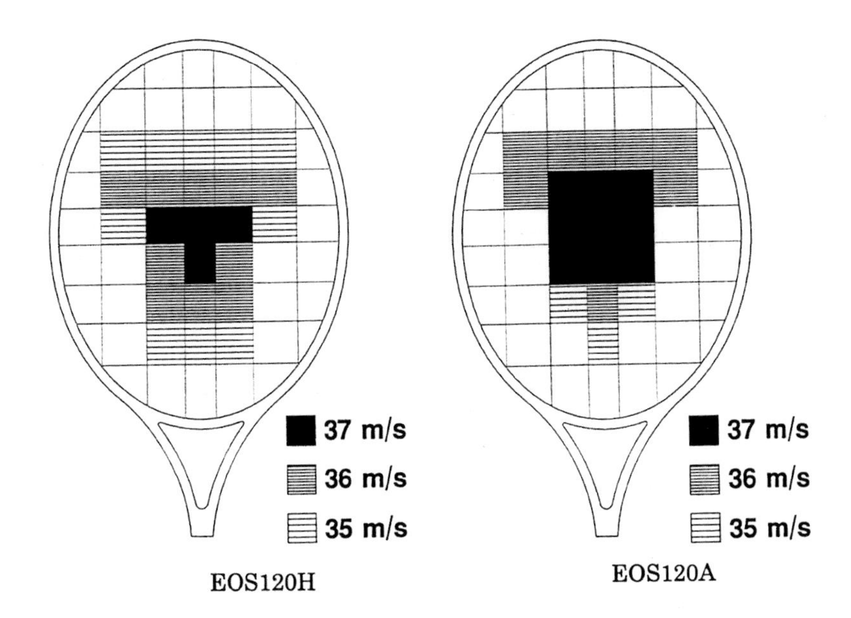

Fig. 7 Predicted sweet area with respect to the post-impact ball velocity V_B (V_{Bo} = 10 m/s, N_s = 56.9 Nm).

$$e = -\frac{(V_B - V_{Ro})}{(V_{BO} - V_{Ro})} \tag{3}$$

Figure 3 shows the predicted rebound power coefficient e when a player hits the ball at the longitudinal axis and off the longitudinal axis (V_{Bo} = 10 m/s, N_s = 56.9 Nm). Figure 4 shows the sweet area with respect to the rebound power coefficient e.

The post-impact ball velocity V_B could estimate the power of the racket when a player hits the ball. V_B can be expressed as Eq.(5).

$$V_B = -V_{Ro}e + V_{Ro}(1+e) \tag{4}$$

Figure 5 shows the predicted pre-impact racket head velocity V_{Ro} (N_s = 56.9 Nm). The head velocity V_{Ro} of super light racket is higher than that of a conventional racket.

Figure 6 shows the predicted post-impact ball velocity V_B when a player hits the ball at the longitudinal axis and off the longitudinal axis of the racket (V_{Bo} = 10 m/s, N_s = 56.9 Nm). Figure 7 is the predicted sweet area with respect to the post-impact ball velocity V_B (V_{Bo} = 10 m/s, N_s = 56.9 Nm).

It is seen that the super-light racket is wider than that of a conventionally mass distributed racket with the sweet area in terms of the post-impact ball velocity or power. However, it also showed that the post-impact ball velocity of the former is lower than that of the latter when a player hit the ball at the near side and off the longitudinal axis of the racket head.

CONCLUSIONS

This paper investigates the prediction of the tennis racket performance in terms of the sweet area where the post-impact ball velocity is higher when a player strikes a ball. The prediction is based on the impact analysis by using the experimental identification of the racket and ball with a simple swing model.

The result of the comparison between the two super-large sized rackets with different mass and mass distribution showed that the sweet area of a super-light racket is wider than that of a conventionally mass distributed racket. However, it also showed that the post-impact ball velocity of the former is lower than that of the latter when a player hit the ball at the near side and off the longitudinal axis of the racket head.

ACKNOWLEDGEMENT

This work was supported by a Grant-in-Aid for Science Research(B) of the Ministry of Education, and Culture of Japan, and a part of this work was also supported by the High-Tech Research Center of Saitama Institute of Technology.

REFERENCES

Kawazoe, Y. (1989) Dynamics and computer aided design of tennis racket. *Proc. Int. Symp. on Advanced Computers for Dynamics and Design'89*, 243-248.

Kawazoe, Y. (1992) Impact phenomena between racket and ball during tennis stroke, *Theoretical and Applied Mechanics*, Vol.41, 3-13.

Kawazoe, Y. (1993), Coefficient of restitution between a ball and a tennis racket, *Theoretical and Applied Mechanics*, Vol.42, 197-208.

Kawazoe, Y. (1994) Effects of string pre-tension on Impact between ball and racket in Tennis, *Theoretical and Applied Mechanics*, Vol.43, 223-232.

Kawazoe, Y. (1994) Computer aided prediction of the vibration and rebound velocity Characteristics of tennis rackets with various physical properties, *Science and Racket Sports*, pp.134-139.E & FN SPON.

Kawazoe,Y. and Kanda,Y. (1997), Analysis of impact phenomena in a tennis ball-racket system (Effects of frame vibrations and optimum racket design), *JSME International Journal, Series C*, Vol.40, No.1, 9-16.

Kawazoe,Y. (1997) Experimental identification of hand-held tennis racket characteristics and prediction of rebound ball velocity at impact, *Theoretical and Applied Mechanics*, Vol.46, 165-176.

Kawazoe,Y. (1997) Performance prediction of tennis rackets with materials of the wood and the modern composites, *5th Japan Int. SAMPE Sympo.& Exhibition,* pp.1323-1328.

Kawazoe, Y. (1999) Performance prediction of tennis rackets with different materials of the wood and the super-light / high-rigidity composite (mechanism of the difference in terms of the post-impact ball velocity), Proc. of 6th Japan Int. SAMPE Symp, (Tokyo, Japan), p.783-786.

On tennis equipment, impact simulation and stroke precision

F. Casolo, V. Lorenzi
Dipartimento Sistemi di Trasporto e Movimentazione, Politecnico di Milano, Italy
H. Sasahara
Department of General Education, Hiroshima University of Economics, Japan

ABSTRACT: A system of simulation-programs for the analysis of tennis equipment behaviour has been developed. The models has been validated using static and dynamic tests. Both the "quasi static" stringing phase and the dynamic of the rebound has been simulated. The impact duration is a key factor to understand ball-racket interaction and to predict ball trajectory. String tension and ball stiffness plays a fundamental role with respect to the contact time. Relevant distances of the impact point from the racket longitudinal axis cause unacceptable errors in rebound trajectories. Player skill may be related to the feeling of the consequences of the racket rotation during the impact. This approach, notwithstanding its computational cost, is suitable to predict the consequences of geometrical or mechanical changes of the equipment with regards to the rebound phase.

INTRODUCTION

How to invert the present negative trend of the audience to tennis matches is still an open question. Tennis federations and tennis player organizations, from different points of view, argue on the usefulness of introducing changes in tennis rules. The point of the debate is to find modifications that can make the match more interesting without affecting the tennis players' skills too much. The new rules will involve the equipment because its evolution is the main cause of the balls' velocity increase and the latter is strictly related, for instance, to the high rate of aces scored during a match.

A profitable approach to the question uses the mechanical characterization of the tennis equipment and the analysis of the influence on the performance of kinematics, dynamic and structural parameters of the ball-racket-player system. Dynamic simulation of the strokes may be helpful, not only to point out the optimal values of the parameters to be changed, but also to study how the player may adapt his skills to the new rules.

Obviously, the first step of this process is to set up a model of the system. The present work deals with this first phase and shows how simulation models can highlight some aspects of tennis mechanics which, in some cases, are not well known. Various kinds of mathematical models can be used for tennis simulation: their adequacy depends on the main aim of the research. For tennis player dynamics we use chains of rigid bodies while for racket, strings, balls and impact surfaces we take in models made of deformable bodies..

Here we deal with racket and ball dynamics, taking into account the player's role - which can only be passive during the impact phase (3 - 5 ms) – simply by means of an equivalent mass, evaluated as it is described in a previous work (Casolo, Ruggieri 1991), and connected to the racket handle.

Despite the scarcity of scientific works concerning tennis balls, the knowledge of the ball behavior during the impact is fundamental for the accurate design, analysis and simulation of the racket, and also for a correct understanding of the player action.

We set up a simulation system which allows us to predict ball rebound, actions transmitted to the player, ball and racket vibration components after rebound and other parameters such as impact duration. At present, the validated routines allow us to simulate impacts at any location within the racket oval but only for ball (input) trajectories perpendicular to the strings plane.

To build this system we developed some independent routines which manage only a portion of the simulation and can generally be validated separately on laboratory experimental results.

The preliminary simulation routines, based on simple models with "concentrated parameters", are quite fast and manageable even with personal computers, but they are too approximate, mostly because they do not take into account the mass distribution within the system components -ball and racket-; therefore, the last versions of most routines are now based on finite element continuous models, which keep the program quite far from being a "real time" simulation system.

MATERIALS AND METHODS

RACKET AND STRINGS

A racket frame FE model has been developed using 3D beam elements. These are straight elements, with two nodes (and an optional third node to define the element orientation around its longitudinal axis) having six degrees of freedom (three translations and three rotations) per node.

Element parameters consist of cross-sectional properties (section moment of inertia with respect to section main axes, width and height of the section - used to evaluate maximum and minimum stresses) and material properties (Young's Modulus, Poisson's ratio and density).

The racket geometry has been defined parametrically in order to simplify the set up of the model of any desired frame shape. The material properties and inertial parameters of the frames have been obtained by dissecting the rackets, analyzing the material distribution within the sections and working specimens for tensile tests.

The frame model has been validated by means of experimental static flexion tests and modal analysis: a good agreement has been found between the model and experimental eigenfrequencies and mode shapes.

This frame model has been used, for example, to investigate the evolution of stress and strain during the stringing phase and to evaluate the final string tension.

To this aim the whole stringing procedure has been simulated modeling strings as 3D spar (two nodes, 3 degrees of freedom - translations-per node) with appropriate material and section properties. The oval has been constrained in six points, according to the constraints of the stringing machine (Scaglia sport tnmc191), using contact elements - four circular profiles prevent frame displacement in the direction of the centre of the oval and two in the opposite direction.

Then the stringing procedure has been followed. The effect of frame deformation on string tension has been taken into account during all the procedure, updating nodal coordinates at each step, according to the computed displacements, thus performing a nonlinear large displacements analysis. In this phase each string has been considered independent by the other ones. Fig.1 shows frame deformation (with an amplified scale for the displacements) during the stringing procedure.

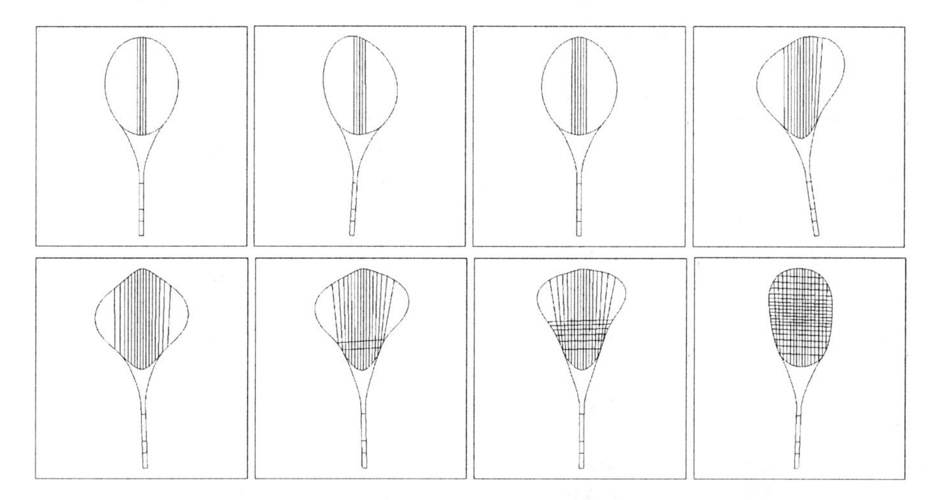

Fig. 1 Deformed frame during the stringing phases (the displacements are enlarged).

Resulting string tensions at the end of the process are not uniform: stringing the racket, for example at a theoretical constant tension of 240N for all the strings, gives final string tension ranging from 200N to 260N. Thus the procedure adopted for the simulation, which corresponds to the sequence suggested by the racket producer, shows that the central strings perpendicular to the long axis of the racket have a lower tension.

Deflection tests have been also performed on the strings with the oval constrained, following the same procedure used in experimental tests and obtaining coherent results. When transverse loading is applied, as in this case, strings are linked together at their crossover.

Modal analysis has then been performed on the stringed frame, taking into account the strings' tension. Fig 2a shows the results of a simulation on a free racket and Fig.2b show the results of the same analysis on the same racket but adding the player equivalent mass to the racket handle. These frequencies values are near to the ones measured by means of strain gauges on a free racket with the same mass. The above mentioned analysis has been performed in order to verify the consistency of the model with some experimental data, obtained with a mechanical model designed for the analysis of the racket rebound; however, a more refined model of the racket-player system, built for a precise study of the vibrations, would require to take into account the non rigid connection between the player equivalent mass and the racket handle.

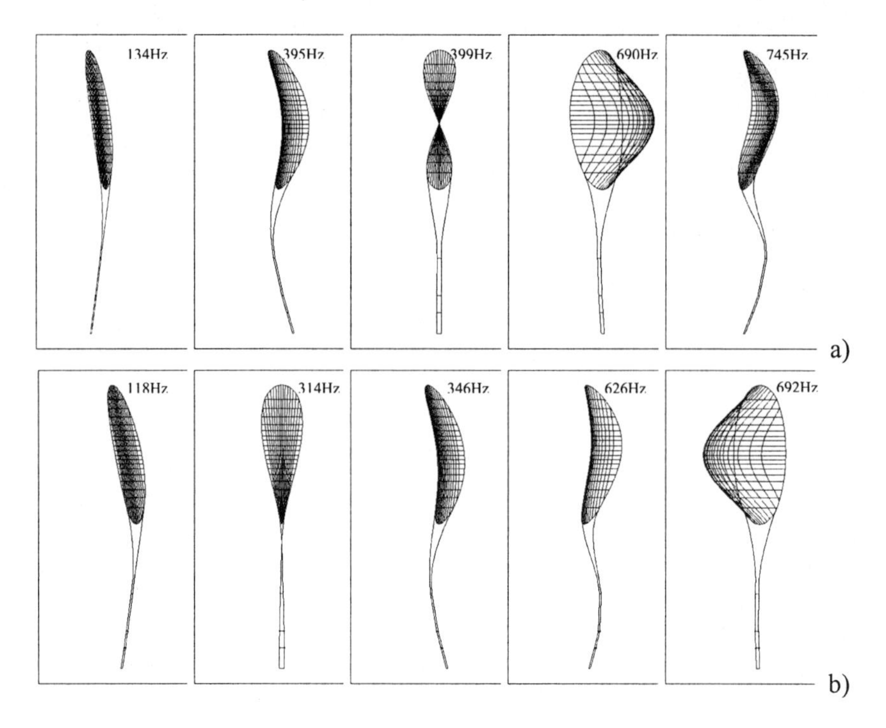

Fig. 2 a, b Frequencies and mode shapes of the first 5 modes of a free racket (up) and of a free racket with added mass (down).

THE BALL

The efficiency of the ball during the rebound in tennis is quite poor and is strictly related to its maximal deformation. Therefore internal pressure and quality of the ball plays a fundamental role in the shot.

The FE model of the ball consists of 3D brick elements with 20 nodes with three d.o.f. per node. Rubber characteristics have been extrapolated by mechanical tests on a hydraulic machine. The hysteresis of the material (hyper-elastic Mooney-Rivling kind) has been introduced by means of an equivalent material damping. Also the covering felt has been considered.

The influence of internal pressure on ball stiffness has also been taken into account: during the ball deformation the internal volume is monitored and the resulting pressure is replaced at each time step.

The ball model has been tested both statically, by compression tests on the hydraulic machine, and dynamically, by means of rebound tests on a rigid wall, using balls with different internal pressure, for a range of velocity from 5 to 55 m/s.
In the same conditions the rebound coefficient of restitution has also been calculated by means of the simulation model.

Most of results have also been reached by using an axial-symmetric model, which can be handled quite easily even if some ball vibration modes cannot be evaluated.

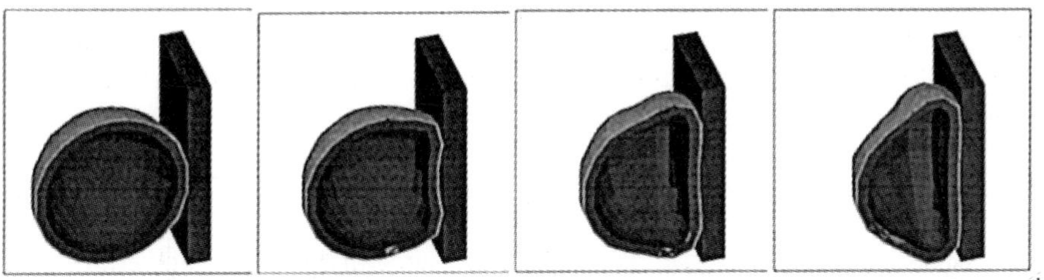

Fig. 3 F.E. simulation of ball impact against the rigid wall. Ball deformation: the 4th frame is 1.25 milliseconds after the impact, the ball pre-impact velocity is 15 m/s

IMPACT ON THE RACKET

The last step of the work is the simulation of tennis strokes varying the impact point, the ball velocity and the string tension. For this aim, contact elements have been added between strings and ball surface. This implies a nonlinear transient dynamical analysis, quite heavy in computational terms.

The ball-racket model has been validated by means of rebound experimental tests in which the rackets, with or without the "player equivalent mass", are automatically released just before the ball hits the string plane. Therefore it is possible to analyze the effects on ball trajectory and velocity of many parameters. The parameters that may be changed are structural and geometric, such as material proprieties of the components, their shape and dimension, but they are also related to the player ability, as for instance the precision of the impact location within the oval.

It is also possible to speculate on rackets' easie of use and precision. For clarity, let us look at the ball-racket system during the impact, superimposing to the whole system a constant velocity equal, in value, but with the opposite direction, to that of one of the two components before the impact. In this way we can easily take all consideration on impact kinematics, if the racket does not rotate before the impact. Therefore we can reduce all strokes, serve included, to the same scheme where, before the impact, the racket does not move, and the ball velocity is equal to the relative velocity. This approach is very profitable for many reasons and, for instance, it allows us to simplify the experimental system for the rebound tests.

Thus, the results of our simulation, which concerns the impact of a ball against a not moving handled racket, may be extrapolated to the other rebound conditions (without spin).

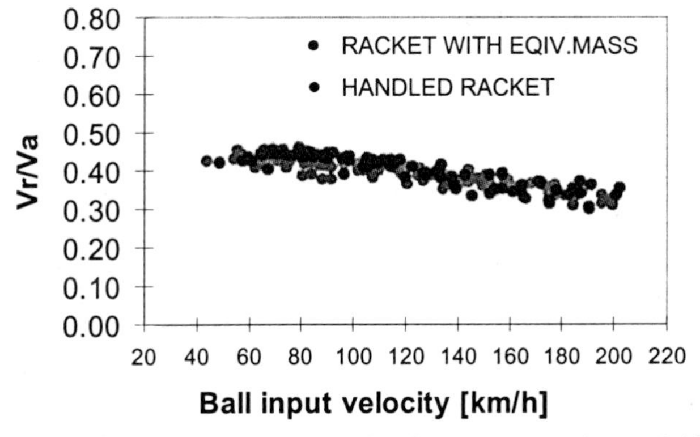

Fig. 4 Rebound laboratory tests: A COR of a handled racket and of the same racket with the "player equivalent mass" on the handle.

COMFORT, "EASINESS " AND PRECISION OF A RACKET

Tennis journals often describe rackets as easy, comfortable or precise. While the feeling of comfort is mainly related to the vibrations transmitted to the arm, the definition of the other two qualities is arguable and probably they have some connections.

There are many factors that make a racket easier with respect to another: it is well known, for instance that, in an off-centre impact, a bigger racket, having a larger moment of inertia with respect to its longitudinal axis, rotates less around that axis than a standard one.

Thus, we may say that, by a certain point of view, the big racket is easy because it reduces the trajectory deflection produced by hitting the oval far from the longitudinal axis. Furthermore, the same phenomena makes the ball trajectory rotate out, moving on the plane perpendicular to the oval and containing the racket longitudinal axis, because, even for a precise hit at the centre of the oval, the action is applied far from the racket centre of rotation, which is generally, near to the handle.

Even in this case a high moment of inertia with respect to the transversal axis reduces the trajectory deflection. Nevertheless this high moment of inertia make the racket less manoeuvrable and this may be in contrast with the concept of easy racket. Many other mechanical parameters of the ball-racket-player system play a role in the phenomena above described, we think that an overall analysis of the causes of the ball-trajectory changes during the impact is also an investigation on the "racket precision".

In our preliminary simulations of the ball rebound, we assume that the racket string plane is oriented perpendicularly to the ball trajectory before the impact, and that the ball has no spin.

The following graphs display the position of the centre of the ball cavity - easier to compute but not exactly coincident to the ball centre of mass – for ten milliseconds, starting just before the impact on the strings. The racket is free in the space and the player equivalent mass is linked to the handle. Two stringing tension are considered in the tests (240N and 330N).

We assume an absolute reference frame located -before the impact- at the lower end of the oval, with Z axis directed orthogonal to the string plane, Y on the racket longitudinal axis and X perpendicular to Z-X plane and directed out of the racket. Looking to an impact 50mm off-center we may notice that the output trajectory of the ball is considerably rotated both for X and Y direction. The big displacement in Y positive direction means that the counterbalancing effect, due to the different lengths of the string segments (to their different inclination during ball sinking and consequently to the inclination of the bisecting line of the comprised angle) is relatively very low.

Lowering strings' tension all three component of the displacement increases: Z because it produces less ball deformation and consequently less energy loss; while Y and Z mainly because ball-racket contact time increases and during this time the racket rotate both in ZX and in ZY planes.

While the deflection of the trajectory in the direction (X) of the ball offset is obvious, (Fig. 5) probably less obvious is the deflection in Y direction which for the lower strings' tension is quite high (in the example displayed in Fig.6 is 8°).

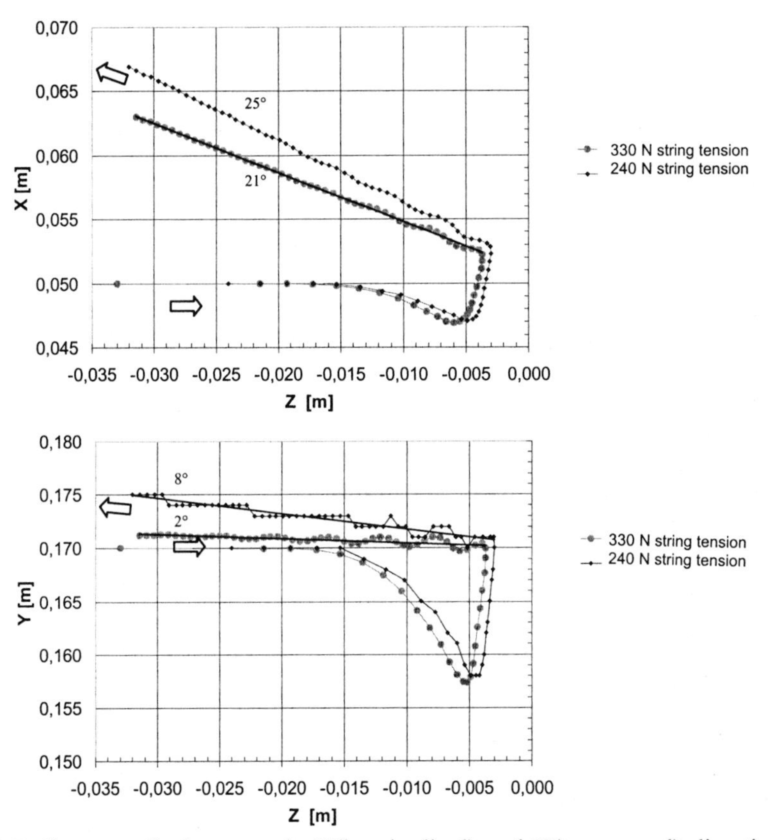

Fig. 5, 6 Ball centre displacement in X(longitudinal) and Y(transversal) direction, with respect to Z displacement (perpendicular to the string plane before the impact) for the 15 milliseconds following the impact and a 50mm impact offset in X direction. The string tension are 240N and 330N for graphs x24, y24 and x33, y33 respectively

Therefore, looking to the tremendous error originated by 50mm offset, we can easily figure out that the precision of hitting on the racket longitudinal axis is an essential characteristic of a tennis player and no racket or stringing procedure may correct such a relevant (lateral) offset error.

On the other hand, in case of perfectly centered shots, while for tighter-strings the trajectory deflection in Y direction is almost negligible, looser string plates cause a sensible deflection in that direction. Repeating the same simulation of fig.5 and 6, but without X offset, we obtain trajectory rotation Y direction of 0,1° and 1° for tighter and looser strings respectively. A one-degree error on the trajectory is huge for a professional player -it may correspond to 0.3 meters on the court.

The bases of the assertion that "tighter strings allow a better control on the ball" are now clear. Other mechanical factors may produce similar errors. However, the key element, for limiting the problem, is the reduction of the ball-racket contact time. The above considerations concerns racket precision, but we may also deduce something on the player skill. If a player do not use extremely tight strings, for saving power, he can't avoid trajectory changes at the impact, thus he must keep into account this occurrence in advance. This is a reason why the full knowledge of . rackets characteristics is very important and why professional player have so many troubles changing their equipment. Moreover we must remember that ball stiffness has the same influence on the rebound, therefore changes in this parameter may also interfere on player skill.

CONCLUDING REMARKS

With the described system of programs we can handle some different tennis situations involving the rebound phase. We can predict, for instance, how modification of the mechanical parameters may affect the stroke, consequently, it is possible to help the player to plan his action, taking into account - or choosing when is possible - racket stiffness and mass distribution, frame shape, string tension, as well as ball pressure and size.

Moreover, the modal analysis can identify the location of the nodes of the racket's main vibration. This point is one of the so called "sweet points" because, for a ball impact on it, the vibration associated with the corresponding mode shape is not excited. From our preliminary tests it seems that many professional tennis players instinctively choose to hit the ball at this point. When the same point is also near to the center of percussion with respect to the handle, the impact force transmitted to the hand is also minimized.

Concerning the string tension, we must pay attention to the stringing procedure. The final string tension is generally quite different from the planned one, because of the oval deformation. Our preliminary tests, simulating standard stringing sequence at equal tension, show that, in most cases, the resulting tension of the transverse strings is lower than that of the longitudinal strings; this kind of error is not a problem because it may improve the cooperation of the strings during the impact because both short and long strings have the same camber. With the same approach, new stringing schemes can be easily tested before racket prototyping.

Interesting results can be obtained, with extensive rebound simulation, with regards to racket precision: we have seen in case of centered impact that ball trajectory after rebound is shifted towards the racket tip, because of racquet backward rotation during the finite time of contact. Strings have no direct effect because their action is symmetric on the ball.

ACKNOWLEDGMENT

We thank Scaglia Sport(Milano) for the support and the electronic stringing machine.

REFERENCES

Brody, H. (1987)Tennis science for tennis players. *University of Pennsylvania Press.*

Casolo, F., Ruggieri, G. (1991) Dynamics analysis of ball-racket impact. Meccanica, 26:67.

Elliott, Blanksby, Ellis, (1980) *Vibration and rebound velocity..*, *Res.Quart.Ex.* Sp. 51.

Kawazoe, Y., (1992) Impact phenomena between racket and ball during tennis stroke, *Theoretical and Applied Mechanics*, 41:3-13.

Prediction of the impact shock vibrations of the player's wrist joint: comparison between two super large sized rackets with different frame mass distribution

Y. Kawazoe
Department of Mechanical Engineering, Saitama Institute of Technology, Saitama, Japan
K. Yoshinari
Shirayuri Women's College, Tokyo, Japan

ABSTRACT: This paper investigates the tennis racket performance in terms of feel or comfort. It predicts the effect of the mass and mass distribution of super-large sized rackets on the impact shock vibrations of the racket handle and the player's wrist joint when a player hits a flat forehand drive. The prediction is based on the identification of the racket characteristics, the damping of the racket-arm system, equivalent mass of the player's arm system and the approximate nonlinear impact analysis in tennis. The result of the comparison between the two super-large sized rackets with different mass and mass distribution shows that the shock vibration of the super-light racket is much larger than that of the conventional weight balanced type racket. It also shows that the sweet area in terms of the shock vibration shifts from the center to the topside on the racket face with a super-light racket compared to the conventional weight balanced type racket.

INTRODUCTION

The implementation of material composites has led to increased flexibility in the design and production of sporting goods. The increased freedom has enabled manufacturers to tailor goods to match the different physical characteristics and techniques of users. However, ball and racket impact in tennis is an instantaneous non-linear phenomenon creating frame vibrations and large deformations in the ball/string system in the racket. The problem is further complicated by the involvement of humans in the actual strokes. Therefore, there are many unknown factors involved in the mechanisms explaining how the racket frame influences the racket capabilities.

This paper investigates the tennis racket performance in terms of the feel or comfort. It predicts the effect of the mass and mass distribution of super-large sized rackets on the impact shock vibrations of the racket handle and the player's wrist joint when a player hits a flat forehand drive. The prediction is based on the identification of the racket characteristics, the damping of the racket-arm system, equivalent mass of the player's arm system and the approximate nonlinear impact analysis in tennis.

The racket, called EOS120A, is employed as a representative example of a super-light racket (mass: 292 g including the weight of strings, the center of gravity L_G: 363 mm from the butt end), while the racket called EOS120H is selected as a representative of a conventional weight and weight balanced racket (349 g, L_G: 363 mm). They are the super-large sized rackets made of carbon graphite with a head size of 120 square inches (Kawazoe 2000).

PREDICTION OF SHOCK ACCELERATIONS TRANSMITTED TO THE ARM JOINT FROM A RACKET IN THE IMPACT

IMPACT MODEL FOR THE PREDICTION OF SHOCK FORCE AT THE ARM JOINTS

Figure 1 shows an experiment where a male tournament player hits a flat forehand drive and Fig.2 shows the locations of attached accelerometers at the wrist joint and the elbow joint in the experiment. Figure 3 shows an impact model for the prediction of shock forces transmitted to the arm joints from a racket. The impact force S_0 at P_0 causes a shock force S_1 on the player's hand P_1, a shock force S_2 on the elbow P_2, and finally a shock force S_3 on the player's shoulder P_3 during the impact at which the player hits the ball with his racket.

Since the intensity of the impulse decreases with the distance from the point of impact with the ball, it can be assumed that the shoulder does not basically alter its velocity, despite the presence of the shock force S_3. Generally speaking, the shock forces S_0, S_1, S_2, and S_3 , which are mainly responsible for the sudden changes in velocity that take place in the brief interval of time considered, is one order of magnitude higher than the other forces in play during the same interval; consequently the gravity force and muscular action are not taken into account. Accordingly, we consider the racket to be freely hinged to the forearm of the player, the forearm being freely hinged to the arm and the arm freely hinged to the player's body. We can deduce that the inertia effect of the arm and the forearm can be attributed to a mass M_H concentrated in the hand; therefore the analysis of impact between ball and racket can be carried out by assuming that the racket is free in space, as long as the mass M_H is applied at point P_1 of the hand grip. If the impact force S_0 between a ball and the racket is given when the ball hits the racket, the shock force S_1 can be obtained (Casolo 1991, Kawazoe et al 2000).

DERIVATION OF THE RESTITUTION COEFFICIENT AND THE IMPACT FORCE BETWEEN A BALL AND A RACKET

The vibration characteristics of a racket can be identified using the experimental modal analysis (Kawazoe 1989, 1990) and the racket vibrations can be simulated by applying the approximate impact force-time curve to the hitting portion on the string face of the identified vibration model of the racket. When the impact force component of k-th mode frequency f_k in the frequency region applies to the point j on the racket face, the amplitude X_{ijk} of k-th mode component at point i can be derived using the residue r_{ijk} of k-th mode between arbitrary point i and j (Kawazoe, 1993).

Fig. 1 Experiment where a male player hits flat forehand drive.

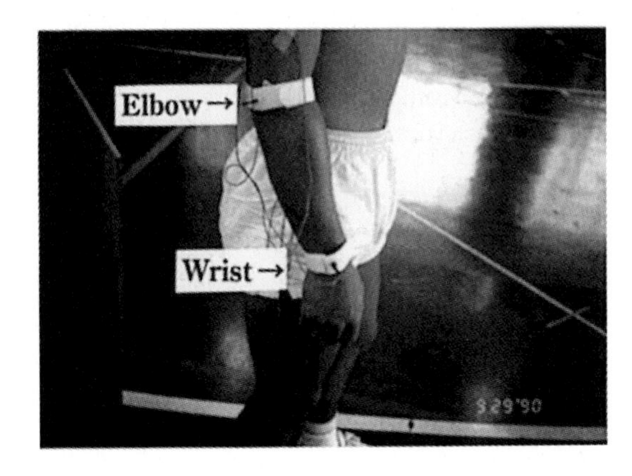

Fig. 2 Accelerometers attached at the wrist and the elbow.

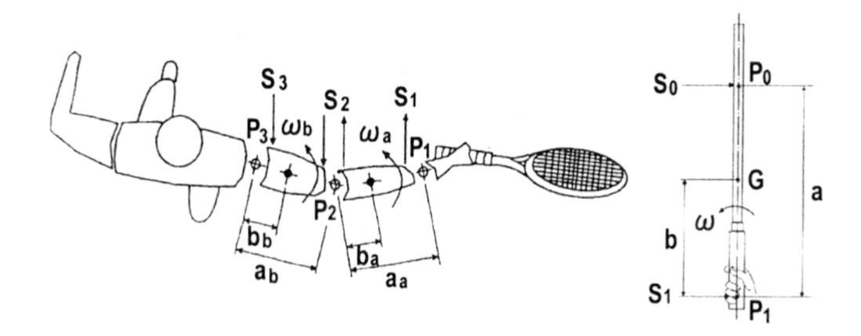

Fig. 3 Impact model for the prediction of the shock force transmitted to the arm joints from a racket.

The energy loss due to the racket vibration induced by impact can be derived from the amplitude distribution of the vibration velocity and the mass distribution along a racket frame when an impact location on the string face and the impact velocity are given.

The coefficient of restitution e_r (COR) can be derived considering the energy loss E during impact. The main sources of energy loss is E_1 due to racket vibrations as well as E_2 due to the instantaneous large deformation of a ball and strings (Kawazoe 1993). Furthermore, the force-time curve of impact between a ball and a racket considering the vibrations of a racket frame can be approximated as

$$S_o(t) = S_{O\,max} sin\left(\frac{\pi t}{T_c}\right)(0 \le t \le T_c) \tag{1}$$

where

$$S_{O\,max} = \frac{\pi}{2T_c}(V_{BO} - V_{Ro})(1+e_r)\frac{m_B}{(1+m_B/M_r)} \tag{2}$$

The contact time T_c during impact can be determined against the pre-impact velocity (V_{BO} - V_{Ro}) between a ball and a racket assuming the contact time to be half the natural period of a whole system composed of the mass m_B of a ball, the equivalent stiffness K_{GB} of ball/strings, and the reduced mass M_r of the racket.

PREDICTION OF SHOCK ACCELERATIONS TRANSMITTED TO THE ARM JOINT FROM A RACKET IN THE IMPACT

The shock acceleration $A_{nv}(t)$ at the hand grip considering the equivalent mass M_H of the arm system can be represented as

$$A_{nv}(t) = S_o(t)\left\{\frac{1}{M_R + M_H} - \frac{aX}{I_G}\right\} \tag{3}$$

where X denotes the distance between the center of mass of the racket-arm system and the location of hand grip, a the distance between the center of mass of racket-arm system and the impact location of the racket, I the moment of inertia around the center of mass of racket-arm system, respectively. The maximum shock force $S_{1\,max}$ transmitted to a wrist joint corresponds to the maximum impact force $S_{0\,max}$.

Figure 4 shows the predicted maximum shock acceleration at the grip portion of (a) a freely suspended racket, (b) a hand-held racket.

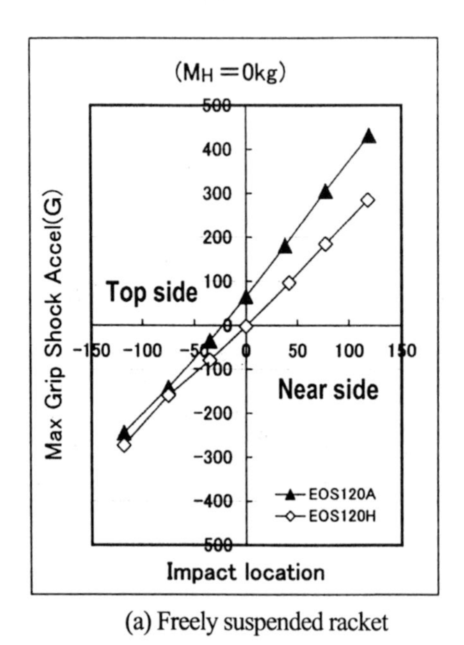

(a) Freely suspended racket

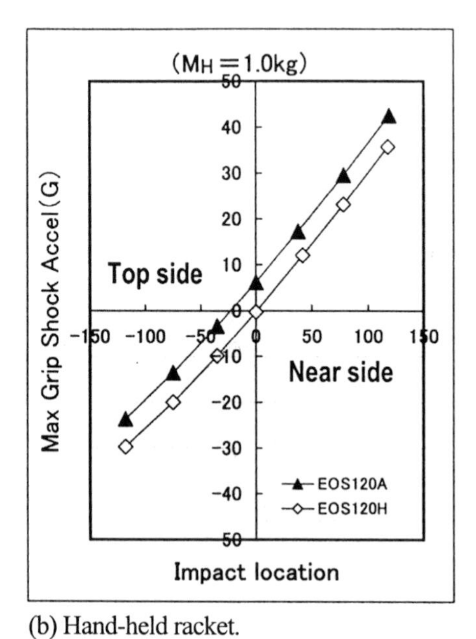

(b) Hand-held racket.

Fig. 4 Predicted maximum shock acceleration at the grip (impact velocity: 30 m/s).

PREDICTION OF VIBRATION COMPONENTS AT THE RACKET HANDLE AND THE WRIST JOINT

The vibration acceleration component of the *k*-th mode at the location *i* of handgrip is represented as

$$A_{ijk}(t) = -(2\pi f_k)^2 r_{ijk} S_{oj}(2\pi f_k) e^{-2\pi f_k \zeta_k t} \sin 2\pi f_k t \qquad (4)$$

where j denotes the impact location between ball and racket on the string face, ζ the damping ratio of k-th mode, $S_{0j}(2\pi f_k)$ the fourier spectrum of Eq.(1). Figure 5 shows the predicted vibration amplitude components during impact at the grip 70 mm from the grip end, where the first four vibration modes of freely suspended rackets are considered. The vibration of the super-light racket at the grip is much larger than that of the conventional weight balanced type racket. It is because the location of grip (70 mm from the grip end) is more apart from the location of node on the handle of the first mode of super-light racket than that of the conventional weight balanced type racket.

It also shows that the sweet area with respect to the vibration is located around 30-mm topside from the center on the racket face with a super-light racket.

PREDICTION OF THE WAVEFORMS OF SHOCK VIBRATIONS AT THE GRIP

The summation of Eq.(3) and Eq.(4) represents the shock vibrations at the handgrip. Figure 6 shows the predicted waveform of the shock vibrations at the grip on comparing the two freely suspended rackets with different weight and weight balance when a ball strikes the various locations on the string face. The impact velocity between the ball and the racket is 30 m/s. The shock vibration of the super-light racket at the grip is much larger than that of the conventional weight balanced type racket.

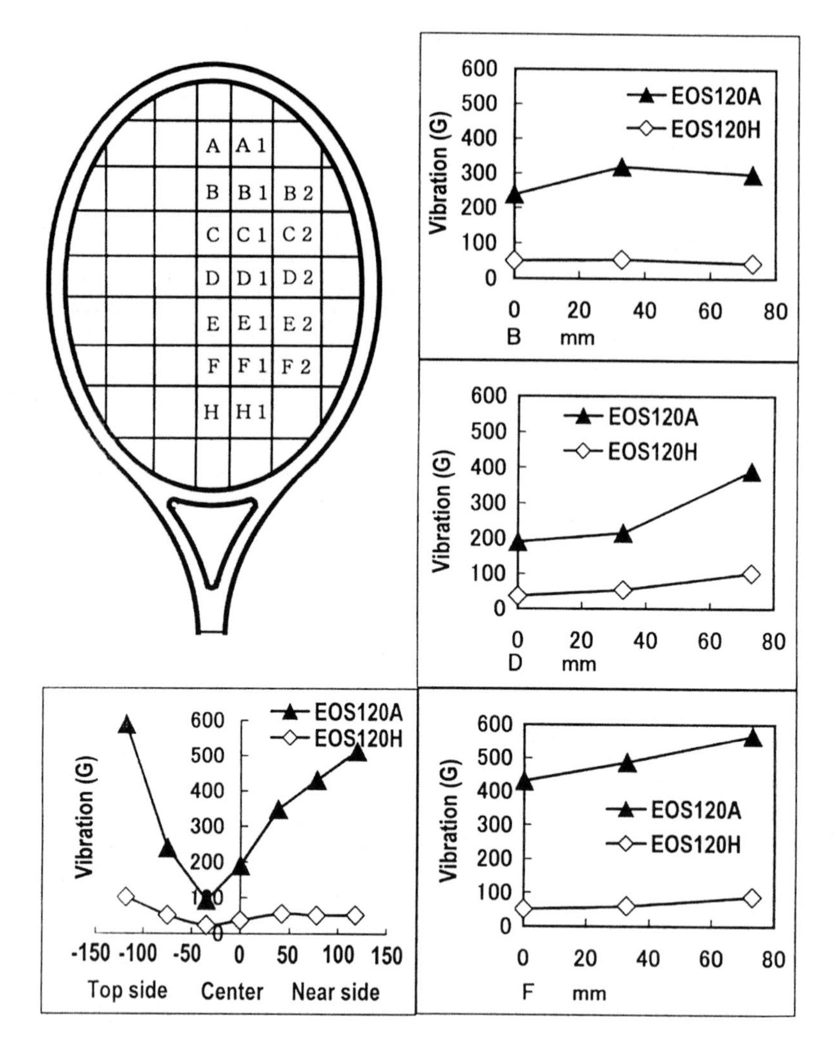

Fig. 5 Predicted vibration components at the grip (impact velocity: 30 m/s).

PREDICTION OF THE WAVEFORMS OF SHOCK VIBRATIONS AT THE WRIST JOINT

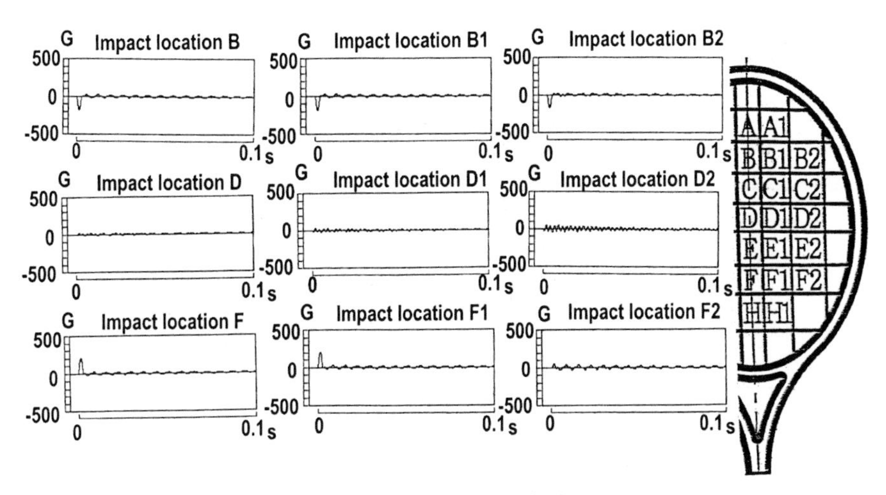

(a) Racket EOS120H: 349 g

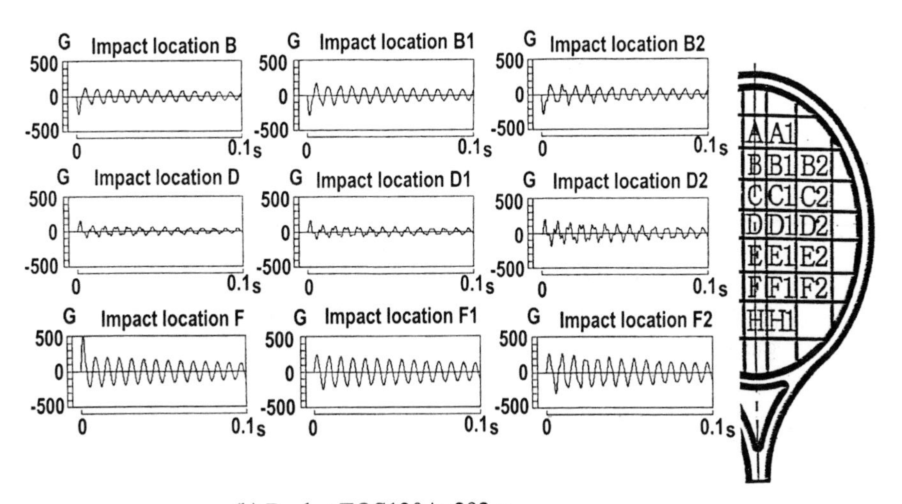

(b) Racket EOS120A: 292 g

Fig. 6 Predicted waveform of the shock vibrations at the grip on comparing the two freely-suspended rackets with different weight and weight balance when a ball strikes the various locations on the string face. The impact velocity between the ball and the racket is 30 m/s.

The predicted waveform of the shock vibrations at the wrist joint agrees fairly well with the measured ones during actual forehand stroke by a player (Kawazoe et al 1997, 2000). Figure 7 shows the predicted shock vibrations of a wrist joint. The damping ratio of a hand-held racket in the actual impact is estimated as about 2.5 times that of the one identified by the experimental modal analysis with small vibration amplitude. Furthermore, the damping of the waveform at the wrist joint was 3 times that at the grip portion of the racket handle. The shock vibrations of super-light racket are much larger than those of conventional weighted and weight balanced racket; the conventional weighted and weight balanced super- large racket is predicted to be very comfortable when the ball is hit with it.

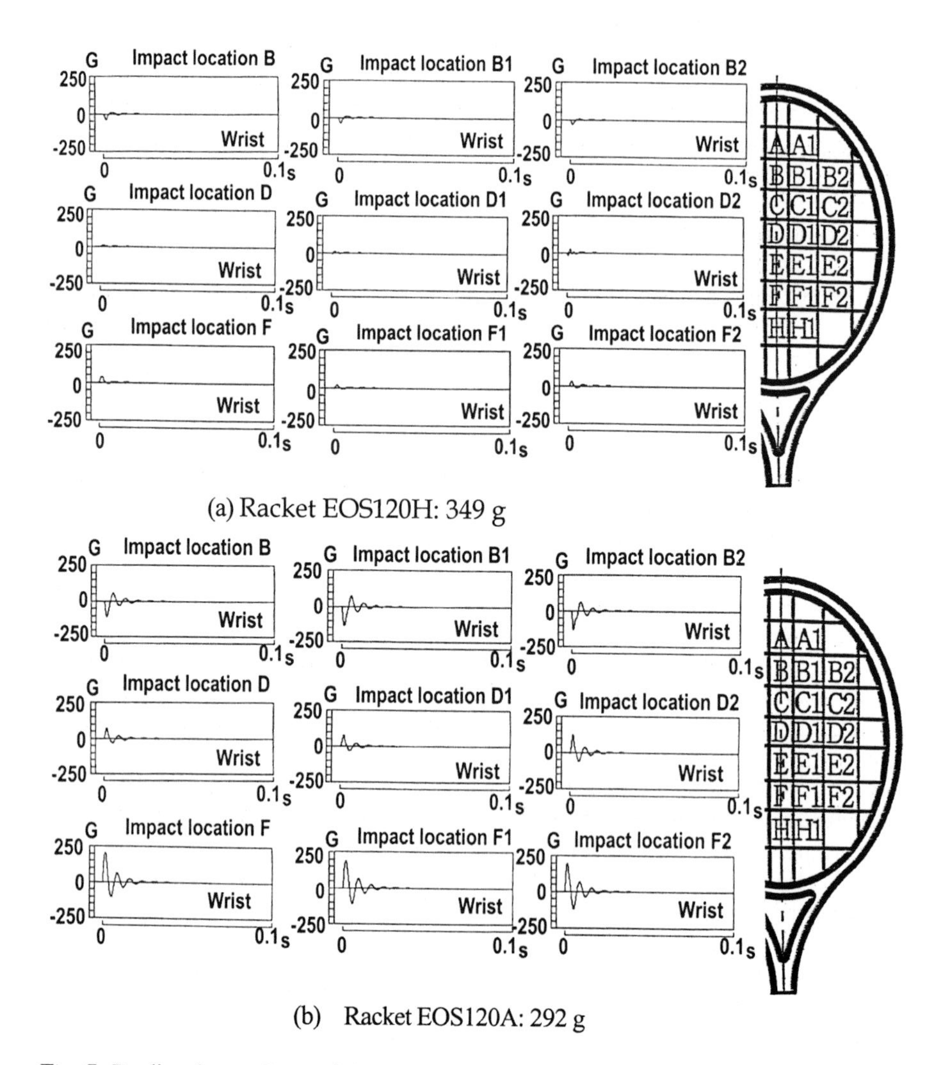

(a) Racket EOS120H: 349 g

(b) Racket EOS120A: 292 g

Fig. 7 Predicted waveform of the shock vibrations of the player's wrist joint on comparing the two rackets with with different weight and weight balance when a ball strikes the various locations on the string face. The impact velocity between the ball and the racket is 30 m/s.

CONCLUSIONS

This paper has investigated tennis racket performance in terms of the feel or comfort. It predicted the effect of the mass and mass distribution of super-large sized rackets on the impact shock vibrations of the racket handle and the player's wrist joint when a player hits flat forehand drive. The prediction is based on the identification of the racket characteristics, the damping of the racket-arm system, equivalent mass of the player's arm system and the approximate nonlinear impact analysis in tennis. The result of the comparison between the two super-large sized rackets with different mass and mass distribution shows that the shock vibration of the super-light racket is much larger than that of the conventional weight balanced type racket. It is seen that the vibration of the super-light racket at the grip is much larger than that of the conventional weight balanced type racket. This is because the location of the grip (70 mm from the grip end) is further from the location of the node on the handle of the first mode of super-light racket than that of the conventional weight balanced type racket. It is also seen that the sweet area with respect to the vibration is located around 30-mm topside from the center on the racket face with a super-light racket.

ACKNOWLEDGEMENT

This work was supported by a Grant-in-Aid for Science Research(B) of the Ministry of Education, and Culture of Japan, and a part of this work was also supported by the High-Tech Research Center of Saitama Institute of Technology.

REFERENCES

Casolo,F. & Ruggieri,G.(1991) Dynamic analysis of the ball-racket impact in the game of tennis, *Meccanica*,Vol.24, pp.501-504.

Kawazoe,Y.(1989) Dynamics and computer aided design of tennis racket. *Proc. Int. Sym po. on Advanced Computers for Dynamics and Design'89*, pp.243-248.

Kawazoe,Y.(1992) Impact phenomena between racket and ball during tennis stroke, *The oretical and Applied Mechanics*,Vol.41, pp.3-13.

Kawazoe,Y.(1993) Coefficient of restitution between a ball and a tennis racket, *Theoretic al and Applied Mechanics*, Vol.42, pp.197-208.

Kawazoe,Y.(1994) Effects of String Pre-tension on Impact between Ball and Racket in Tennis, *Theoretical and Applied Mechanics*, Vol.43, pp.223-232.

Kawazoe,Y.(1994) Computer Aided Prediction of the Vibration and Rebound Velocity Characteristics of Tennis Rackets with Various Physical Properties, *Science and Rack et Sports*,pp.134-139.E & FN SPON.

Kawazoe,Y.(1997) Experimental Identification of Hand-held Tennis Racket Characteristics and Prediction of Rebound Ball Velocity at Impact, *Theoretical and Applied Mechanics*, Vol.46, 165-176.

Kawazoe, Y., Tomosue, R. & Miura, A.(1997) Impact shock vibrations of the wrist and the elbow in the tennis forehand drive: remarks on the measured wave forms cosidering the racket physical properties, *Proc. of Int. Conf. on New Frontiers in Biomechanical Engineering*,pp.285-288.

Kawazoe, Y. (1997) Performance Prediction of Tennis Rackets with Materials of the Wood and the Modern Composites, *5th Japan Int. SAMPE Sympo.& Exhibition*, pp.1323-1328.

Kawazoe, Y. and Kanda, Y.(1997), Analysis of impact phenomena in a tennis ball-racket system (Effects of frame vibrations and optimum racket design), *JSME International Journal, Series C*, Vol.40, No.1, 9-16.

Kawazoe, Y., Tomosue, R., Yoshinari, K. and Casolo, F. (2000), Prediction of impact shock vibrations of a racket grip and a player's wrist jpint in the tennis forehand drive, *3rd Int. Conf. on the Engineering of Sport*, to be presented.

Development of a racket with an "Impact Shock Protection System"

T. Iwatsubo
Department of Mechanical Engineering, Kobe University, Japan
Y. Kanemitsu, S. Sakagami, T. Yamaguchi
Sumitomo Rubber Industries, Ltd., Sports Goods Research Department, Japan

ABSTRACT: In order to reduce the impact acceleration at a player's elbow, we have developed a dynamic damper which we call "Impact Shock Protection System (ISPS)". We simulated the vibration mode of a tennis racket by FEM, and when we found out which vibration mode was, then we determined where to hold the racket in order to reduce the impact acceleration at a player's elbow. With the result, we selected the location of ISPS on the frame and tuned the weight of ISPS to the frequency of the vibration. This ISPS located in the suitable position on the tennis racket frame can reduce impact acceleration at a player's elbow. The player can feel the reduction of impact shock with ISPS.

INTRODUCTION

The reduction of tennis racket weight goes on along with the improvement of materials and molding technologies. A player expects a racket as light as possible, to improve of his or her play (an earlier reaction and a faster swing speed of a service). As a result, tennis rackets weighted less than 200g are beginning to appear in the market. On the other hand, we are concerned about the increase in the number of injury cases, e.g. tennis elbow, by reducing the racket weight. As a countermeasure, tennis racket manufactures are promoting the development of rackets that absorb shock and vibration. We prototyped an effective dynamic damper as a method that absorb shock and vibration in this research. The person who released the racket with a dynamic damper first in the market was Lacoste F. R. (1995). Lacoste's dynamic damper is attached to the grip end and it is designed as the 1st bending vibration mode is absorbed.

The racket does not have only the 1st bending vibration mode but also the 2nd, the 3rd and high dimension. Furthermore, there is the distortion vibration mode that results by off center impact. It is said that the frequency of vibration that a human being can feel is under 1kHz. Therefore the human being does not feel the vibration of the 3rd over.

At first, we investigated the impact point on the string area, to find out the vibration mode that occurs when a human being hits a ball with a stroke. Although it changes by the specifications of rackets (sweet spot, balance, the moment of inertia, etc.), we confirmed that the impact point is in the range of ±5cm from the center of a string area.(Fig. 1) We found that mainly the 2nd bending vibration mode are occurred when we struck a ball at the center, and the 1st distortion vibration mode are occurred when we struck a ball at the off center, by the FEM simulation (Yamaguchi et al. 1995).

In this research, by reducing the 2nd bending vibration mode and the 1st distortion vibration mode, we tried to design an effective dynamic damper in order to absorb the shock and the vibration. To confirm an effect, we measured the acceleration of a racket and player's elbow at the time of a ball impact.

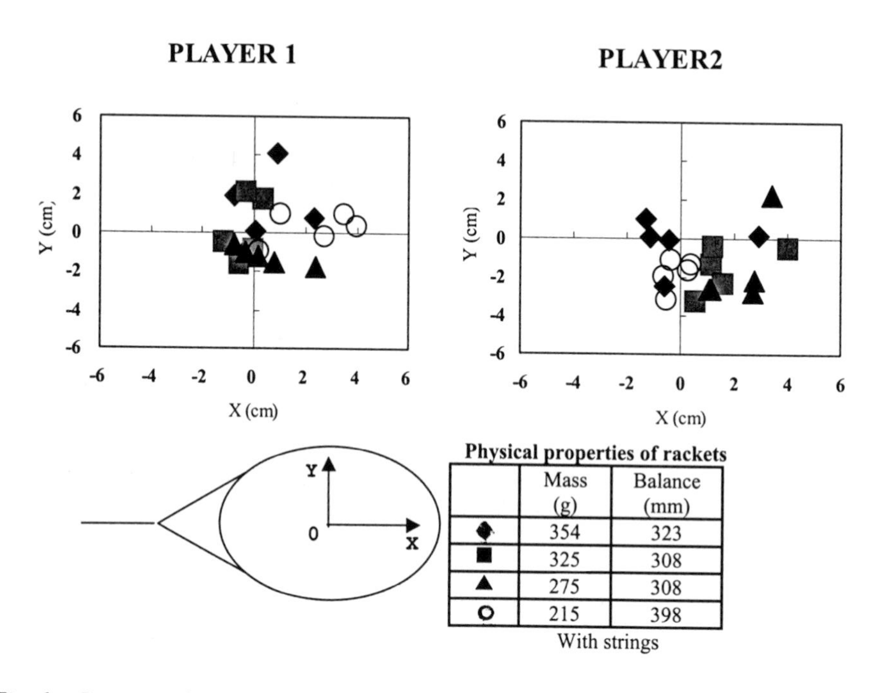

Physical properties of rackets

	Mass (g)	Balance (mm)
◆	354	323
■	325	308
▲	275	308
○	215	398

Fig. 1 Impact point of ball on the string area

PROTOTYPING

The specifications of the base model racket that we used to the evaluation are shown in Table 1. The dimension of a damper and attachment detail drawing is shown in Fig. 2. We used the rubber with three kinds of Young's modulus for the viscoelastic part of a dynamic damper, and made the mass part of stainless steel and lead 8-12g, and adhered the mass part to rubber. The Young's modulus of rubber, the weight of a mass and the attachment position to a racket are shown in Table 2. We considered that an attachment position fits antinode of the vibration mode.(Fig. 3) For example, the attachment position of prototype A is the antinode of the 1st bending vibration mode and the 1st distortion vibration mode. The attachment position of prototypes B, C and D is the antinode of the 2nd bending vibration mode. The Young's modulus of the rubber is calculated with the compressed displacement of specimen (30x30x8mm) at 196N by the compression test. The cross head speed is 500 mm/min.

Table 1 Specifications of a racket

Model	DUNLOP PRO30Lady Tour
Length (inch)	27
Face size (square inch)	110
Mass (g) *	272
Balance (mm) *	333

*without string

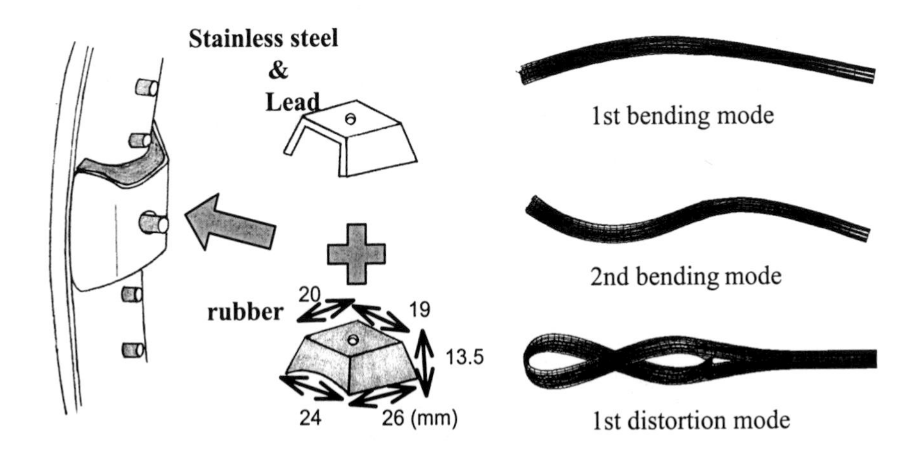

Fig. 2 Dimension & attachment detail of a damper

Fig.3 Vibration mode of a racket

Table 2 Young's modulus of rubber, the weight of mass and the attachment position

	Young's modulus of rubber (MPa)	Weight of mass (g)	Position
A	0.48	12	A
B	0.48	8	B
C	1.74	8	B
D	5.34	8	B

Young's modulus = W/(Ax(δL/Lo))
W: Load 196N
A: Cross-section area
δL: Displacement
Lo: The length of a specimen

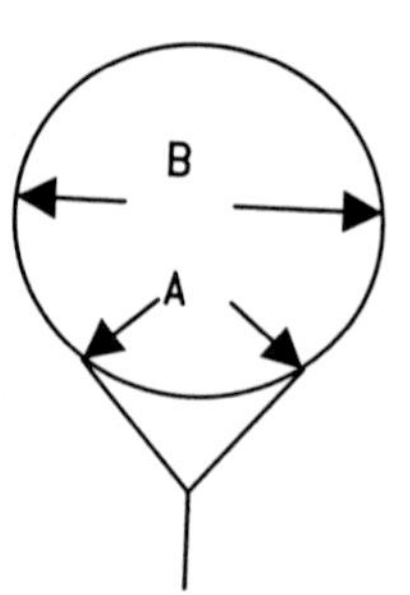

EVALUATION

Measurement of natural frequency and damping ratio

We measured the damping ratio of a racket, to confirm the vibration inhibition effect of a dynamic damper. In order to measured damping ratio at the position where each vibration mode is occurred, we attached an acceleration pickup and the impacted by an impact hammer as shown in Fig. 4. The racket is a free support.

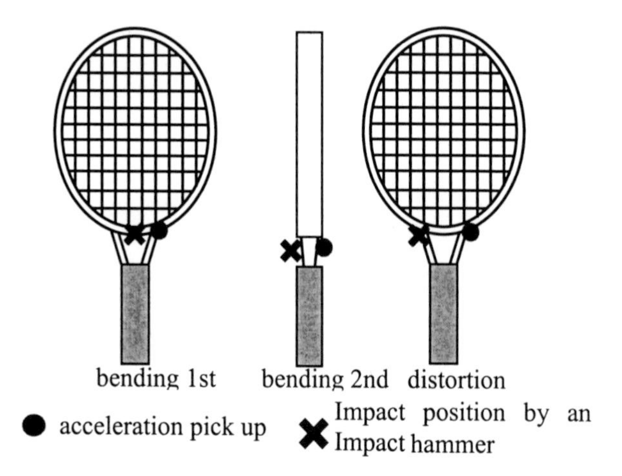

Fig. 4 Measurement method of vibration

Sensory test

Eight players in the middle and high level class tested four kinds of prototypes. They evaluated 5 stages of 5 point full marks about the smallness of the shock that given to an arm, the paucity of vibration and the width of a sweet area.

Measurement of acceleration that given to a racket and the elbow

We supposed that the bigger the acceleration is, the worse the influence to the elbow is, we tried to measure the acceleration that given to a racket and the elbow of a human being at impact. We attached an acceleration pick up to the shaft of a racket and the elbow of a male player. He hit the ball whose speed was about 6m/sec. We evaluated three market products, lightweight prototype and prototypes B and C with the dynamic damper. Those weight are 220-270g (without strings) to confirm the influence of a damper effect and racket weight. We pickup the height of the first acceleration peak at ball impact.

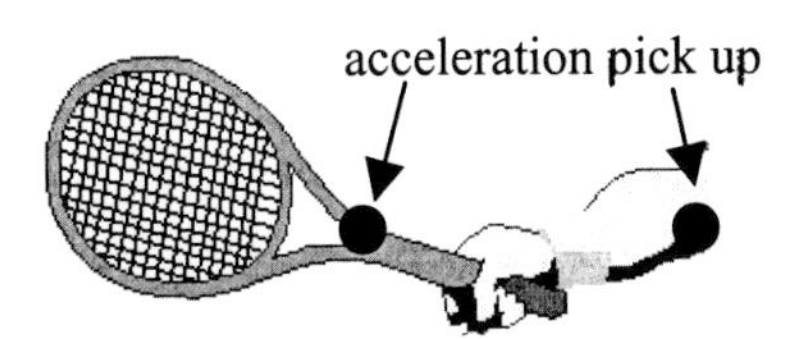

Fig. 5 Position of acceleration pickup

RESULTS AND DISCUSSIONS

Results of natural frequency and damping ratio

The natural frequency, damping ratio, weight and balance of our damper prototyping rackets are shown in Table 3. Frequency response function of prototype C is shown in Fig. 6. The dotted line shows result without the dynamic damper and the continuous line shows result with the dynamic damper. The peak of the natural frequency in the 2nd bending mode is not clear. The damping ratio is very high with 4.564%. Because we attached a damper in the position where is the antinode of the 2nd bending mode (B Table 2), the natural frequency of a damper is fitting with one of a racket and a damper is absorbing the kinetic energy of a racket.

Table 3 Results of natural frequency and damping ratio

	Bending				distortion	
	1st		2nd		1st	
	Freq. (Hz)	Damping ratio (%)	Freq. (Hz)	Damping ratio (%)	Freq. (Hz)	Damping ratio (%)
A	131	1.608	425	0.980	342	3.008
B	140	0.891	401	1.420	378	0.606
C	141	0.579	366	4.564	370	1.167
D	141	0.678	390	0.631	373	0.466
Grip end	160	2.820	440	0.330	404	0.480

In the case of prototype A, we tried to locate the dynamic damper at the position of the antinode of the 1st bending mode (A, Table2) and used low Young's modulus rubber in order to shift the natural frequency of the damper. The damping ratio of the natural frequency in the 1st bending mode becomes high. Furthermore, because the position of A is also the antinode of the 1st distortion mode, the damping ratio of the 1st distortion frequency was enlarged very much.

Although the rubber Young's modulus of prototype B is the same as one of prototype A, we are able to enlarge the damping ratio of the natural frequency in the 2nd bending mode of a racket. Because we fit an attachment position of the dynamic damper to B and changed the dynamic damper weight, in order to adjust the natural frequency of a damper.

For prototype D, a damper is attached in the same position as B and C, but the natural frequency of a racket and a damper are not fitting because the Young's modulus of rubber is high. Therefore, the damping ratio of all the natural frequency is small.

Even though the configuration (XL IMPEDANCE2) differs from prototypes, we put the data of the racket that attached a damper to the grip end. The damping ratio of the 1st bending natural frequency is very high, but there is not the effect to the 1st distortion natural frequency.

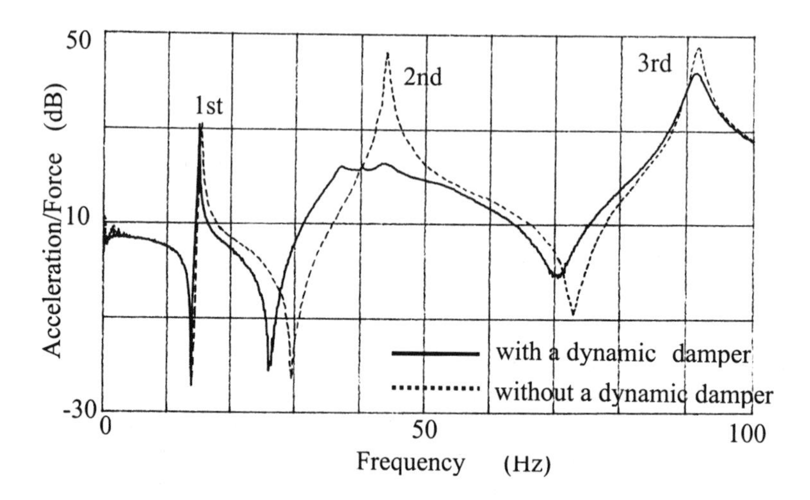

Fig.6 Frequency response function of prototype C

From the above result, we can say that a dynamic damper can absorb the aimed vibration mode of the racket by adjusting the Young's modulus of rubber, weight and the attachment position. Also, we understood that it has an effect in the distortion vibration mode in addition to the bending vibration mode, by attaching a damper in the string area of the frame.

Results of sensory test

The average ratings of the smallness of shock, the paucity of vibration and the width of a sweet area are shown in Table 4. The ratings of prototypes B and C with effective damper in the 2nd bending mode are high. In the case of prototype A and grip end damper, the rating is low, even though the damping ratio of the natural frequency in the 1st bending mode is high. It is conceivable as a cause that hitting point of a human being concentrates almost at the center of a string face and the occurred mode is the 2nd bending natural frequency. On the other hand, prototypes A and C received higher evaluation ratings for sweet area. They have larger damping ratio of the natural frequency in the 1st distortion mode. The thing that players can feel rather small vibration caused by distortion in case of off-center hitting is suggested to result the fact that they give higher evaluation and feel the sweet area wide.

Table 4 Results of sensory test (The average of 5 point full marks)

	Shock	Vibration	Sweet area
	big 1-2-3-4-5 small	many 1-2-3-4-5 few	narrow 1-2-3-4-5 wide
A	3.25	3.43	3.40
B	3.75	3.86	3.20
C	3.67	3.67	3.60
D	2.88	2.57	3.25
Grip end	2.75	2.50	3.00

From the above results, in the viewpoint of the shock absorption characteristics of a racket, we can say that it is more effective to absorb the vibration of the 2nd bending mode rather than the 1st bending mode. Furthermore, we can say that players feel sweet area wider when the distortion vibration mode is absorbed.

Results of acceleration that given to a racket and the elbow

The relationship between the acceleration given to a racket and the weight of a racket and the relationship between the acceleration given to the elbow and the weight of a racket are shown in Fig. 7. The lighter the racket is, the bigger it's acceleration becomes. Although it is not so evident as a racket, the same trend can be seen in the results of the acceleration of the elbow. The acceleration of prototype C with a damper effect is smaller than D without a damper effect. Therefore, we can say that the appropriate damper can absorb the shock.

However, we think that the factor that influences the acceleration given to a racket and the elbow is not only the weight of a racket. We will examine the shock including the factor of the balance, toughness and repulsion of a racket, and the change of player's swing. We want to find the conditions of the optimum racket to players and the racket that does not cause injury.

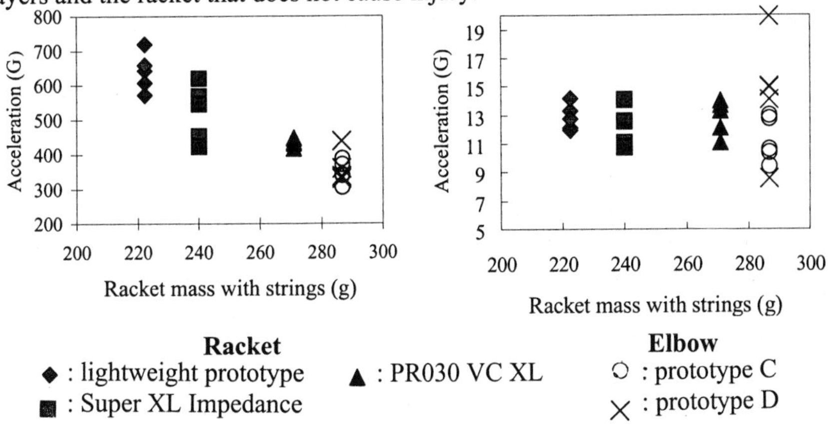

Fig.7 Acceleration of racket and elbow

CONCLUSIONS

(1) By adjusting the Young's modulus of rubber, weight and the attachment position we can absorb the aimed vibration. The dynamic damper has an effect to absorb the distortion vibration in addition to the bending vibration, because it is attached in the strings area of a frame. It differs from the damper that is attached to the grip end.
(2) In sensory test, the shock & vibration rating of prototype with the enlarged damping ratio of the 2nd bending natural frequency is good. Because the human being hits a ball mainly at the center of the string face and the 2nd bending vibration mode of the rackets are excited. It is more effective to absorb the 2nd bending vibration rather than the 1st bending vibration.

As for the prototype with the enlarged damping ratio of the distortion, players feel the sweet area wide.
(3) The acceleration given to a racket and the elbow is bigger at ball impact when the racket is lighter. An appropriate dynamic damper can absorb the acceleration of a racket and the elbow.

REFERENCES

Lacoste, F.R., JP Patent No.888259

Yamaguchi, T., Yamamoto, T., Tsunoda, M. and Iwatsubo, T. (1995) A study for optimum design of tennis racket, *Trans. Japan Soc. Mech. Eng. (in Japanese)*, C, Vol. 61, No. 588, pp.3355-3360.

One size fits all? Sensitivity to moment of inertia information from tennis rackets in children and adults

S. Beak, K. Davids & S. Bennett
Department of Exercise and Sport Science, Manchester Metropolitan University, UK

ABSTRACT: Are children and adults sensitive to moment of inertia from wielded tennis rackets? This question was examined by investigating the ability to perceive moment of inertia information of 6 tennis rackets in 3 groups varying in age and tennis experience. Data revealed that all 3 groups displayed sensitivity to small changes in moment of inerta from the wielded rackets. The implication is that initial sensitivity to haptic information can be exploited by learners to select appropriate tennis rackets for skill acquisition. Manufacturers need to increase the choice of rackets available to adults and children by varying the range of moment of inertia.

INTRODUCTION

In Ecological Psychology, there is growing interest in the role of haptic information to support the manipulation of objects and implements in the environment (e.g., see Solomon & Turvey, 1988; Turvey, 1996; Beak et al., in press). The prevalence of manipulative movements in everyday life led Turvey et al. (1998) to argue that "the role of dynamic touch in the control of manipulatory activity may be both more continuous and fundamental than that of vision" (p.35). Solomon and Turvey (1988) demonstrated that, when rod wielding, performers were sensitive to haptic information provided by the invariant, moment of inertia (I). Moment of inertia is an object's resistance to being turned/ rotated and can be picked up during wielding or hefting activities. Previous studies concerned with haptic invariants, have tended to examine a non cross-sectional population of participants, usually young adults performing a task such as rod wielding (Bingham et al., 1989; Solomon & Turvey, 1988). Typically, the adult participants were required to perceive spatial characteristics of a rod and not its utility for action.

An interesting issue raised by Solomon and Turvey (1988), is whether the findings generalize to sport performance. An important task constraint in many sports is the ability to wield implements in order to effect some change to the environment. For example, bats and rackets are wielded to hit a moving ball in ball games, oars are manipulated to row a boat in water and foils are aimed at an opponent's trunk in fencing. In tennis, learners often need to be able to judge the appropriateness of one racket over another. Haptic information available when wielding a racket is picked up by the nervous system and perceived in terms of relevance for striking a ball.

Although previous research demonstrated the sensitivity of adults to the moment of inertia of wielded implements (see Turvey, 1996), there have been few attempts to examine the implications for the design of tennis rackets. One study by Carello and colleagues recently compared the ability of expert adult tennis players to novice adults in perceiving the length of a tennis racket as well as it's sweet spot (Carello, Thout, Anderson & Turvey, 1999). Using visual manipulations, they found that novices were as adept as experts in perceiving the location of a racket's sweet spot and key spatial characteristics, such as length, and that these perceptions were constrained by the inertial properties of the racket.

The lack of expert-novice differences in that study poses the question whether humans have an innate sensitivity to moment of inertia information, which can be exploited for selecting appropriate tennis rackets for practice. A major difficulty for junior players is that, during development, as the properties of the learner change, the perceptual information relevant for assembling movement task solutions also changes. During development, organismic constraints (growth, strength, postural control etc.) change, along with the individual's action capabilities. Sensitivity to haptic information would aid in the task of choosing relevant rackets for practice over time.

This paper examines the relationship between sensitivity and attunement to haptic information as constraints on perceiving the functional utility of a tennis racket to strike a ball. It was predicted that any differences in the perception-action systems of the participants, due to experience and development, would be observed through changes in preferred choice of racket. Any effect of condition would also be seen through changes in perceptual judgements on each racket.

METHOD

PARTICIPANTS

Three groups participated in the study (for all groups n = 10): an inexperienced child group (mean age = 10 yrs), an inexperienced adult group (mean age = 28 yrs.) and an experienced adult group (mean age = 22 yrs.). All participants in the inexperienced groups were selected on the basis of no previous coaching in any racket sports. The inexperienced child group was specifically selected for their limited experience in wielding a racket. The experienced adult group were required to be University 1st team standard or above with at least 5 years competitive experience.

APPARATUS

The experiment required 6 identical rackets with equal moments of inertia. These were donated by a well-known racket manufacturer (Prince, Inc., USA) whose junior "*Rad*" range are marketed as purpose-designed rackets for use by 6-12 year olds. To coincide with the age range of the inexperienced child group the rackets used throughout the study were Rad 10s. Each racket was constructed of aluminum with length 63.5 cm, and mass 0.252 kg.

Although initially having equal moments of inertia, each racket was manipulated to produce different overall moments of inertia. In order to precisely control this experimental manipulation, an external mass of 0.05 kg was added to each of the 6 rackets along its longitudinal axis at intervals of 10 cm. This value was calculated so that the overall mass of each racket (0.302 kg) was similar to the Rad 12 (0.307 kg) i.e., the racket designed and marketed for a 12 year-old. By keeping the racket dimensions within Prince's junior range the children would not be forced to choose rackets they would normally discard due to perceived heaviness (i.e., as with an adult racket). The decision to manipulate the mass distribution instead of the mass itself was primarily due to the effect it would have in changing the overall moment of inertia (I) of the racket. From equation 1), distance r, has a larger effect on I because its value is squared. The external mass was covered in black tape to secure it to the racket and to prevent any biasing through knowledge of its value. The calculations for moment of inertia of the six racket systems (I_{RS}) were based upon the formula:

(1) $I_{RS} = I_{butt} + m.r^2$

where I_{butt} = moment of inertia of the racket about its butt end, m = mass of attached weight (kg) and r = position of weight from butt[1].

The moments of inertia for the 6 new racket systems were then calculated to be 0.0334, 0.0349, 0.0375, 0.0410, 0.0456 and 0.0512 kgm^2 for rackets 1-6 respectively.

EXPERIMENTAL PROCEDURE

Each participant wielded the rackets during an individual testing session lasting approximately 10 minutes. There were 2 randomly-ordered wielding conditions: visual and non-visual. Participants wielded each of the 6 rackets separately, and were asked to judge which racket would allow them to strike a sponge tennis ball to a maximum distance. In a forced-choice paradigm, participants indicated their preferred 3 rackets which they judged as optimal for such an action. Participants were allowed to wield each racket as many times as required. The rackets were presented in a random order to counterbalance any possible sequence effects. Grip position was standardized throughout, with the participants holding the racket at the very end of its handle. Participants were not instructed how to wield, nor given a specific length of time in which to complete the procedure. The 2 conditions were separated by a 5-minute rest period. In the visual condition, the participants were able to see the rackets. In the non-visual condition, vision of the rackets was occluded by the use of a large screen with a window. A sheet was draped over the screen so that the participant could not see either his/her arm nor the racket.

During two separate pilot studies, the reliability of the preferred choice of rackets for both the inexperienced child and inexperienced adult group was examined. The groups repeated the described procedure at two separate data collection periods, one week apart under two different conditions. For each participant, a mean preferred moment of inertia was calculated using the weighted method employed by Bingham et al. (1989). A 2 x 2 (Condition by Time) repeated measures analysis of variance was conducted on the data from both groups.

---

[1] The moment of inertia of the racket about its butt end (I_{butt}) was provided by Prince Inc. The value was equal to 0.0328 kgm^2. The positions for each weight from the butt end of the racket were 0.106 m, 0.206 m, 0.306 m, 0.406 m, 0.506 m, 0.606 m for rackets 1-6 respectively.

A significant main effect of time was reported for the child group, ($F(1, 9) = 16.24$, ($p < 0.01$), but no main effect of condition nor time by condition interaction was found. Further analysis revealed simple effects for condition at both the initial time period, $F(1, 9) = 11.24$ and one week later, $F(1, 9) = 11.00$, (both $p<0.01$). However there were no differences between performance in each condition across time. The analysis of the inexperienced adult group data also indicated no main effect of time, $F(1, 9) = 0.227$ ($p>0.05$). To examine the spread of variability across the 1-week period, the standard deviation scores were calculated from the 3 preferred rackets and then subjected to a 2-way (Time by Condition) ANOVA with repeated measures on both factors.

A significant main effect of time was found for the child group, $F(1, 9) = 17.09$, ($p<0.01$), but no main effect of condition, nor time by condition interaction was found. Simple effects analysis revealed a significant difference between conditions at the initial stage $F(1, 9) = 7.95$ ($p< 0.05$), but no effect of time. No significant main effects of time or condition were found for the adult group. These results suggest that both the inexperienced children and adult groups consistently selected rackets that yielded the same degree of variability across the 1-week period. This finding indicates that their perceived choices were not randomly selected and may be considered reliable. Based on this evidence of reliability, there were no obvious reasons to suspect that the preferences of the experienced adults would be unreliable.

RESULTS

After the mean preferred moment of inertia was calculated for each participant in each group, the data were subjected to a two-way, 2 x 3 (Condition by Group) ANOVA with repeated measures on the condition factor. Main effects were found for condition $F(1, 27) = 7.37$, and group $F(2, 27) = 3.50$ (both $p< 0.05$). Simple effects analysis revealed significant differences between the visual and non-visual conditions for the child group, $F(1, 27) = 6.056$, and the experienced adult group, $F(1, 27) = 6.079$, (both $p< 0.05$) but not for the experienced adults. Between-group significant differences were found in both the visual, $F(2, 27) = 25.5$, and the non-visual conditions, $F(2, 27) = 27.1$, (both $p< 0.01$). Follow-up Tukey tests revealed differences in the visual condition, between the child group and the experienced adult group, and between the inexperienced adult group and the experienced adult group, (both $p< 0.05$). Under the non-visual condition, significant differences were reported between the child and inexperienced adult groups, and between the child and experienced adult groups (both $p< 0.01$). Figure 1 illustrates an apparent trend in the data with the two adult groups preferring those rackets with higher moment of inertia values (i.e. those rackets that have similar values to those marketed as adult rackets and are generally perceived to be ‘heavier’). The children demonstrated a preference for values at the lower end of the inertia scale. This finding concurs with the dimensions of the manipulated rackets and the age range that similar rackets are marketed towards.

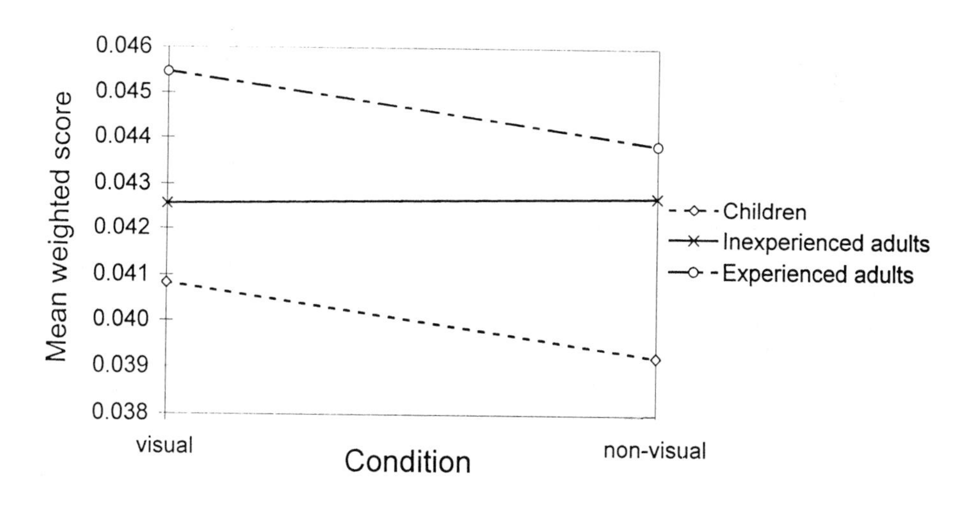

Fig. 1 Mean weighted scores for each group within both visual and non-visual conditions.

This trend is also observed in figures 2(a-c). Both adult groups showed a distinct preference towards the top end of the inertia scale. The child group demonstrated a greater degree of intra-individual variability in racket choice compared to the adults. The experienced adult group (fig. 2c) demonstrated the lowest levels of intra-individual variability of all by picking consecutive rackets along the inertia scale, resulting in a very tight clustering of choices. This trend was evident under both conditions.

Levels of inter-individual variability were found to be highest within the child group, although there was also some variability evident in both adult groups (see for example participant 9 in the experienced adult group, figure 2c). Inter-individual levels of variability increased under the non-visual condition for both the adult groups, but appeared to decrease slightly in the child group (i.e., the children's preferred choices became increasingly clustered when vision was occluded). To statistically examine the spread of variability within and between each group, the standard deviation scores were calculated from the 3 preferred rackets. These data were subjected to a 2-way (Group by Condition) ANOVA with repeated measures on the condition factor. No main effects of group or condition were found, but an interaction was evident $F(1,27) = 4.09$ ($p< 0.05$). Simple effects analysis revealed an effect of the visual condition between groups, $F(2,27) = 7.259$, ($p< 0.05$), but no effect of the non-visual condition. Also found was a simple effect of condition within the child group. Further Post Hoc analysis using the Tukey test revealed differences between the child group and the inexperienced adult group under the visual condition, and also between the child group and the experienced adult group again under the visual condition.

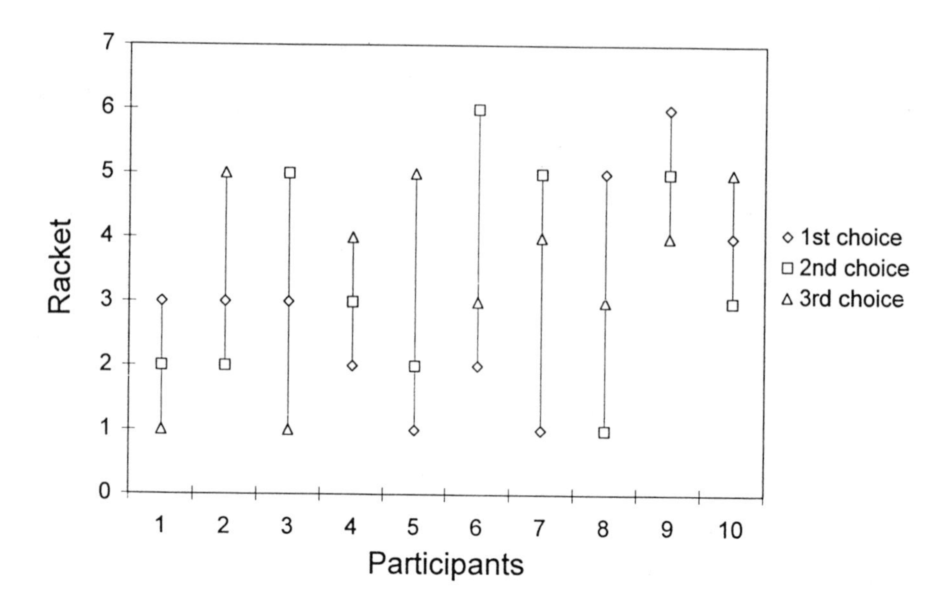

Fig. 2a The data illustrate the children's individual choice of rackets under the visual condition.

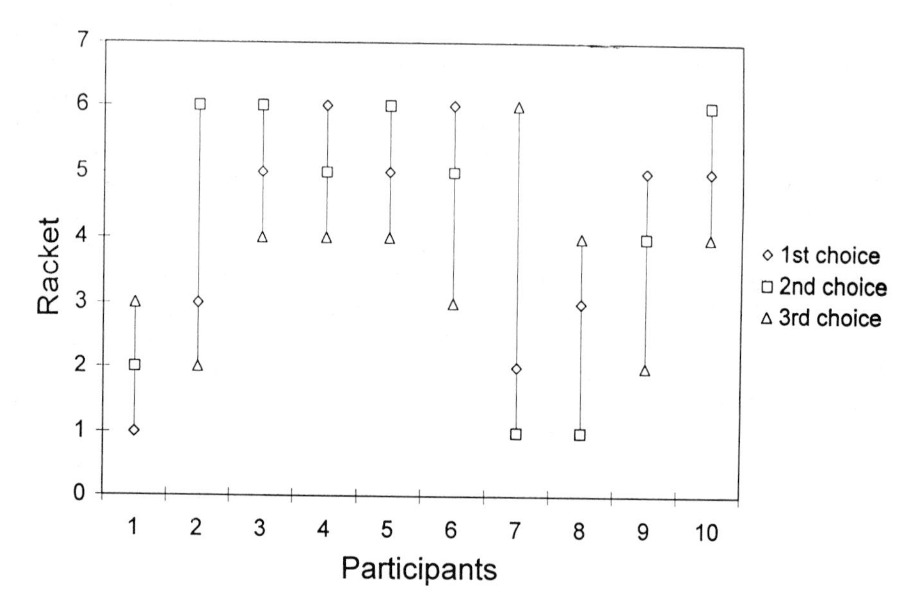

Fig. 2b The data illustrate the inexperienced adults' individual choice of rackets under the visual condition.

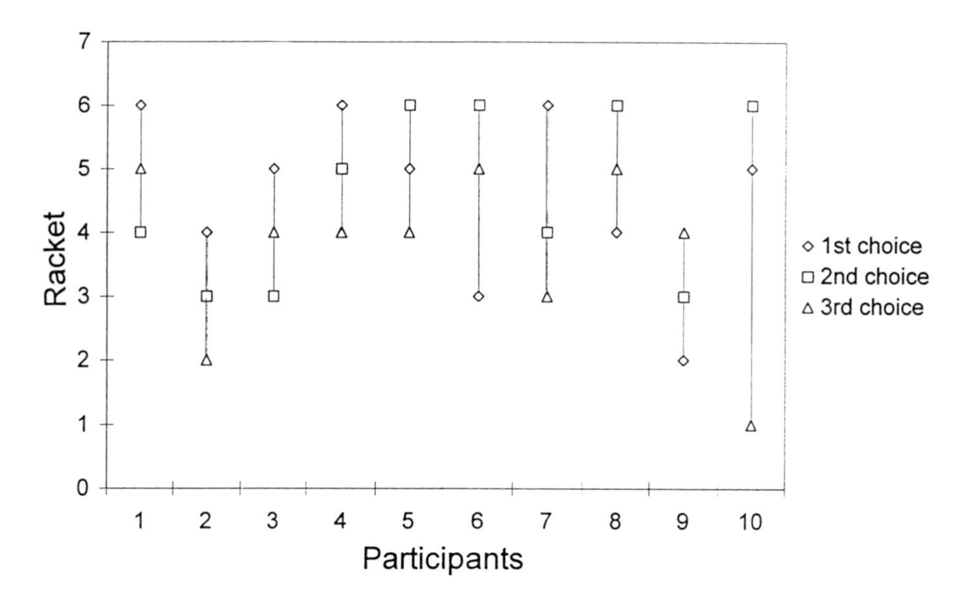

Fig. 2c The data illustrate the experienced adults' individual choice of rackets under the visual condition.

DISCUSSION

Results revealed that all three groups were sensitive to small changes in moment of inertia information from the wielded rackets. Data supported the findings of Solomon and Turvey (1988) and Bingham et al. (1989) on the significance of haptic information for perceiving characteristics of implements and objects during manipulation. In addition, differences were identified between the three group's preferences for rackets considered optimal for striking a ball to a maximum distance.

In the presence of both visual and haptic information, children experienced a conflict in informational constraints, creating greater variability in choices made. Without prior experience in racket wielding for striking, they seemed unable to integrate both visual and haptic information from manipulating the rackets. Without vision there was a greater clustering effect in the children's choices. These results appear to contest the view of Rock and Harris (1967) that, when vision and touch provide contradictory information, perception is dominated by visual information. The results from the present study suggest that preferences of the children were more definite and demonstrated less variability when vision was occluded.

In contrast, the experienced adults showed slightly more variability in choice of rackets when vision was removed. Each available source provides *different* information about the properties of the rackets being wielded, permitting a more global understanding of what each racket affords the actor. If a valuable source of information is removed or unavailable, then only part of the overall picture can be perceived and understood. In contrast, the perceptual systems of the inexperienced children may not have been developed enough at this stage to select relevant information from the total pattern available. Therefore, when a part of the information pattern was removed (i.e., in the non-visual condition), the children found the remaining information easier to interpret.

The present study extends previous understanding through experimental manipulations of developmental status and task-specific experience. The sensitivity of each group to haptic information clearly altered as a function of age and experience. For the children, the task was relatively novel since they had no specific prior experience on which to base preferences. More intra- and inter-individual variability was predicted, and observed, in the choices made by the children. It is possible that they were sensitive to moment of inertia information on the basis of haptic experiences during early infancy and childhood. However, without specific experience in striking a ball with a racket, they were unable to map the haptic information gained through wielding onto the movement dynamics of each racket, and confidently judge which afforded a maximal strike. In this respect, their lack of experience and understanding about the information of the rackets constrained the perception-action system.

These findings have strong implications for skill acquisition. If novices are to become skilled at picking up haptic information sources, to which they are sensitive, they need to be exposed to a variety of tennis rackets. When introduced to new tasks learners should be given pleny of opportunities to explore all the possibilities offered by the equipment, the current developmental status of their bodies, and their immediate environment.

From figures 2a-c, it can be seen that the inexperienced adults demonstrated less inter-individual variability than the children, but not as much as the experienced adult group. This developmental difference may be explained through the greater opportunities that adults typically have for general wielding experiences. Non-specific experiences of wielding and perceiving for action may have allowed the adult perceptual systems to have become better attuned to the availability of haptic information from implements. So, when only haptic information was available to the inexperienced adults, they simply relied upon more general previous experiences to guide their actions and choices.

To summarise, these data support previous work suggesting that children and adults have a high level of sensitivity to the haptic invariant moment of inertia. This sensitivity can be exploited during wielding to help learners select appropriate rackets. Manufacturers need to re-consider the limited range of moment inertia values currently used to design tennis rackets. Future work needs to examine whether more experienced performers show greater sensitivity to changes in one of the two components of moment of inertia (m (mass) and r^2 (distance of m from the point of rotation), compared to less-experienced counterparts. Finally, the data imply that coaches and teachers need to provide a broad range of implements to match the wide range of organismic constraints amongst learners, during skill acquisition. In this sense, one size does not fit all!

REFERENCES

Beak, S., Davids, K. & Bennett, S.J. (2000). Child's play: Children's sensitivity to haptic information in perceiving affordances of rackets for striking a ball. In J.Clark & J.Humphrey (Eds.), Motor Development: Research and Reviews, V.4. New York: AMS Press.

Bingham, G., Schmidt, R.C., & Rosenblum, L. (1989). Hefting for a maximum distance throw: A smart perceptual mechanism. *Journal of Experimental Psychology: Human Perception and Performance,* **15**, 507-528.

Carello, C., Thout, S., Anderson, K.L., & Turvey, M.T. (1999). Perceiving the sweet spot. *Perception,* **28**, 307-320.

Rock, I. and Harris, C.S. (1967). Vision and touch. *Scientific American,* **216**, 96-104.

Savelsbergh, G., Wimmers, R., van der Kamp, J. and Davids, K. (1999). The development of movement control and coordination. In M.L. Genta, B. Hopkins, and A.F. Kalverboer (Eds.), *Basic issues in Developmental Biopsychology.* Dordrecht:Kluwer Academic Publishers.

Solomon, H.Y. & Turvey, M.T. (1988). Haptically perceiving the distances reachable with hand-held objects. *Journal of Experimental Psychology: Human Perception and Performance,* **14**, 404-427.

Turvey, M.T. (1996). Dynamic touch. *American Psychologist,* **51**, 1134-1152.

Dynamic properties of tennis strings

R. Cross
Physics Department, University of Sydney, Sydney, Australia

ABSTRACT: Measurements are presented on elastic and frictional properties of different tennis strings. Theoretical and experimental results show that the most desirable properties of a good string are high elasticity and a high coefficient of friction between the strings and the ball.

LABORATORY TESTS

Most tennis players are agreed that natural gut is the best string to use in a racket, although it is usually too expensive for the average player, and it is not as durable as synthetic strings. Many professional players actually prefer synthetic strings, despite the generally inferior elastic properties of synthetics. Regardless of the type of string, elite players describe new strings as having a "crisper" feel than old strings, and they describe old strings as being lifeless or lacking the power of new strings. With this in mind, a number of laboratory tests were undertaken to compare the properties of a Babolat natural gut string with an Isospeed Professional synthetic string, claimed by some players to be comparable to or better than natural gut. Eighteen other strings were also tested.

Measurements of elongation vs tension were made using an Instron machine. In this test, a string was clamped at each end by means of metal jaws 300 mm apart and stretched at a rate of 100 mm/min to a tension of up to 700 N. Results are shown in Fig. 1. The important part of these curves is the change in elongation in the region from about 200 to 350 N, since this determines the increase in string plane stiffness during a high speed impact. Within this range, natural gut had the highest elasticity of all the strings tested. The Isospeed string was slightly less elastic, while all other strings were significantly less elastic, as illustrated by the Klip XL (nylon) string. The stiffness of the string plane for small displacements is proportional to the string tension and is independent of the actual elongation at that tension. Consequently, a racket strung with steel strings would have the same initial string plane stiffness as one strung with gut, if the string tension were the same and the displacement remained small. This is not the case for a fast serve or smash where the strings may be displaced by 30 mm or more, in which case there is a significant increase in tension and string plane stiffness due to the additional stretch of the strings. The result is a decrease in the impact duration, an increase in the force transmitted to the arm and an increase in the amplitude of frame vibrations, leading to differences in the "feel" of different strings.

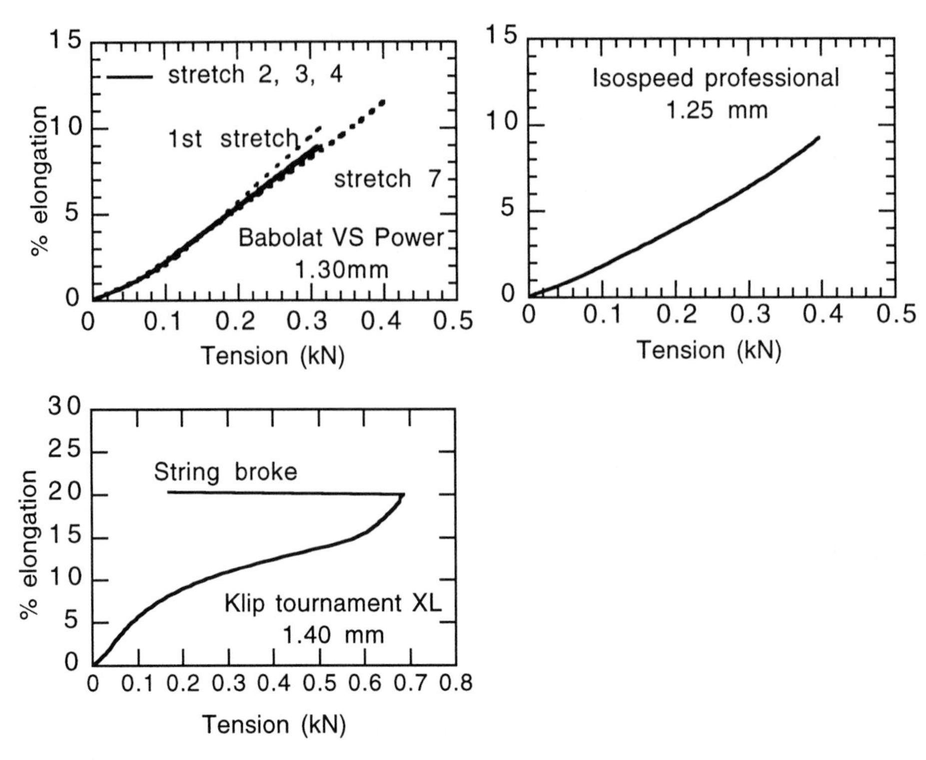

Fig. 1. Percentage elongation vs string tension for three different strings.

A second test involved stretching a string in a rigid metal frame, using a load cell to monitor the tension. This apparatus (Cross, 2000a) was designed to monitor changes in string tension with time and with repeated impacts, while leaving the length of the string fixed. The string was clamped in metal jaws separated by a distance of 340 mm prior to stretching, and then stretched to a tension of 250 N. The tension was then allowed to decrease for a period of one hour before applying an impulsive load to the string. The impulse was applied by allowing a mass of 0.29 kg to impact at low speed, 2.5 ms^{-1}, at the centre of the string at at right angles to the string. The mass was mounted at the end of a light wood beam and allowed to swing into the string as a pendulum or hammer. Losses in the string were estimated from the rebound speed of the hammer.

Under normal conditions, the strings of a racket experience a peak transverse force of up to about 1500 N. Such a force, acting on a ball of mass 57 gm over a period of about 5 ms, is required to change its velocity from +30 ms^{-1} to -30 ms^{-1}. The force is distributed over all the strings, but if one assumes that the force is shared mainly by five mains and five cross strings, then the peak force on each string is about 150 N. In the impact tests, the 0.29 kg mass changed its velocity from +2.5 ms^{-1} to about -2.4 ms^{-1} over a period of ~ 30 ms, giving a peak force of about 130 N on the single string. The impact duration is longer than normal, but this has the advantage of simulating the cumulative effect of a number of impacts each of duration 5 ms.

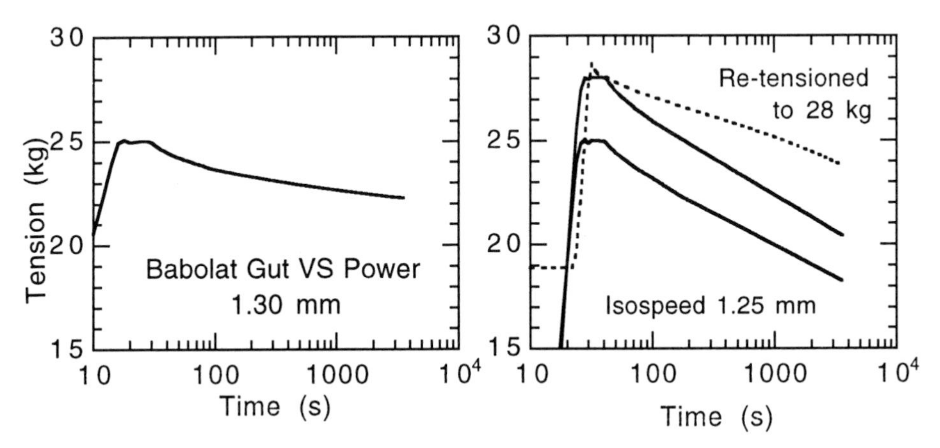

Fig. 2. Tension vs log(time) for two different strings.

Tension loss with time, due to stress relaxation, is shown in Fig. 2. Each string was tensioned to 25 kg, held at this value for about 10 s, and then clamped at a fixed length. The tension dropped rapidly at first, and then more slowly as time progressed. When the tension is plotted vs log(time), the result is linear after the first 100 s, even over periods of several days. Natural gut holds tension better than any other string tested, but Isospeed dropped faster than any other string. If a string is re-tensioned after this test, the tension then drops at a significantly lower rate. Figure 2 shows this effect for a fresh Isospeed string tensioned to 28 kg, then re-tensioned to 28 kg after one hour. This can be repeated many times, and the rate of tension loss decreases each time the string is re-tensioned. One can model this effect in terms of the gradual breaking of bonds between long chain molecules, where the weakest bonds break first.

IMPACT RESULTS

When a string is subject to an impact, the tension rises during the impact and then falls to a value lower than that before the impact. This effect is shown in Fig. 3 for a series of 10 impacts about 40 s apart. The data was digitised at a rate of one point every 2 seconds, so the increase in tension during each impact was not properly recorded, apart from the occasional single point captured during an impact. The string tension and the string displacement were also monitored on a storage oscilloscope. The increase in tension (ΔT), the impact duration (τ) and the maximum string displacement (y) varied by only a few percent from the first to the last impact. The average values for the three strings tested, as well as a steel piano wire (at a reduced hammer speed 2.0 ms^{-1}), are shown in Table 1. The % elongation prior to impact, at a tension of 25 kg, is also shown.

All three tennis strings were tested after relaxing from 25 kg for an hour, so they were tested at slightly different tensions. However, the transverse stiffness of each string was approximately the same, as indicated by the values of τ and y. The Klip string was somewhat stiffer, with a smaller τ and y due to the relatively large increase in string tension during each impact. The stiffness of this string could be reduced by reducing the tension, but the subsequent increase in tension during an impact is then *larger* than the 18.5 kg shown in Table 1. The impact duration can be extended by reducing the tension, but the force waveform is then more non--linear with a narrower peak. Of all 20 strings tested, Babolat gut had the smallest increase in tension, followed by Isospeed. Steel is far too stiff at large displacements to use as a tennis string. Soft strings "feel" better since the peak force on the strings is relatively low and since a long duration impact does not excite frame vibrations as efficiently as a short duration impact.

Table 1 Impact parameters

String	Diameter	Elongation	ΔT	τ	y
Babolat gut	1.30 mm	8.4%	10.0 kg	35 ms	30.6 mm
Isospeed	1.25 mm	4.2 %	15.0 kg	35 ms	30.8 mm
Klip nylon	1.40 mm	9.7%	18.5 kg	32 ms	27.6 mm
Piano wire	1.00 mm	0.2 %	62.0 kg	18 ms	16.1 mm

The ratio of the rebound to incident speed of the hammer was 0.96 ± 0.01 for all strings regardless of the number of impacts. The nylon string was tested with 200 impacts. Despite the severity of each impact and the significant loss in tension after 200 impacts, the hammer still rebounded at 95% of the incident speed. This result is surprising since it is commonly assumed that strings lose resiliance with age and use.

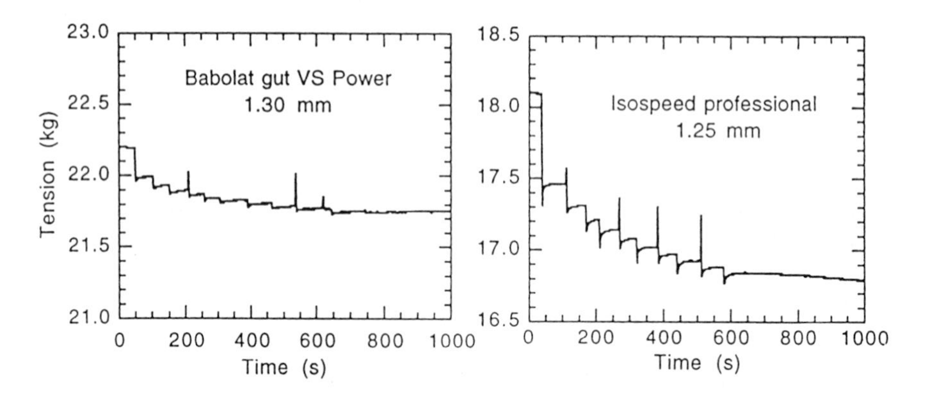

Fig. 3 Tension vs time when a string is subject to 10 impacts.

RESULTS WITH A STEEL BALL

Additional tests were made by dropping a 760 gm steel ball (a boule) from heights up to 2.4 m onto the strings of a racket. The racket head was rigidly clamped to a massive timber block resting on a solid timber floor. The ball was released from an electromagnet to land precisely at various points on the string plane, and the bounce height and angle were measured either using a laser beam (at drop heights < 1.5 m) or by eye against a grid for drop heights > 1.5 m. The kinetic energy of the steel ball, when dropped from a height of 2.4 m, was equivalent to that of a tennis ball incident at 24 ms^{-1}.

Several different rackets strung at various tensions were tested this way. It was found that the ball rebounds at 95 ±2% of the incident speed regardless of whether the strings are new or very old or worn, and regardless of the drop height or impact energy. This result is consistent with the impact tests on a single string and it shows that energy losses in the strings are negligible. However, the results are completely at odds with the alleged loss of power in well--used strings as reported by players.

THEORETICAL CONSIDERATIONS

It is well known that racket power can be increased by reducing string tension since the collision is softer and less energy is dissipated in the ball. However, recent calculations show that this effect is essentially negligible (Cross, 2000b; 2000c). The magnitude of the effect can be estimated by modelling the ball as a mass m_1 connected to a spring of spring constant k_1, and modelling the racket as a mass m_2 connected to a spring (ie the strings) of spring constant k_2. An impact of the ball on concrete can be modelled with $k_1 = 4 \times 10^4$ Nm^{-1} and a coefficient of restitution (COR) $e_1 = 0.75$. If the ball impacts on the strings of a head-clamped racket, and if there is no energy dissipation in the strings, then the COR increases to a value e given by

$$e^2 = \left(k_1 + k_2 e_1^2\right) / \left(k_1 + k_2\right)$$

For example, if $k_2 = 3 \times 10^4$ Nm^{-1} then $e = 0.90$, and if $k_2 = 4 \times 10^4$ Nm^{-1} then $e = 0.88$. The COR is unaltered if the head is not clamped, provided the ball impacts at the vibration node in the centre of the strings. In this case, a ball incident at speed v_1 on a stationary racket will rebound at speed $v_2 = e_A v_1$ where

$$e_A = \left(e m_2 - m_1\right) / \left(m_1 + m_2\right)$$

is the apparent coefficient of restitution, typically about 0.4 for an impact near the centre of the strings. Alternatively, if the racket is incident at speed v_R on a stationary ball, then the ball will be served at speed

$$v_{out} = \left(1 + e_A\right) v_R$$

The effective mass of the racket at the centre of the strings is typically about three times larger than the mass of the ball. If $m_2 = 3m_1$, then v_{out}/v_R increases from 1.413 to 1.426 when k_2 is reduced from 4×10^4 to 3×10^4 Nm^{-1}. Since k_2 is directly proportional to the string tension, it can be seen that a 25% decrease in string tension results in an increase of only 0.9% in the ball speed.

It appears likely, therefore, that any loss in string power must be associated with a reduction in the transverse component of the ball velocity, rather than a decrease in the normal component. The dynamics in a direction perpendicular to the string plane depend on the COR. In a direction parallel to the string plane, the dynamics depend on both the COR and the coefficient of sliding friction, μ_S, between the ball and the strings. If μ_S is small, then ball will tend to skid and rebound at a low angle, similar to the rebound off a grass surface. If μ_S is large, the ball will rebound at a larger angle, similar to the rebound off a clay surface. Detailed calculations of this effect

(Cross, 2000d) show that the ball trajectory does not depend strongly on μ_S when $\mu_S > 0.3$ but there is a large change in the trajectory when $\mu_S < 0.3$, as shown in Fig. 4. When $\mu_S < 0.3$, the ball slides along the strings without rolling. When $\mu_S > 0.3$, there is usually sufficient friction for the ball to start rolling before it rebounds off the strings. The coefficient of rolling friction is very small, and has no significant effect on the parallel dynamics.

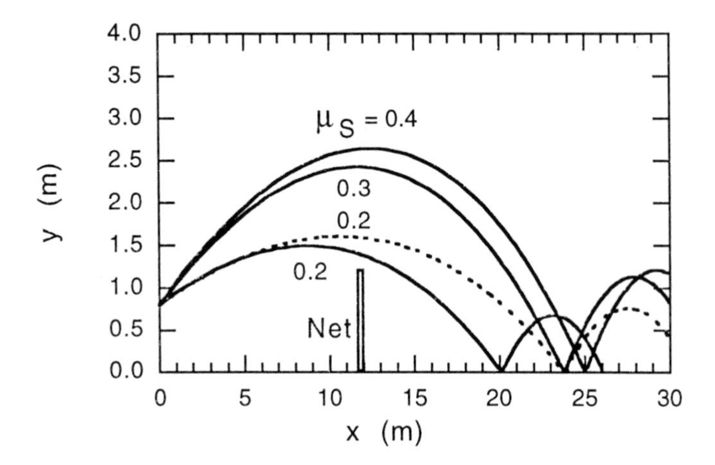

Fig. 4 Ball trajectories for a topspin forehand. The ball is incident horizontally at 10 ms^{-1}, striking a racket moving at 24 ms^{-1} and rising upwards at 45°, with $\mu_S = 0.2$, 0.3 or 0.4 (solid lines). The dashed curve is for a racket speed 32.3 ms^{-1}, with $\mu_S = 0.2$. In each case, the racket head is tilted forwards 5°.

The parameters in Fig. 4 were chosen so that the ball lands on the opponent's baseline when $\mu_S = 0.3$ and when the racket speed $v_R = 24$ ms^{-1}. If $\mu_S = 0.4$, then the ball would land 1.23 m beyond the baseline. If μ_S drops to 0.2, then the ball would land 3.7 m short of the baseline since the ball rebounds from the racket at a relatively small angle. In the latter case the player could correct the stroke, for example by increasing the racket speed to $v_R = 32.3$ ms^{-1}, in order for the ball to land on the opponent's baseline, as shown by the dashed trajectory in Fig. 4. A player is likely to conclude, if μ_S drops below 0.3, that the strings are not as responsive or that the strings have lost power. This conclusion will be reinforced if the player needs to hit the ball much harder to achieve the same range as strings with $\mu_S = 0.3$. A more severe test of string friction is provided by the topspin lob, shown in Fig. 5 for a case where the racket rises upwards at 60° to the horizontal at a speed $v_R = 25$ ms^{-1} with the head vertical. The ball is incident at 10 ms^{-1} and at an angle of 20° down from the horizontal. In this example, the ball reaches a height of 3 m only when $\mu_S > 0.37$.

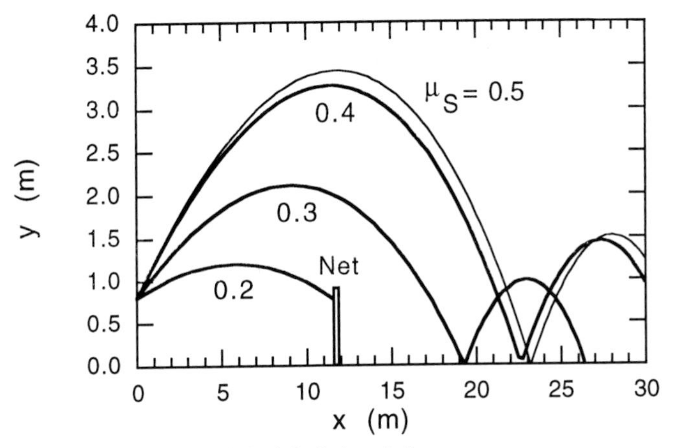

Fig. 5 Topspin lob, with μ_S = 0.2, 0.3, 0.4 or 0.5.

FRICTION MEASUREMENTS

A simple experiment was undertaken to determine the coefficient of sliding friction between a ball and the strings of a racket, at relatively low values of the normal reaction force. Using masses up to 10 kg held lightly on top of a new tennis ball to prevent the mass toppling, and a spring balance to measure the force needed to drag the ball at constant speed across the strings, it was found that μ_S varied from 0.27 to 0.42 for five different rackets, with Isospeed having the highest value of μ_S. By comparison, it was found that μ_S = 0.8 for a tennis ball sliding on rubber, 0.54 when sliding on the Sydney Olympic Games tennis courts, 0.25 on a low pile carpet, 0.13 on a smooth wood surface and 0.68 on a fine grain emery paper (grit size ~ 0.02 mm).

A number of individual strings were also tested, by wrapping three turns around a 66 mm diameter, horizontal cylinder covered with tennis cloth. By hanging a 200 gm mass on one end and pulling the mass at constant speed by a force applied on the other end, the relative coefficients of sliding friction of different strings could be determined. For example, a force of 3.5 kg was required using teflon coated wire, a force of 4 kg was needed for the natural gut string and a force of 11 kg was needed for the Isospeed string. All other strings tested required a force between 3 and 4 kg. The force is exponentially rather than linearly proportional to the number of turns and to the coefficient of friction in this experiment. The results of this test are in fact consistent with the measurements of μ_S since $e^4 \cong 11/0.2$ and $e^3 \cong 4/0.2$. These results imply that μ_S is determined by the surface properties of a string rather than the overall roughness of the string plane. This was confirmed by applying a thin film of epoxy and fine grit to the strings of a racket, which had the effect of increasing μ_S from 0.4 to 0.6.

CONCLUSIONS

A comparison of various tennis strings showed that the two most highly rated strings, natural gut and Isospeed professional, are more elastic than other strings. By itself, this result seems to indicate that the most desirable feature of a string is high elasticity at string tensions in the range from about 200 to 350 N. It was also found that natural gut holds its tension better than any other string tested, but Isospeed was the worst string in this respect, dropping from 25 kg to 21 kg in only 10 minutes. This result suggests that string tension is not as important to a player as is commonly assumed. The increase in racket power at low string tension cannot explain the popularity of Isospeed, since this effect was found to be too small to be of any significance.

Apart from the fact that spaghetti strings are banned by the ITF, since they can be used to impart excessive spin on a ball, very little attention has been paid to the significance of string friction. Calculations of the ball trajectory show that the performance of a string deteriorates rapidly if μ_S drops below about 0.3. The rebound speed of the ball does not change significantly, but the rebound angle drops, with the result that the ball lands short of the target. This is likely to be interpreted by the player as a loss of racket power. Measurements show that μ_S is typically between 0.3 and 0.4. Isospeed had the largest value of μ_S, which probably accounts for its popularity. A thin film of epoxy and fine grit on the strings increases μ_S considerably.

REFERENCES

Brody, H. (1979) Physics of the tennis racket. *Am. J. Phys.* **47**, 482-487.

Cross, R. (2000a) Physical properties of tennis strings. *3rd ISEA International Conference on the Engineering of Sport*, Sydney. in press.

Cross, R. (2000b) Effects of string tension and frame stiffness on racket performance. *Sports Engineering* in press.

Cross, R. (2000c) The coefficient of restitution for collisions of happy balls, unhappy balls and tennis balls. *Am J. Phys.* in press.

Cross, R. (2000d) Effects of friction between the ball and strings in tennis. *Sports Engineering* submitted for publication.

Methods to determine the aerodynamic forces acting on tennis balls in flight

S.G. Chadwick & S.J. Haake
Department of Mechanical Engineering, University of Sheffield, UK

ABSTRACT: The aim of this study is to assess the methods available to measure the aerodynamic forces acting on a tennis ball during flight. The assessment covered a range of spin rates, velocities and angles, but also looked at the effect of improving the quality of source data and smoothing techniques. The methods assessed incorporate wind tunnel techniques, projecting devices, digital photography and motion analysis techniques. Given the equipment available, a wind tunnel technique utilising a force balance was found to return the best quality results for drag forces, whilst a wind tunnel technique utilising a specially designed spin drop device and digital photography should return the best results for lift forces.

INTRODUCTION

The trajectory of a tennis ball is determined by the gravitational and aerodynamic forces acting on it during its flight. This study was designed to find the drag and lift forces acting on a tennis ball during its flight, and discuss the methods used to best achieve it.

The drag and lift forces are a function of the ball characteristics and the fluid through which it passes. Equations 1 and 2 show the relationship between drag/lift forces and drag/lift coefficients,

$$F_D = \tfrac{1}{2} \rho V_R^{\,2} A C_D \qquad (1)$$

$$F_L = \tfrac{1}{2} \rho V_R^{\,2} A C_L \qquad (2)$$

where, F_D is the drag force
F_L is the lift force
ρ is the density of the fluid within which the ball is projected
V_R is the velocity of the fluid
A is the projected area of the ball
C_D is the drag coefficient
C_L is the lift coefficient

The average serve speed of a top male competitor can be up to 58ms^{-1} (130 mph) and may be struck with very little spin. In contrast, a drop shot can be struck with a much reduced velocity and spin rates in excess of 3000 rpm.

Aerodynamic studies on tennis balls are limited, generally related to specific shot types (Stepanek, 1988). There have been two main methods used in previous work: one involving dropping a ball through the airflow of a wind tunnel (Davies, 1949); the second utilising a 3 component wind tunnel balance, generally using a larger model ball (Bearman and Harvey, 1974). A further method used the 3-dimensional trajectory of a volleyball as projected by a human (Deprá, 1998). It is important that the object being tested remains a true representation of the original and obviously the best method of this is to use the original unmodified version. Each of the methods mentioned has been adapted to find the C_D and in some cases C_L of tennis balls.

THREE COMPONENT WIND TUNNEL BALANCE

There have been two methods adopted for use in this study, many of the characteristics of the test method are the same, differing mainly in measurement techniques and wind tunnel attributes. Testing involves the use of a 3 component wind tunnel balance capable of measuring drag force, lift force and pitching moment. Thus far, however, only the force due to drag had been studied.

The balls are attached to the 3 component wind tunnel balance via a sting. Whilst the sting is different in each test, the purpose and design criteria are the same. The sting protrudes from the rear of the ball and is rigidly connected ensuring that all forces exerted are translated to the 3 component wind tunnel balance. Shrouding has been utilised to minimise drag forces exerted on non-essential sting components.

A digital manometer attached to the wind tunnel was used to give the dynamic pressure across the working section which was used to calculate the wind velocity in the working section.

There have been two wind tunnels and two measurement techniques used for this study. There are no constraints regarding which measurement technique is used in which wind tunnel and the set-ups used are described below.

HIGH SPEED WIND TUNNEL

The high speed wind tunnel is rated with a maximum wind velocity of approximately 61ms^{-1}, with a working section of 1.65m x 1.2m. The wind velocity was increased from zero to approximately 61ms^{-1} in equal increments giving 22 readings over the complete range. Force is translated from the sting assembly via fine wires, weighted down to ensure tension.

Displacement transducers are used to monitor any out of balance effects, the output shown on a digital voltage meter. A moving mass is used to regain the balance point, the distance moved by the mass being calibrated against drag force.

LOW SPEED WIND TUNNEL

The low speed wind tunnel is rated with a maximum wind velocity of approximately $26ms^{-1}$, with a working section of 715 mm x 512 mm. The wind velocity was increased from zero to approximately $26ms^{-1}$ in equal increments giving 12 readings over the complete range. Force is translated from the sting assembly via a rigid 25 mm diameter bar.

Displacement transducers are used to monitor the force applied to the sting assembly, with the output shown on a digital voltmeter. The voltage difference is directly calibrated to the drag force.

RESULTS

The 3 component wind tunnel balances are calibrated prior to experimentation, hence the values obtained can easily be converted to drag forces. Figure 1 shows results obtained for a 6" diameter smooth ball the aerodynamic properties of which are well known. The results for this ball match the results obtained by Achenbach (1974) which implies that this method is suitable. Figure 1 also shows the results obtained in the high speed wind tunnel for three balls with different surface properties. It can be seen that the ball with the fluffed nap has a higher C_D than the one with a normal nap, which in turn has a higher C_D than the ball with a shaved nap.

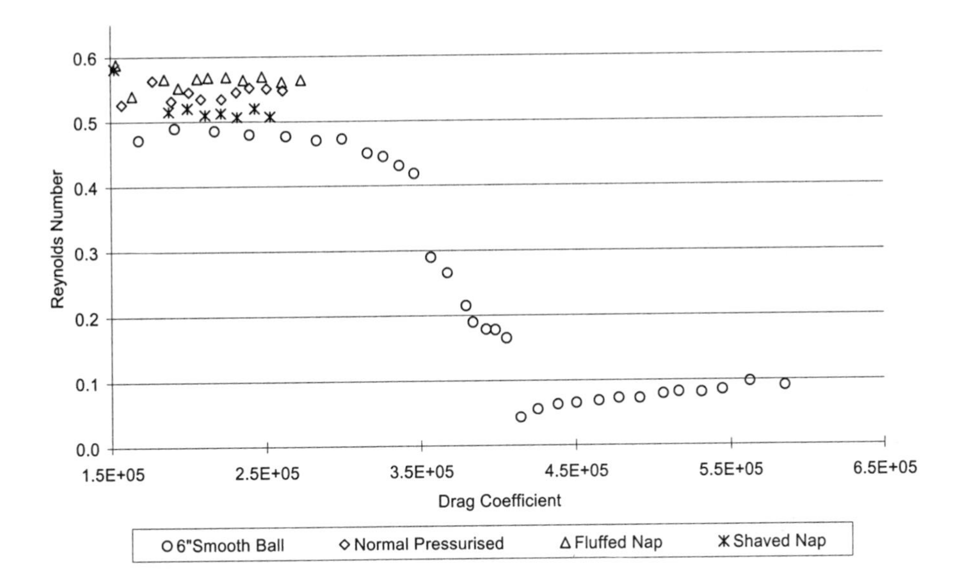

Fig. 1 Results of drag coefficient versus Reynolds number for 4 different balls in the high speed wind tunnel

When using the high speed wind tunnel, the results show some scatter at low velocities and with increased wind speeds the C_D becomes more stable. Given this fact, results from the high speed wind tunnel are limited to those above a Reynolds number of 1×10^5, whereas in the low speed wind tunnel useful results are those above a Reynolds number of 6×10^4. These are not necessarily limitations of the wind tunnels, more the method by which the C_D is being deduced. The method used in the high speed wind tunnel has typical errors in the range of 4.8%, and errors are reduced to less than 2% when using the method in the low speed wind tunnel.

The trajectory of an object is affected both by drag and lift, but thus far only C_D has been obtained. Whilst it is possible to find C_L using a 3 component wind tunnel balance, the equipment required is both large and intricate, which could lead to increased subtraction errors and high development costs. It was decided that a more suitable way forward would be to obtain results for both C_D and C_L using a method of dropping a spinning ball through the moving air stream of a wind tunnel.

USING TRAJECTORIES TO CALCULATE C_D AND C_L

The method of dropping a ball through the moving air of a wind tunnel utilises a dropping device and a method of capturing images. The wind tunnel has a working section measuring 715 mm deep; 512 mm high; and 910 mm long. The dropping device utilises pneumatic pistons to hold the ball in place, whilst a 24V motor rated at 10,000 rpm applies spin to the ball. Rotational speed is varied using a motor controller while a digital voltage readout allows a calibration against spin rate.

At the desired spin rate the pistons are activated to release the ball. Wind velocity through the working section was increased in equal increments to a maximum of approximately $20 ms^{-1}$. A selection of rotational speeds, ω, and wind speeds, v, were chosen to give a range of values ω/v.

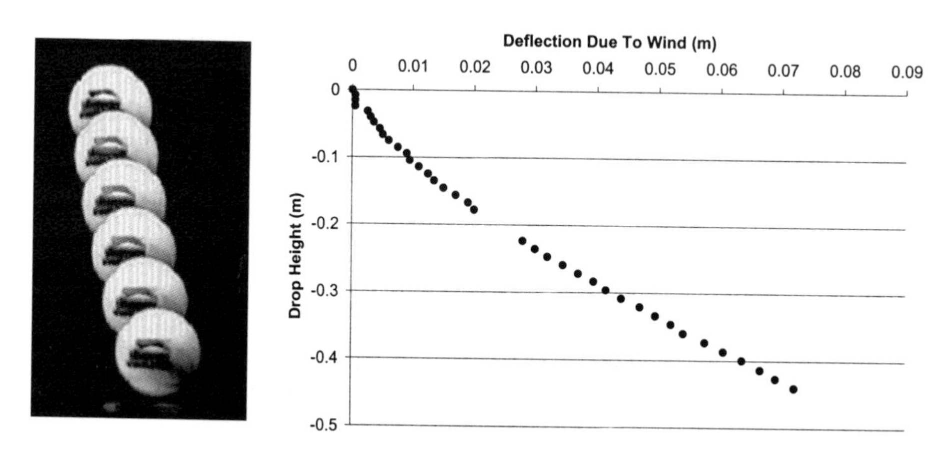

Fig. 2 **Left**: A series of images from a typical ball drop with no spin captured using a KODAK Motioncorder high speed video camera, airflow travelling from left to right. **Right**: The resulting digitised trajectory.

The motion of the ball during flight was captured digitally using a KODAK Motioncorder high speed video camera at a frame rate of 240 frames per second and a shutter speed of 1/10,000 second. The frames containing the flight of the ball were transferred from the Motioncorder and captured digitally on a computer, creating a collection of images over the duration of the flight. A typical set of images for a non-spinning ball drop is shown in Fig. 2. Motion analysis software was used to obtain the co-ordinates of the ball at several points for each trajectory. The co-ordinates were then used to calculate speeds, spins and angles at all points along the trajectory.

Figure 3 shows the force diagram developed to describe the motion of a rotating object passing through a flow of air. The equations of motion for the given scenario can then be defined from the force diagram.

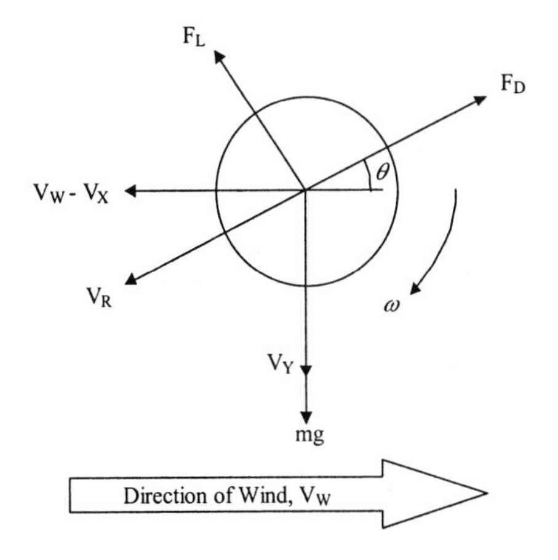

Fig. 3 Force diagram for an object falling through a moving air stream.

Hence,

$$m\frac{d(V_W - V_X)}{dt} = F_L \sin\theta - F_D \cos\theta \qquad (3)$$

$$m\frac{dV_Y}{dt} = mg - F_L \cos\theta - F_D \sin\theta \qquad (4)$$

$$\tan\theta = \frac{V_Y}{(V_W - V_X)} \qquad (5)$$

A computational trajectory model has been developed to assess the various methods used to find C_D and C_L, formed using the equations of motion. A selection of the methods assessed and the algorithms used are as follows.

METHOD 1 - manipulating the equations of motion to give C_D and C_L.

$$C_D = \frac{2m}{\rho V_R{}^2 A(dt)}\left[g(dt) - d(V_Y) - \frac{d(V_W - V_X)}{\tan\theta}\right] \qquad (6)$$

$$\Leftrightarrow C_L = \frac{2md(V_W - V_X)}{\rho V_R^{\ 2} A(dt)\sin\theta} + \frac{C_D}{\tan\theta} \tag{7}$$

where, V_W is the wind tunnel air stream velocity
V_Y and V_X is the vertical and horizontal components of the ball velocity

METHOD 2 - possibly the most common method used to find C_D and C_L was that utilised by Davies (1949) and later by Stepanek (1988). The analysis uses the total deflection of the ball due to wind and spin for a known drop height. Calculations are simplified with the use of a modified force diagram and the assumption that acceleration is constant. The initial velocity is assumed to be zero as the object is dropped from within the wind tunnel. A modification to the calculations was required to allow for a drop from outside the wind tunnel.

$$C_D = \frac{4m}{\rho V_R^{\ 2} A T^2}\left(Deflection - (V_{Xi} \times T)\right) \tag{8}$$

$$C_L = \frac{2}{\rho V_R^{\ 2} A}\left[\left(\frac{2m}{T^2}\left((V_{Yi} \times T) + DropHeight\right)\right) + mg\right] \tag{9}$$

where, T is the total time taken for the drop
V_{Xi} and V_{Yi} are the initial velocities in the x and y direction respectively

For all analyses an ideal trajectory was used, calculated using the equations of motion with a C_D of 0.55. The trajectory was modified to incorporate the effect of spin, starting with $C_L = 0$ denoting zero spin, then increasing it to ± 0.2 simulating clockwise and anti-clockwise spin. To further simulate an actual drop, an initial velocity was introduced, where the ratio of V_{Yi} with V_{Xi} can also be used to modify the incoming angle.

Figure 2 shows that the source data is far from perfect, especially when considering a change in velocity and hence further modification was required. To simulate real data, scatter was applied to the perfect data as a tolerance. The effect of the scatter was then reduced using a 'best fit' polynomial function applied to the drop and deflection data against time.

RESULTS

Figure 4 shows the calculated C_D and C_L values using ideal theoretical data for a trajectory with: incoming velocities of 1.8 ms^{-1} vertical and 0.09 ms^{-1} horizontal; wind velocity of 11.5 ms^{-1}; C_L equal to 0.2; a scatter of 0.037% drop and 0.264% deflection; and a smoothing polynomial. Several different orders of polynomial have been used to find the best approximation of the trajectory. The results obtained using method 1 return a good correlation with the values of input C_D and C_L for all polynomials used, however, higher order polynomials appear to induce small errors. Method 2 returns results with a clear offset from the input C_D and C_L, and as the method uses only the start and end points of the trajectory, the smoothing function makes little or no difference.

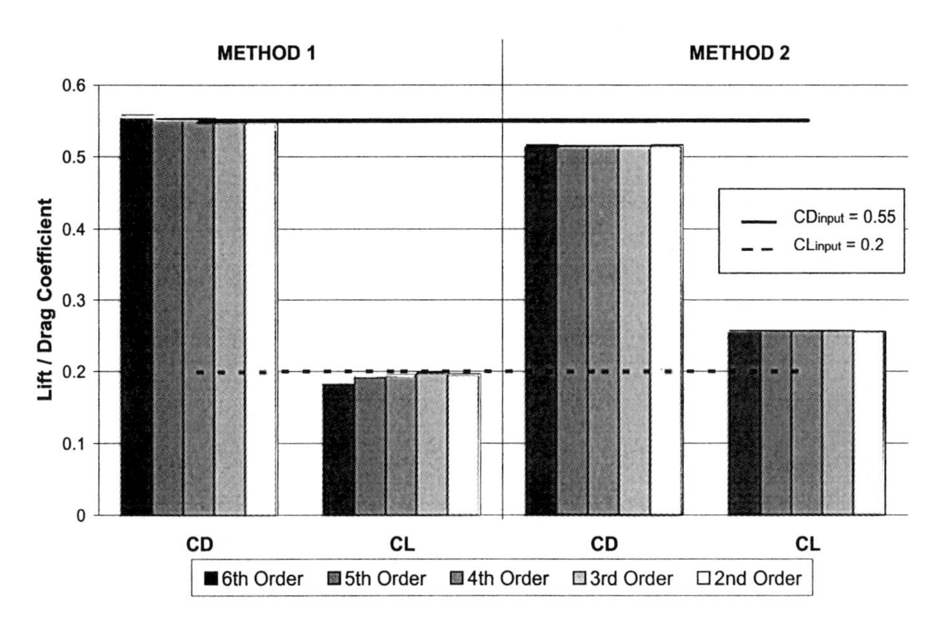

Fig. 4 Results obtained for C_D and C_L using polynomial smoothing for an ideal trajectory with 0.037% drop and 0.264% deflection scatter introduced.

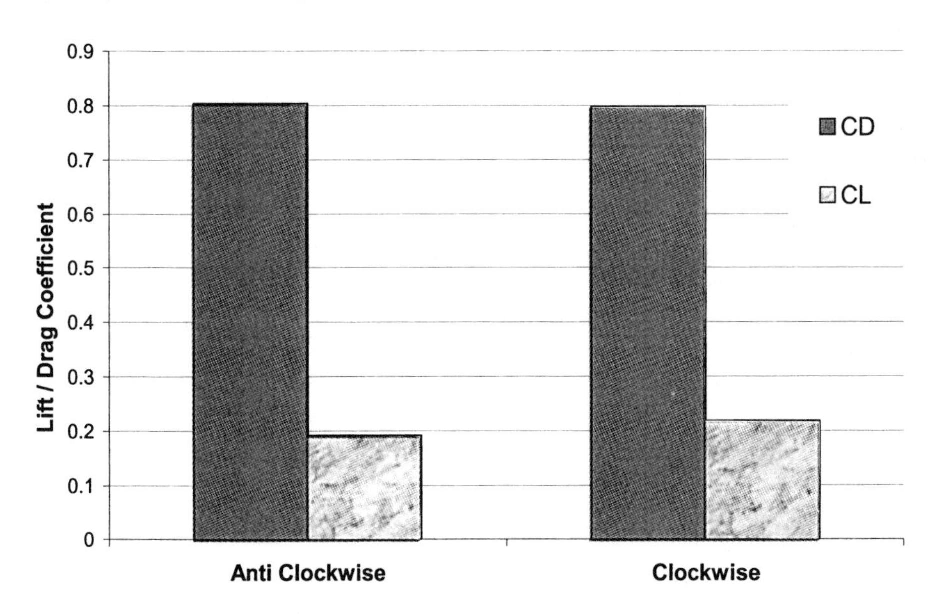

Fig. 5 Results obtained for C_D and C_L using polynomial smoothing for real data for a ball dropped with ±1600 rpm through an air stream moving at 11.6 ms^{-1}.

Figure 5 shows the calculated C_D and C_L values obtained using method 2 for an actual ball dropped through an air stream with wind velocity of 11.6 ms^{-1} and a spin rate of 1600 rpm both anti-clockwise and clockwise. Stepanek (1988) found results of approximately 0.73 and 0.25 for C_D and C_L respectively which compare well with these shown here. Results using method 1, however, were certainly not so good.

CONCLUSIONS

From the two methods discussed it is clear that the results obtained for C_D using the 3 component wind tunnel balance are both the easiest to obtain and the most accurate. Using this method to find C_L is possible, however the materials required for the sting using the present set-up would be large and heavy, possibly leading to substantial subtraction errors. It is feasible that a load cell could be designed and developed specifically for the purpose of finding the forces on a tennis ball.

The method of dropping a ball through the moving air of a wind tunnel was found to return suitable results for both C_D and C_L providing the source data was of good quality. Two methods have been discussed in this paper, and whilst method 1 was found to be superior to method 2 when using perfect data, errors were apparent when using real data. Further work is required to develop this method and reduce the errors incurred.

ACKNOWLEDGEMENT

The authors gratefully acknowledge the funding of this project by the International Tennis Federation. Thanks also to Dr. Alison Cooke and the technical staff at Cambridge University Aero Department for their help in setting up the project.

REFERENCES

Bearman, P. W. & Harvey, J. K. (1976) Golf Ball Aerodynamics. *Aeronautical Quarterly*, **27**, 112-122

Davies, John M. (1949) The Aerodynamics of Golf Balls. *Journal of Applied Physics*, **20**, 821-828.

Deprá, P. (1998) Fluid Mechanics Analysis in Volleyball Services. *Proceedings XVI International Symposium on Biomechanics in Sports, pub UVK–Universitätsverlag Konstanz GmbH, Germany*

Stepanek, A. (1988) The Aerodynamics of Tennis Balls - The Top Spin Lob. *American Journal of Physics*, **56**, 138-142.

Achenbach, E. (1974) The Effects of Surface Roughness and Tunnel Blockage on the Flow Past Spheres. *Journal of Fluid Mechanics*, **65**, 113-125

Tennis science collaboration between NASA and Cislunar Aerospace

J.M. Pallis
Cislunar Aerospace, Inc., Napa, California, USA
R.D. Mehta
National Aeronautics and Space Administration, Moffett Field, CA, USA

ABSTRACT: The National Aeronautics and Space Administration (NASA) and Cislunar Aerospace, Inc. have developed an Internet based education project designed to instruct students and educators on sport science. Using tennis as its theme, the materials provide an interactive study of the basic aerodynamics, physics, mathematics, motion capture and analysis techniques and biomechanics of sports. The researchers conduct experiments and report results on ball/court interaction, wind tunnel tests, serve racquet head speed, ball speed and spin, computational fluid dynamics (CFD) simulations, and professional player biomechanics such as footwork and body flex. Wind tunnel experiments have included air flow visualizations of top and underspin balls as well as tests to determine the drag of a variety of tennis balls including the new larger tennis balls.

INTRODUCTION

The "Aerodynamics in Space Technology" and "Sport Science" projects are two collaborations between NASA and Cislunar Aerospace, Inc. Supported by NASA's Learning Technology project and the Fluid Mechanics Laboratory, these projects serve as part of NASA's education outreach mission to communicate science and introduce Internet technology into grade K-12 classrooms.

Using tennis as its theme, the projects provide an interactive study of sports science designed to teach students basic aerodynamics, physics, mathematics, motion capture and analysis techniques and biomechanics. Research questions and experiments are defined by a team of aeronautical engineers, sports scientists, and educators. Students learn about tennis science, sport science and engineering careers, follow the progress of the project, interact with researchers and participate in on-line activities such as web chats and Internet video conferences through the project's web site (http://wings.ucdavis.edu/Tennis). As part of the project, the team conducted experiments and reported results on ball speed, ball spin, ball/court interaction, wind tunnel tests, CFD and player biomechanics. The project showcases computational and experimental methods developed by NASA and other researchers and has made use of new generation high-speed digital cameras to follow equipment and player motion. The project team accumulated data of top players from professional tennis tournaments such as the US Open.

WIND TUNNEL TESTS

Flow Visualization Studies

An 11" diameter tennis ball (Wilson novelty ball) was used as the model in a 48" X 32" test section of an open-circuit wind tunnel at NASA Ames. The larger model was used so that the test could be conducted at lower flow velocities, making it easier to visualize the flow patterns. The testing procedure involves matching the Reynolds numbers commonly encountered on a tennis court. The Reynolds number (Re) is defined as, $Re = Ud/\nu$, where U is the ball velocity, d its diameter and ν is the kinematic viscosity of air. Therefore, for a given Re, the ball (or flow) velocity can be lower if the ball diameter is increased. The flow visualization technique consisted of injecting smoke into the flow (ahead of the ball) and then observing the flow over the ball by illuminating the smoke particles using a light sheet (Fig. 1).

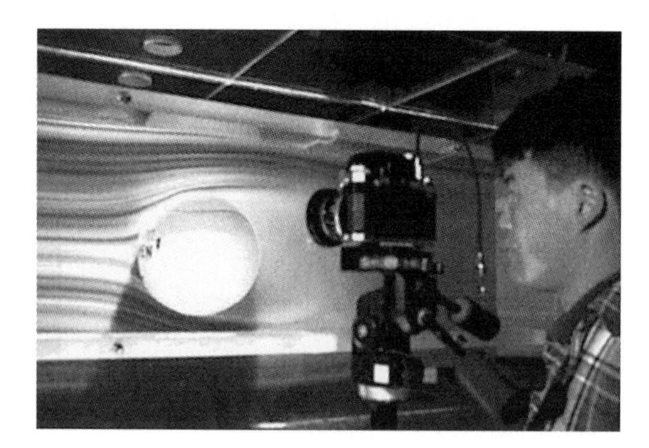

Fig. 1 Wind tunnel test with smoke blown over an 11" diameter tennis ball.

The ball was filled with a rigid polyurethane foam, which preserved the ball's shape and provided a means to attach the ball to a support structure. The model was bonded to a 1" steel rod that was attached to a support outside of the wind tunnel. A crank was mounted at the end of the steel rod, which allowed the ball to be spun. The flow visualization studies were conducted at flow velocities between 20 mph to 35 mph; this corresponded to a Re range of 180,000 to 300,000 and standard-sized tennis ball velocities of 90 to 150 mph.

The first tests were conducted with the ball stationary (not spinning). The first observation at the lower test velocities was that the boundary layers over the top and bottom of the ball separated relatively early, at about 90 degrees from the front stagnation point (Fig. 2). This normally implies that the boundary layers are laminar (transition to a turbulent state is yet to occur). However, on increasing the wind tunnel velocity (even up to the maximum), no significant changes were observed, much to our surprise. At some point, we expected the assumed laminar boundary layers to undergo transition and this would be evidenced by a sudden rearward movement of the separation points, thus leading to a smaller wake and less drag.

Fig. 2 Wind tunnel test over an 11" diameter tennis ball with no spin.

So the new conclusion was that the flow over the ball was in fact post-critical (turbulent boundary layer separation) over the whole Re range tested. Although we had expected the felt to affect the critical Re at which transition occurs, it seemed as though the felt was a more effective boundary layer trip than we had expected. The fact that the boundary layer separation over the top and bottom of the non-spinning ball is symmetric leading to a horizontal wake was, of course, expected since a side force (upward or downward) is not expected in this case. In one series, the ball orientation was altered to check for seam effects, but none were noted. Although ball seam orientation affects the flight and trajectory of other sports balls, these effects were not observed on the tennis ball. Unlike cricket balls and baseballs (Mehta 1985), the seam on a tennis ball is indented and the cover surface is very rough obscuring any seam effects.

Fig. 3 Wind tunnel test over an 11" diameter tennis ball with topspin.

In the second round of testing, spin was imparted to the ball by turning the crank that was attached to the ball support. In Fig. 3, the ball is spun in a counter-clockwise direction to simulate a ball with top spin while in Fig. 4 the ball is spun in a clockwise direction (underspin). A variety of spin rates were tested, equivalent to 1000 to 3000 rpm on a standard sized tennis ball. In Fig. 3, the boundary layer separates earlier at the top of the ball compared to the bottom. This results in an upward deflection of the wake behind the ball, and following Newton's 3rd Law of Motion, implies a downward (Magnus) force acting on it which would make it drop faster than a non-spinning ball. On the other hand, in Fig. 4 the wake is deflected downwards which means that the ball has an upwards Magnus force acting on it that would make it fly further than a non-spinning ball. The fact that the Magnus force occurred in a direction that was expected (positive Magnus force) over the whole speed range further confirmed that the flow over the ball was post-critical with turbulent boundary layer separation.

Fig. 4 Wind tunnel test over an 11" diameter tennis ball with underspin.

Drag Measurements

A second series of wind tunnel tests were conducted to determine the drag coefficient of several types of tennis balls over a range of Re. The drag coefficient is defined as, $C_d = D/(0.5\rho U^2 A)$, where D is the ball drag, ρ is the air density, U is the flow velocity and A is the ball projected area. Six balls were tested in this first phase of the experiments: a smooth ball (Plexiglas sphere), a new Wilson US Open ball, a used Wilson US Open ball, a bald (no felt) tennis ball, and two of the new larger diameter tennis balls (a 6.5% and a 8.5% larger ball both made by Wilson).

An open-circuit wind tunnel with a 15" by 15" test section (NASA Ames) was used. Each ball was mounted on a sting attached to a symmetric airfoil shaped strut. The strut was attached to a reaction torque cell. To damp vibrations from this assembly, the bottom of the strut had an extension attached to it, which protruded through the test section floor and was immersed in a small container of oil. The vibrations are mainly a result of unsteady forces induced by vortex shedding from the ball. The tare of the system was measured to account for the drag of the sting and strut assembly. In order to include additional effects caused by vortex shedding and turbulence in the ball wake, a tare ball was held in place through a mount from the side of the wind tunnel, but not touching the sting. The drag measurements were made over a Re range of 100,000 to 300,000 which corresponds to ball velocities of about 50 mph to 150 mph (Fig. 5).

In order to verify the accuracy of the whole measurement system, the first test was conducted on a smooth Plexiglas sphere with about the same diameter as a standard tennis ball. The results (Fig. 5) compare very well with the classic data of Achenbach (1972). A C_d of about 0.5 with a slight increase with Re initially, as the laminar boundary layer separation point creeps upstream, followed by a slight decrease at the higher Re as boundary layer transition is about to occur, are all evident in these results. Transition, evidenced by a sudden drop in C_d occurs at a Re of about 300,000 for a smooth sphere (too high for the present set-up). The turbulent boundary layer, by virtue of its more energetic state, is better able to withstand the adverse pressure gradient over the ball and separation is therefore delayed resulting in the sudden drop in $C_{d.}$ Premature transition can occur (at a lower Re) if the wind tunnel flow quality is inadequate (high turbulence levels) or if the model vibrates a lot; the fact that it does not occur here is again evidence of a suitable experimental set up.

Transition does occur at a Re of about 140,000 on the bald ball which was installed with the "seam" (a small step where the two halves of the ball are joined together) perpendicular to the flow. Beyond Re = 200,000 the Cd starts to increase again as the turbulent boundary layer is thickened by the seam and the separation line moves forward. Beyond Re = 280,00 or so, the Cd would be expected to level off at a near constant value of about 0.4. This is the regime the tennis ball data are in, only at relatively low Re, corresponding to ball velocities normally encountered in tennis.

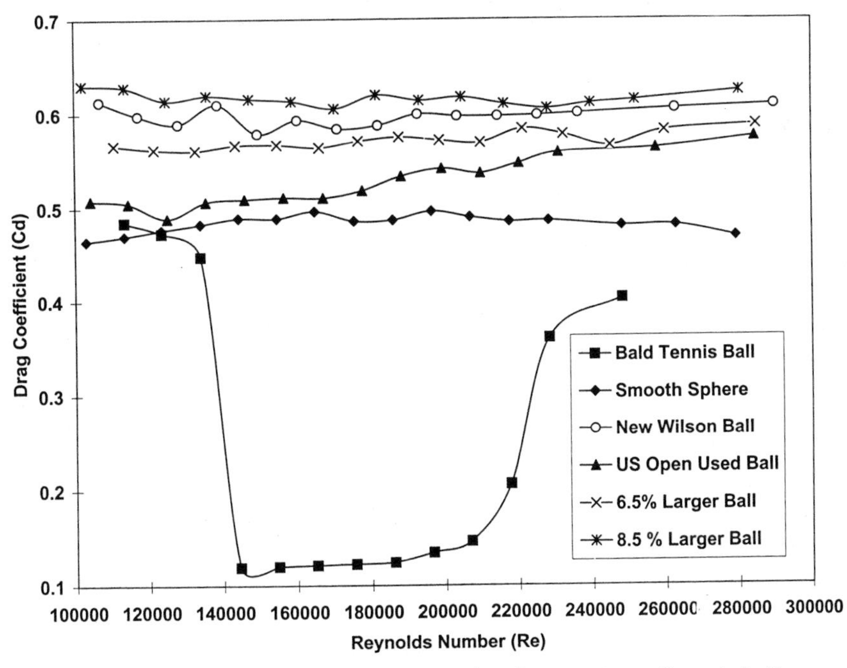

Fig. 5 Drag Coefficient versus Reynolds Number for a variety of tennis balls.

The results for the new standard tennis ball show a C_d of about 0.6 over the whole Re range. The data for the 8.5% larger ball shows a slightly higher C_d of about 0.62 while that for the 6.5% larger ball is slightly lower ($C_d \approx 0.57$). The main idea behind introducing a larger ball is to decrease the maximum ball velocity since the drag is proportional to the ball projected area. This is assuming that the C_d for the larger ball is about the same, or at least not lower, since then the desired effect of higher drag is clearly offset. The two larger balls looked quite new, although according to the supplier (USTA), they had been used in play for about 30 minutes. Also, the felt on these two balls had less nap than a standard ball, making the felt less thick. These results clearly show that the condition and type of felt can affect the C_d of a tennis ball, and not always in the expected direction. This effect of the felt is further borne out in the results for the used ball (used in the 1997 US Open for either 7 or 9 games). At the lower Re, the C_d for the used ball is just slightly higher than that for the smooth sphere. However, it increases with Re and at the highest Re, it is only slightly lower than that for a new standard ball. As the Re increases, the worn felt starts to thicken the turbulent boundary layer and the separation line moves upstream, similar to the effect observed for the bald ball at the higher Re.

COMPUTATIONAL FLUID DYNAMICS (CFD)

CFD is a numerical technique used to simulate fluid motion over objects. The methods have been used extensively in the design and analysis of air and spacecraft. The techniques allow scientists, engineers and designers to use computers to calculate the same flow forces and other physical characteristics obtained from inserting a model in a wind or water tunnel, such as separation, vortex shedding, areas of low or high pressure and forces. Team members used the NASA computer code OVERFLOW to simulate the flow of air over a tennis ball and calculate the lift, drag, velocity and pressure around a spinning tennis ball. Team researchers modified the NASA software to emulate "spinning". Note the asymmetric velocity distribution on the spinning ball in Fig. 6. Although the goal was to introduce students to basic CFD concepts, one of the unexpected benefits of the project was NASA's use of the team's modifications for a rotating heart pump simulation.

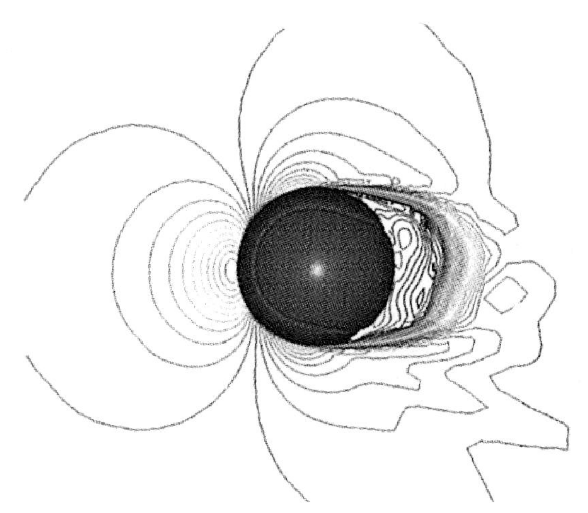

Fig. 6 Velocity contours of a 120mph tennis ball spinning at 1000rpm.

BALL/COURT INTERACTION

The pre- and post- ball bounce velocity, spin, angles of incidence and reflection, trajectory, rebound height and distance were calculated for a hard, green clay, red clay and grass court located at the USTA Player Development Center. Using a ball machine, which controlled both spin and speed, ball flight was captured using a high-speed digital camera

Wilson US Open balls were used on all four court surfaces. In addition, a Wilson clay court ball was tested on a green clay court, a Roland Garros (French Open) ball on red clay and a Slazenger (Wimbledon) ball on a grass court. The footage very clearly demonstrated that regardless of the type of spin before the bounce, balls rebound with topspin. (Although it is possible to rebound with underspin, this is very rare.)

For flat (no spin), low, medium and heavy topspin, medium and heavy underspin rates, the results quantified the effects created by the friction between balls and the different court surfaces (Table 1). Less ball/court friction resulted in a lower ball rebound angle and faster court. The complete test results are located on the Internet (http://wings.ucdavis.edu/Tennis/Project/bounce-01.html).

Table 1 Wilson US Open Ball - Flat (No Spin) Angle Before and After the Bounce.

Court Type	Angle In	Angle Out	Difference
Green Clay	26.8	37.5	10.7
Red Clay	26.5	37.5	11.0
Hard	23.9	32.9	9.1
Grass	24.9	29.4	4.5

BIOMECHANICS

A study of footwork speed, stances, and jump heights of Chang, Sampras and Agassi is currently underway. A 3D model (skeleton and muscles) of the lower body was created to study leg motions (Fig. 7) and teach students the basic concepts of biomechanics.

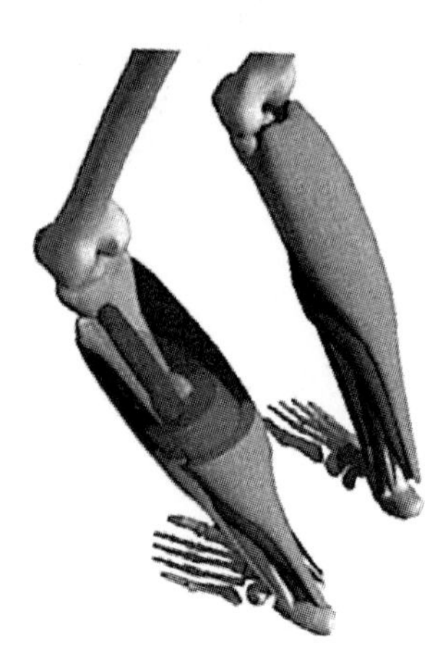

Fig. 7 3D-skeleton and muscle model used to study player biomechanics.

BALL SPEED AND SPIN

Using footage from the 1997 and 1998 Sybase Open, an analysis of ball speed and trajectories was conducted. A series of professional player serve, return, groundstroke and volley and overhead speeds were calculated over the flight path of the ball.

Students learned about the physics laws that govern trajectories. Complete results can be found on the web site

(http://wings.ucdavis.edu/Tennis/Project/speed-01.html).

A summary of professional men and women player ball spin was calculated from the 1997 US Open footage including first and second serve, forehand, backhand, volleys and overhead shots.

RACQUET-HEAD SPEED

Racquet-head tip speed and angles were calculated from toss to post-serve follow-through. Using high speed digital footage (250 frames/sec.) that the team captured at the US Open in 1997 and 1998, first and second serves were analyzed for Sampras, V. Williams, Agassi, Courier (among others). Team members were able to quantify and qualify the practices, speed, body and racquet angles of the best tennis players in the world.

The racquet head tip was marked using a software program which also captured and calculated velocities (Fig. 8). Patterns quickly emerged for flat, slice, and American twist serves. Students were asked to assess "unknown" patterns on their own, as they learned the mathematics and science associated with velocity, acceleration, angles, geometric pattern recognition, and 2D and 3D spatial relationships.

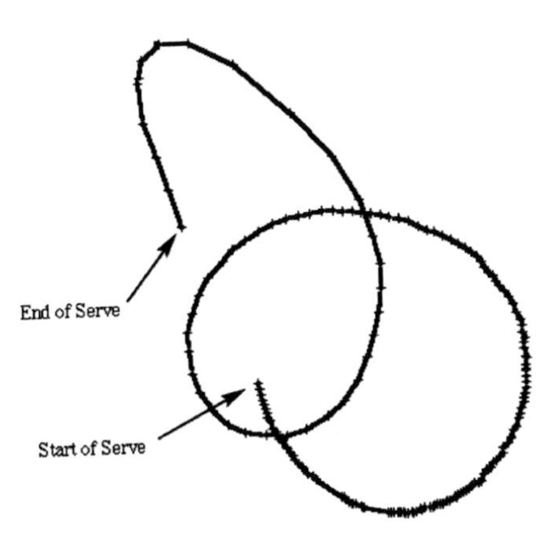

Fig. 8 Racquet-head tip pattern from toss to post-serve follow-through for Sampras second serve.

CONCLUSIONS

A successful Internet based tennis science educational project is followed by students, adults and professionals. The flow visualization studies revealed that the flow over tennis balls is post-critical and that it is the Magnus effect that makes the ball swerve. A subsequent test was conducted to calculate the C_d of several different tennis balls. A standard (Wilson) tennis ball has a C_d of about 0.6 over the velocity range, 50 to 150 mph. The most interesting results are for the used ball; the C_d increases from about 0.5 to 0.6 over the same velocity range. Studies on ball spin, ball speed, ball/court interaction have demonstrated basic tennis sport science concepts. Several biomechanic studies have been proposed including the completion of a full skeletal/muscular model, qualitative and quantitative studies of professional player motions.

ACKNOWLEDGMENTS

This work was supported through several agreements between NASA and Cislunar Aerospace, Inc. (NCC2-9014, NCC2-9010 and SAA2-400190). The footwork and biomechanics studies have been supported in part by a grant from the USTA. The authors wish to thank: the USTA for the usage of the high speed footage from the 1997 and 1998 US Open to conduct the studies and the larger diameter tennis balls; Greg Zilliac, David Yaste and Kent Shiffer from NASA Ames and Jim Pallis from Cislunar Aerospace, Inc. for their assistance during the wind tunnel tests.

REFERENCES

Achenbach, E. (1972) Experiments on the flow past spheres at very high Reynolds Number. *Journal of Fluid Mechanics*, **54**, 565.

Cislunar Aerospace, Inc. (1997-2000), http://wings.ucdavis.edu/Tennis.

Mehta, R.D. (1985) Aerodynamics of Sports Balls. *Annual Review of Fluid Mechanics*, **17**, 151-189.

Aeromechanical and aerodynamic behaviour of tennis balls

T.M.C. Brown
British Sugar, Peterborough, Cambridgeshire, UK
A.J. Cooke
Department of Engineering, University of Cambridge, UK

ABSTRACT: This paper draws together research conducted at Cambridge University Engineering Department, discussing the aeromechanics and aerodynamics of tennis balls. It concludes that initial unsteady motion effects are negligible and that variation in the coefficient of restitution of the ball/racket system might be an effective solution to ball speed reduction during tennis serves. It recommends the specification of a comprehensive design parameter set, with particular reference to the importance of aerodynamic data.

INTRODUCTION

Competitive tennis has flourished since the All England Croquet and Lawn Tennis Club held its first tournament in 1877. The advances of professionalism and racket technologies have combined to bring about higher ball speeds. With the advent of sponsorship, viewing figures are critical to the financial future of the game. Rally length is an important factor for viewing figures. High ball speeds make it difficult to return the ball, especially on serve. This shortens rallies and thus jeopardises viewing figures. Ultimately, this could affect the financial future of the game.

A sport's governing body has a responsibility to monitor and control the development of its sport and to ensure that the needs of all stakeholders in the game are met. Guidelines from the governing bodies to the manufacturers on equipment design often do not fully specify the equipment's performance. This is because the engineering issues involved in the equipment development are complex. They require research before full understanding of equipment performance is gained.

The International Tennis Federation was one of the first sports' governing bodies to embark on a systematic programme of research. Research projects on rackets, balls, surfaces, and their associated interactions are being undertaken. Their aim is to provide tools for designing, testing and selecting sports equipment.

This paper draws together research conducted at Cambridge University Engineering Department (CUED), UK, on aspects of tennis ball design during the period 1996-1997. Brown (1997), Cartwright (1997), Nevill (1996) and South (1996) carried out this unpublished work. The aims of this paper are to:

(1) Discuss aspects of the aerodynamics and mechanics of tennis balls that affect performance.
(2) Identify design parameters that define ball behaviour (particularly affecting service speeds).
(3) Highlight needs for future work.

SUMMARY OF CAMBRIDGE REPORTS

Nevill (1996) examined the subject as four, separate regimes.

(1) The Impact – between the racket and the ball that sets the ball in motion.
(2) The Initial, Unsteady Motion – of the ball after the impact as the ball and surrounding fluid motion alter dramatically and thus cannot be approximated by steady state mechanics.
(3) The Quasi-Steady Motion – of the ball and surrounding fluid once they have settled enough to be accurately approximated by steady state mechanics.
(4) The Bounce – The impact between the ball and the surface of the court.

He conducted a study of the mechanics of tennis and examined the factors, in each regime, that affect ball performance with a view to reducing the maximum speed of the serve. This framework has been adopted for summarising the four reports.

THE IMPACT

Nevill stated that the fastest recorded service speed was 62 m/s (a Reynolds number of approximately 270,000). Nevill and South (1996) examined the impact using several techniques; a linear relationship and a Voigt model for ball deformation, conservation of momentum, conservation of energy and consideration of coefficient of restitution. Impact is a function of, the player, the swing, the racket's properties (e.g. moment of inertia, size, string tension), and the ball's properties (e.g. mass, internal pressure, wall thickness and material properties.) Kinetic energy from the swing is transferred to potential energy in the deformation of the ball and strings. Elasticity allows most of the energy to be restored as the kinetic energies of the ball and racket.

Voigt Model for Ball Deformation

Daish (1981) conducted experiments on duration of impact. From his work the duration of impact is approximately 4.3 ms for most tennis shots. This compares well with work by Hatze (1976) who estimated it to be ~ 4 ms. If the ball/racket system was linear (force $\propto$ displacement) then contact time would be constant irrespective of approach velocity. After observing that the ball deformation was not linear for small deformations but only for larger deformations, Nevill investigated the ball deformation using a Voigt model. A Voigt model is one spring in series with a dashpot in parallel with a second spring.

This model cannot be solved analytically due to non-constant coefficients in the O.D.E. Examining the model in two extreme positions is of benefit. For rapid deflections the damper will effectively lock and the model consists of two springs in series. At slower deflection speeds the unit will deflect reducing the apparent stiffness. This explains longer contact times for lower impact velocities. Small contact times/high velocities give large accelerations, forces and deformations. Hysteresis is a measure of the energy lost in the impact. Nevill examined hysteresis using coefficient of restitution e.

Coefficient of Restitution

Using the conservation of momentum and the definition of e, and ignoring any racket rotation, then

$$v' = \frac{M[u(1+e) - ev] + mv}{M + m} \tag{1}$$

v' is the final velocity of the ball, M is the racket mass, m is the ball mass, u is the initial velocity of the racket and v is the initial velocity of the ball. Using ITF data on the permitted rebound height of a tennis ball, the coefficient of restitution of a tennis ball colliding with a rigid surface is 0.73 – 0.76. The fraction of the initial energy remaining after the bounce is e^2. The higher the impact speed the larger the deformation and the more energy will be dissipated as heat. The ratio of the energy stored by the ball to the energy stored by the racket is equal to the ratio of their stiffnesses. The combined ball/racket coefficient of restitution, $e_{b/r}$, is related to the ball coefficient of restitution, e_b, and the ball and racket stiffnesses, k_b and k_r, by:

$$e_{b/r}^{\,2} = \frac{k_b + k_r e_b^{\,2}}{k_b + k_r} \tag{2}$$

THE INITIAL UNSTEADY MOTION

Nevill examined the initial unsteady motion to see what significance this part of the flight had on ball speed. South investigated two forms of initial unsteady motion: viscous and inviscid motion. Nevill investigated 3 forms of unsteadiness: vibration, initial viscous motion and initial inviscid motion.

Vibration

Experience suggests that the ball deformation during impact does not last long. A ball bouncing on the floor after being struck (without spin) seems to bounce without the irregular flight that would be consistent with a non-spherical collision. In order to assess the restoration of the ball to its spherical shape, South (1996) measured pressure variation inside the ball after impact. He estimated that significant large-scale deformation ends about 5 ms after impact. Nevill deduced that the vibrations most probably cease after 28 ms, shortly after the ball has actually left the racket.

Initial inviscid motion

Nevill and South considered a potential "virtual mass" for the ball on leaving the racket, for inclusion in the energy equations. The "virtual mass" consists of the mass of the ball and an effective mass of a region of fluid surrounding it. South deduced mathematically that this effect would be negligible.

Taylor (1942) considered the motion of a body in fluid when subjected to a sudden impulse and calculated how the surrounding fluid resists the motion of the ball and dissipates some of its energy in the form of a sound wave. Nevill and South, drawing on Taylor's work, considered a rigid sphere at rest in a compressible fluid. An assessment was made of the development of the flow around the ball and its affect on drag, force and thus ball speed. Both authors concluded that the sound wave had a minimal effect on the eventual steady state velocity for the quasi-steady regime. Hence, it was suggested by both authors that the inviscid, unsteady motion of a tennis ball does not last for an appreciable length of time and does not substantially affect the velocity of the ball.

Initial viscous motion

Nevill and South investigated the effect of air viscosity on the initial unsteady motion as the boundary layer developed. Complex mathematical models by Batchelor (1967) were considered for boundary layer growth on infinite flat plates and cylinders. By approximating a tennis ball to a cylinder in two dimensions, South achieved an order of magnitude figure for the time to ensure full boundary layer growth. He deduced that, for a ball accelerated to 50 m/s, the steady flow regime was achieved in a time of the order of 3 ms.

QUASI-STEADY MOTION

Work by Brown (1997), Cartwright (1997) and Nevill sought to investigate the effect of quasi-steady motion on ball velocity. The expected range of Reynolds number found in tennis was calculated by Brown to be between 0 and 320,000, i.e. maximum speed of serve at 72 m/s. This maximum is higher that the current service record of 61 m/s because records are average measures only. Nevill used a Matlab trajectory model to extrapolate for more accurate service velocities.

Skin friction drag is due to the shearing fluid in the boundary layer, and is the resultant of all the forces tangential to the body surface. The pressure drag is the resultant of all the forces normal to the body surface. A tennis ball is a bluff body and thus has large pressure drag (associated with the wake), relative to the skin friction component.

The drag coefficient, C_D, is defined by:

$$C_D = D/\tfrac{1}{2}\rho U^2 A \quad (3)$$

where D is the total drag force, U is the velocity of the fluid and A is an area. At very low Reynolds numbers the flow around a smooth sphere is entirely laminar, Liebster (1927), and the drag is mostly due to skin friction drag and is low. As the Reynolds number increases the adverse pressure gradient towards the rear of the sphere becomes too great for the boundary layer to negotiate, thus the flow separates and a large turbulent wake forms and the pressure drag increases accordingly, Achenbach (1972). At still higher Re numbers the transition between laminar and turbulent flow occurs in the boundary layer before separation and the now turbulent boundary layer is able to negotiate more of the adverse pressure gradient before separating. Thus the wake shrinks and the drag falls dramatically. This is known as the transition point. As Re continues to increase, the separation point moves forward and the wake size increases again, thus increasing C_D.

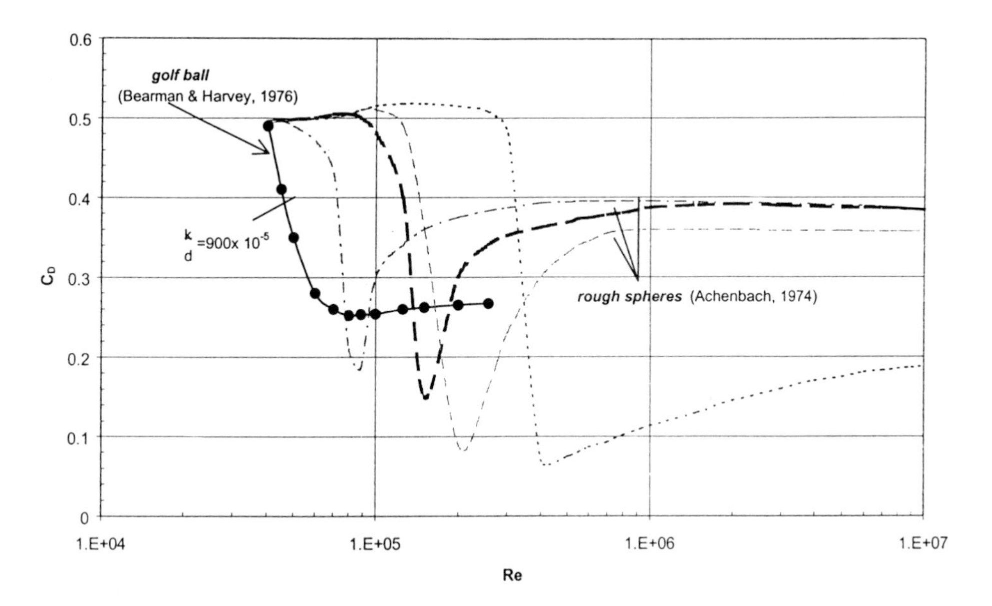

Fig. 1 Graph of C_D V Re, for spheres of differing surface roughness factors.

Since surface roughness (k/d) promotes transition from laminar to turbulent boundary layers, the consequent drag reduction associated with transition occurs at lower Re numbers. However the drag coefficient soon rises again to a higher value than for smooth spheres. In figure 1, k is the average size of roughness and d is the diameter of the sphere. Brown estimated k from ball manufacturer's data on felt thickness. The resulting k/d was 0.0154. The k/d parameter is aerodynamically important as it affects Re critical (Re at which transition occurs).

Brown and Cartwright collected drag coefficient data for numerous spherical models with nodules and seams for comparison with a smooth sphere. The data could not be directly related to tennis.

THE BOUNCE

Nevill briefly considered the ball bounce and spin. This is very similar to the impact section and is chiefly governed by e_b. To allow the ball to bounce high there must be a high e value. That depends on the playing surface and the ball. Spin affects the angle of bounce to the vertical, from the angle of flight, and the velocity after the bounce. This depends on the amount of spin imparted at racket/ball impact and also on the coefficient of friction (μ) between playing surface and ball. Nevill makes the assumption that the μ is insufficient to deviate the ball on service in normal serves. This claim is doubtful and the whole bounce issue needs more research. Bounce must be related to all relevant design parameters that define a ball's characteristics.

DISCUSSION

This section summarises the key findings of the four authors and discusses them alongside other literature.

IMPACT

From the impact work, Equation 1 can be used to investigate how to reduce ball velocity during the serve. However, as mentioned earlier, it does not take into consideration the racket rotation on impact. Racket rotation is considered by Brody (1997) and a more advanced model of hysteresis is developed by Hatze (1992). By assuming that the initial velocity of the ball is zero, Equation 1 becomes:

$$v' = \frac{Mu(1+e)}{M+m} \qquad (4)$$

One can reduce ball velocity on leaving the racket by; reducing u (racket velocity before impact); reducing e; reducing M; increasing m. Changing u depends on the complex relationship between racket properties (mass and first moment of inertia etc) and the torque that a human can impart into their swing. An increase in racket mass, m, has a minimal effect on v'. Nevill, using Daish (1981), believes that increases in M would have a negligible effect, although M directly affects u.

The size and thus effect of M in the denominator of Equation 4 limits the effect of reducing m. A doubling of the mass would mechanically only lead to an 11% reduction in initial ball speed. Using Equation 2 and estimates from Nevill on ball and racket stiffnesses the coefficient of restitution for the ball/racket system is 0.89. Recent work by Brody (1997) measured e as 0.85 for a rigid, clamped racket model. Thus reducing e_b, k_b and increasing k_r all reduce $e_{b/r}$. All are possible. Using Equation 4, Nevill also calculated that, reduction of e to 0.2 would decrease the service velocity by only 35%. The system coefficient of restitution, $e_{b/r}$, is a function of impact velocity, Kotze (2000).

INITIAL UNSTEADY MOTION

If the initial inviscid situation is examined (i.e. the "virtual mass" effect), then by using the appropriate numbers for a tennis ball and the associated conditions the ball slows to a steady state velocity Us equal to 0.999 Uo. The time factor for the exponential slowing of the ball falls to half its initial value in a time of just 66.2 μs.

In the initial viscous situation the ball receives an impulse and accelerates from stationary to its initial post impact velocity rapidly. The Mach number for a tennis ball moving at it initial post impact speed is approximately 0.24, low enough to allow the 'information' about the flow to propagate quickly. Steady state should be reached rapidly. Nevill argues steady state would be reached in less than 10 diameters distance, equivalent to ~10 ms. It is therefore reasonable to assume that all ball deformation, vibration, and initial non-steady motion has concluded within approximately 10ms and thus can be considered negligible. It is best to view, model and analyse tennis ball flight as a quasi-steady motion.

QUASI-STEADY MOTION

The Quasi-steady motion is critical to the ball velocity. Figure 2 taken from Nevill shows the trends of C_D against Re (velocity) for a smooth and rough (k/d = 0.0154) sphere with a diameter similar to a tennis ball and over a range of velocities experienced in the game. Haake et al. (2000) measured the C_D of an actual tennis ball in the range $200,000 < Re < 250,000$ (maximum velocity 62 m/s) and it is constant at approximately 0.55. Further work is needed to explain the differences, and to develop a proven methodology for determining tennis ball roughness in order to relate it to Achenbach's work on smooth spheres. Altering the aerodynamic drag coefficient would have a significant effect on the forces acting on the tennis ball and would slow the ball speeds considerably, especially for the receiver. Aerodynamic data for spinning balls must be collated.

BOUNCE

Any changes to e of the ball must be analysed from the perspective of the bounce. If the ball were to be made smooth then the friction effects must be considered to maintain playing characteristics with and without spin.

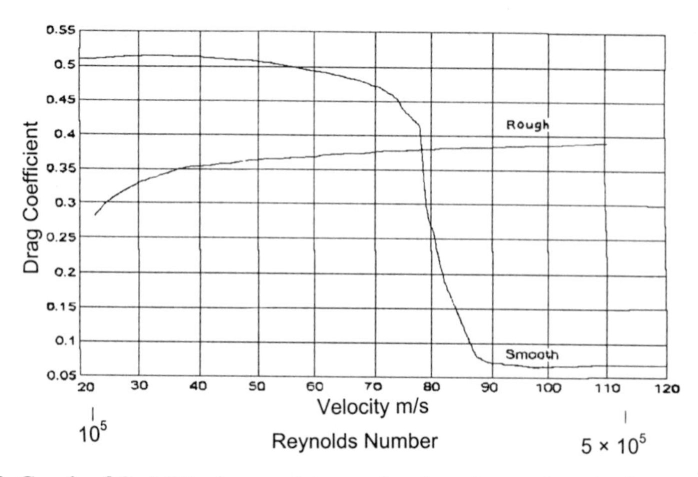

Fig. 2 Graph of C_D V Re for rough/smooth sphere in tennis velocity range.

DESIGN PARAMETERS

As discussed by Cooke (1992) in her work on shuttlecocks, in order to define a projectile's performance a comprehensive set of relevant design parameters must be specified and each one systematically examined and quantified. From the above discussion, key design parameters for consideration when aiming to reduce ball speeds are: coefficients of restitution (both at low and high velocities), drag coefficients, surface roughness, diameter, mass. Coefficients of restitution will require determination of material properties and structural analysis of ball deformation. Also, design parameters can be investigated for rackets and surfaces.

CONCLUSIONS

This paper has drawn together research on the aeromechanics and aerodynamics of tennis balls from four authors at CUED, UK, during the period 1996-97. The flight of the ball was broken down into four areas: impact, initial unsteady motion, quasi-steady motion, and bounce. Two regimes dominate the tennis ball speed; impact and quasi-steady motion. The initial unsteady motion (both inviscid and viscous) can be considered to have negligible effect on the ball speed because of the duration and magnitudes of the forces involved. The whole flight can be modelled as a quasi-steady motion, dominated by pressure drag.

Impact The coefficient of restitution of the ball/racket system was estimated as 0.89, similar to other work, Brody (1997). Reducing $e_{b/r}$ may be an effective solution for reducing initial ball speeds at service. Material properties and ball deformation requires further investigation to fully define $e_{b/r}$.

Quasi steady motion A comprehensive aerodynamic data set for actual tennis balls (with and without spin) is needed. A tested methodology for evaluation of the effects of surface roughness of the tennis ball should be developed and correlated with Achenbach (1972).

It is suggested that further work aims to define and quantify a complete design parameter set for ball, racket and surface. A design parameter set must include drag coefficient, ball diameter and mass, ball surface roughness, coefficients of restitution for ball/racket and ball/playing surface impacts, and material properties of the ball.

ACKNOWLEDGEMENTS

The authors would like to note the contribution to this paper of Mr Nevill, Mr South and Mr Cartwright and also to Professor Ffowcs-Williams and Dr Kimon Roussopoulos, all from CUED. Special thanks must go to Donald Brown for all his faith and love.

REFERENCES

Achenbach E. (1972) *Experiments on the flow past spheres at very high Reynolds numbers*, Journal of Fluid Mechanics, volume 54, part 3, pp565-575.

Achenbach E. (1974) *The effects of surface roughness and tunnel blockage on the flow past spheres*, Journal of Fluid Mechanics, volume 65, part 1, pp 113-125.

Batchelor G. K. (1967) *Fluid Dynamics*.

Brody H. (1997), *The Physics of Tennis III: the ball-racket interaction,* American Journal of Physics vol. 65 (10), 981-987.

Brown T.M.C. (1997) *Aerodynamics of tennis balls*, M.Eng. project report, Cambridge University Engineering Department (CUED).

Cartwright A (1997) *Tennis Ball Aerodynamics*, M.Eng. project report, CUED.

Cooke (1992), *Aerodynamics and Mechanics of Shuttlecocks*, PhD thesis, CUED

Daish C. B. (1981) *The Physics of Ball Games*. Hodder and Stoughton.

Haake S. J., S. G. Chadwick, R. J. Dignall, S. Goodwill (2000), *Engineering tennis – slowing the game down,* Sports Engineering Journal (Haake, S. J. ed.), Volume 3 Issue 2, pub. Blackwell Science, Oxford

Hatze H. (1976), *Forces and duration of impact, and grip tightness during the tennis stroke,* Medicine and Science in Sports vol. 8 (2), 88-95.

Hatze, (1993), *The relationship between coefficient of restitution and energy losses in tennis rackets,* J. App. Bio. Vol. 5, 124-144.

Kotze, J., (2000), *Issues affecting racket power in the tennis serve*, Sports Engineering Journal (Haake, S. J. ed.), Volume 3 Issue 2, pub. Blackwell Science, Oxford

Liebster, H. (1927), *Annual of Physics*: (4) 82, 541.

Nevill N. D. (1996) *The Aeromechanics of Tennis Balls*, M.Eng. project report, CUED.

South N. (1996) *Aeromechanical Behaviour of the Tennis Ball*, M.Eng. project report, CUED.

Taylor G. I. (1942) *Motion of a body in water when subjected to a sudden impulse.*

Analytical modelling of the impact of tennis balls on court surfaces

R.J. Dignall and S.J. Haake
Department of Mechanical Engineering, University of Sheffield, UK

ABSTRACT: An analytical model was created to simulate the bounce of a tennis ball normal to a playing surface, based on a spring and a damper in parallel. Experimental data was used to find stiffness and damping coefficients, and to verify the normal model. Balls were projected between 6.7 and 20 ms^{-1} onto a rigid surface and compared to model predictions. The model was then extended to oblique impacts by introducing a horizontal frictional force. On an acrylic court where the surface is rigid in comparison to the ball, there was a good comparison with experimental data for the outgoing speed, angle and spin. A plot of predicted centre of mass displacement during an oblique impact was an excellent match to experimental co-ordinates.

INTRODUCTION

One criticism of tennis is that it is too fast or is dominated by the serve, and one proposal to slow the game down is to increase the size of the tennis ball. The primary effect is an increased drag but one effect may be an altered bounce of the ball. This paper is concerned with the modelling of tennis ball impacts on court surfaces to enable parameter changes, such as diameter, to be made.

IMPACT MODEL

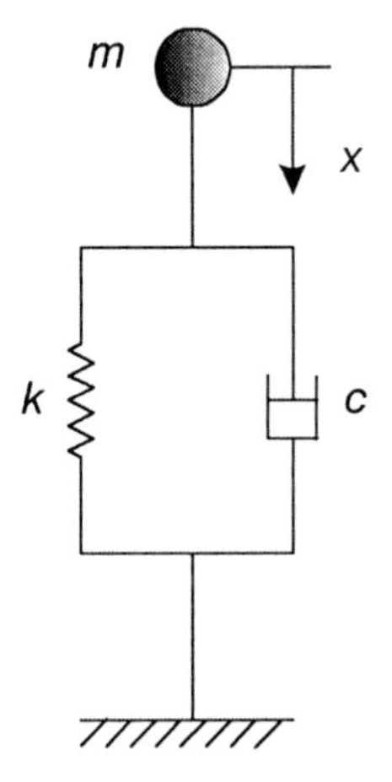

Fig. 1 Representing the ball impact by a model containing a spring and a damper

The simplest model which can be used contains a spring and a damper. The damper was placed in parallel with the spring as shown in Fig. 1. The governing equation for this system is

$$m\ddot{x} + c\dot{x} + kx = 0 \tag{1}$$

Given the boundary condition of $x = 0$ at time $t = 0$, the solution to this equation is

$$x = ae^{-bt} \sin \omega t \tag{2}$$

Differentiating gives

$$\dot{x} = ae^{-bt}\left[\omega \cos \omega t - b \sin \omega t\right] \tag{3}$$

and $$\ddot{x} = ae^{-bt}\left[\left(b^2 - \omega^2\right)\sin \omega t - 2b\omega \cos \omega t\right] \tag{4}$$

The boundary condition $x = 0$ at $t = T_C$ where T_c is the contact time gives $\omega = \dfrac{\pi}{T_C}$.

Equating (3) to the incoming and outgoing velocities gives two more boundary conditions:

$$\dot{x}_{t=0} = V_{in} = a\omega \tag{5}$$

and $$\dot{x}_{t=T_C} = V_{out} = a\omega e^{-bT_C} \tag{6}$$

Thus $$a = \frac{V_{in}}{\omega} = V_{in}\frac{T_C}{\pi} \tag{7}$$

And $$b = -\frac{1}{T_C}\ln\left(\frac{V_{out}}{a\omega}\right) = -\frac{1}{T_C}\ln\left(\frac{V_{out}}{V_{in}}\right) \tag{8}$$

Although the constants *c* and *k* are not necessary for the mathematical modelling, they are useful to give some physical understanding. As for the undamped model,

$$k = m\frac{\pi^2}{T_c^{\,2}} \tag{9}$$

Substituting both $\dot{x}$ and $\ddot{x}$ back into (1) leads to the expression

$$c = 2mb = -\frac{2m}{T_C}\ln\left(\frac{V_{out}}{V_{in}}\right) \tag{10}$$

Thus, the constants a, b and c can be calculated if the incoming and outgoing velocities and the contact time are known.

NORMAL IMPACT EXPERIMENTAL DATA COLLECTION

After pre-compression, pressurised tennis balls were dropped from a height of 100 inches (2.54 m) onto a piezoelectric force plate. This gives an impact speed of around 6.7 ms^{-1}. Light beam timers were used to measure the incoming and rebound velocities and a computer sampled the force output at 60 kHz. Balls were then fired at the force plate using a modified Bola cricket bowling machine to give velocities of between 13 and 20 ms^{-1}.

Table 1 shows that the COR decreases as the impact speed increases. The contact time ranged from 3.7 ms at 20 ms^{-1} to 4.5 ms at 6.7 ms^{-1}, which indicates that the stiffness k varies.

Table 1 Incoming and outgoing velocities

Incoming velocity V_{in} (ms^{-1})	Outgoing velocity V_{out} (ms^{-1})	Coefficient of Restitution (COR)
6.68	5.14	0.77
13.35	9.25	0.69
16.25	10.39	0.64
20.05	12.11	0.60

STATIC STIFFNESS DATA

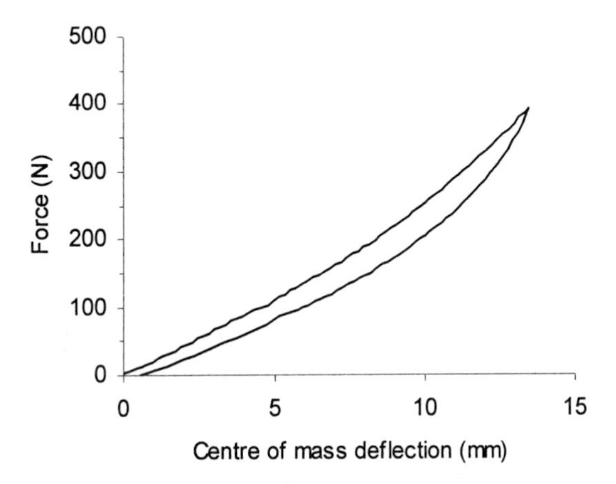

Fig. 2 Graph showing force against deflection of ball centre of mass

Fig. 2 shows a modified static force-deflection curve for a pressurised tennis ball. The data was found using the standard ITF ball compression test, but the ball deformation was halved to give the ball centre of mass deflection. This assumes that the ball deformation was symmetrical. If a cubic polynomial is fitted to the loading curve and differentiated, it gives the following expression for stiffness k related to centre of mass deflection x where x is measured in mm:

$$k = 0.1818x^2 - 0.5616x + 21.97 \tag{11}$$

Therefore the stiffness for zero deflection is 21.97 N/mm or 21 970 N/m.

Table 2 shows the stiffness values found using the experimental contact times from the drop tests and the static stiffness (at zero deflection), as well as the damping coefficients found using the contact times. The stiffness values are shown graphically in Fig. 3 (a), which suggests a linear relationship between stiffness and incoming velocity. Fig. 3 (b) also suggests a linear fit between damping coefficient c and incoming velocity V_{in}.

Table 2 Contact times together with stiffness and damping coefficients

Incoming velocity (ms^{-1})	Contact time (ms)	Stiffness k (kN/m)	Damping c (Ns/m)
static	-	21.97	-
6.68	4.5	27.78	6.59
13.35	3.9	36.99	12.87
20.05	3.7	41.09	15.90

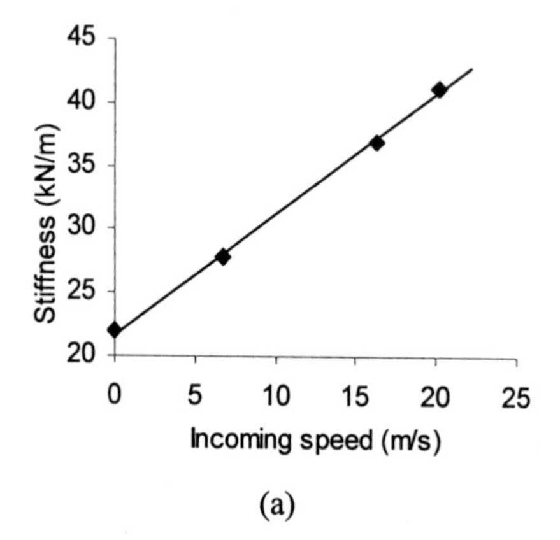

(a)

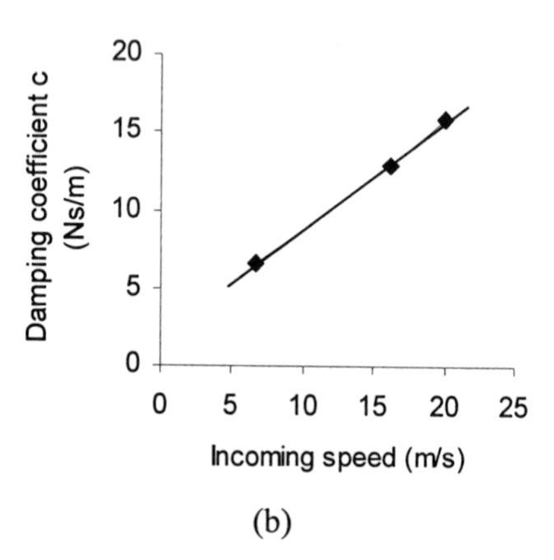

(b)

Fig. 3 The variation of (a) tennis ball stiffness (including static stiffness) and (b) tennis ball damping coefficient against incoming speed

OBLIQUE MODELLING

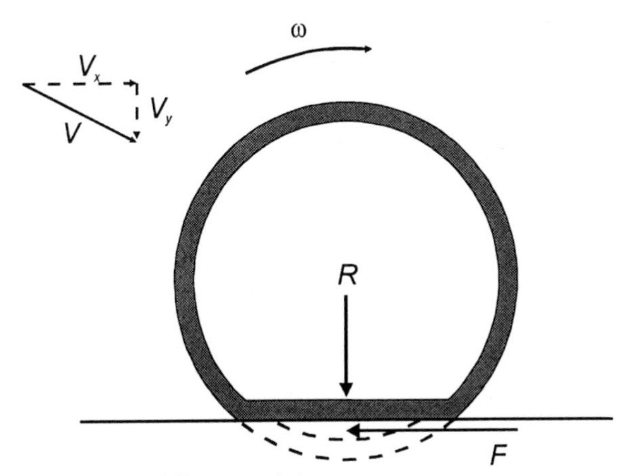

Fig. 5 Forces acting on the oblique model

Fig. 5 shows a ball impacting obliquely on a rigid surface. Its velocity is split into linear components V_x and V_y, together with a rotational component ω (topspin positive). The vertical component is modelled using the normal impact model described previously. The horizontal frictional force F was assumed to be proportional to the vertical reaction force R such that $F = \mu R$. For a force F acting at a radius r, the angular acceleration $\ddot{\theta}$ of a body with moment of inertia I can be calculated using

$$Fr = I\ddot{\theta} \tag{13}$$

For a thin shell of mass m and radius r, $I = \frac{2}{3}mr^2$ and thus (13) can be simplified to give

$$\ddot{\theta} = \frac{3F}{2mr} \tag{14}$$

OBLIQUE IMPACT RESULTS

Pressurised balls were fired obliquely at an acrylic court surface at 35 ms^{-1}. The impact was filmed at 9000 frames per second using a Kodak EktaPro 4540 camera. The coefficient of friction for the court was measured as 0.57 taken on the same court using the ITF proposed standard for a court friction testing method (ITF 1997). Table 3 shows the incoming and outgoing speeds, angles and spins averaged over three impacts. Table 3 also shows the rebound properties found using the oblique model.

Table 3 Comparison of speeds, angles and spins for experimental and predicted results for an oblique impact on an acrylic court surface.

	Incoming			***Outgoing***		
	Speed (ms^{-1})	Angle (to horiz)	Topspin ($rads^{-1}$)	Speed (ms^{-1})	Angle (to horiz)	Topspin ($rads^{-1}$)
Experimental	35.2	16.3	-97.1	24.2	16.6	320
Model				25.2	16.5	344

One of the three impacts was further analysed by looking at the positional information throughout the contact period. The experimental contact time was found to be 4.22 ms, giving 39 consecutive images on the high speed video. The vertical displacement data was modified to allow for the displacement of the centre of mass relative to the top of the ball, caused by the ball deformation. (Dignall, 2000). Fig. 6 shows the position data, plotting the co-ordinates of the centre of mass relative to the start of the impact. As can be seen, there is an excellent agreement between the model and experiment.

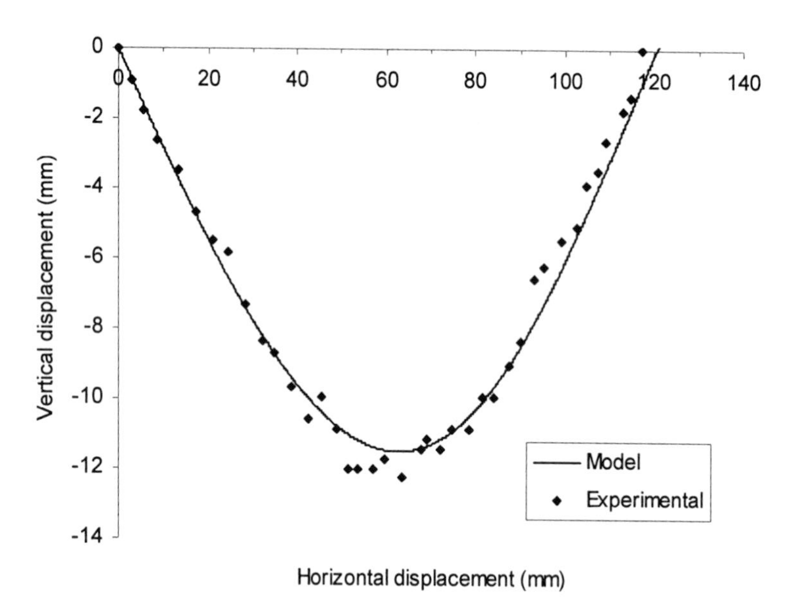

Fig. 6 Position data during impact for an oblique impact on an acrylic court surface

DISCUSSION

One important feature of this work is that Fig. 2 clearly shows that the stiffness is not constant with deflection for a static impact. The values of k predicted will therefore be some measure of average stiffness for an impact. This average stiffness will be increased with velocity by both the viscoelastic nature of the ball and the increased deflections caused by higher speeds.

The linear fit of both c and k with V_{in} suggests that a small number of tests would be sufficient to model the impacts at a range of velocities. Current ITF testing procedures can be used to find static values of k from a ball compression test, and the 100 inch drop test can be used to give the COR for a low speed impact. In fact, for a ball to pass the approval testing procedures, these values are limited to a relatively narrow range. The only additional data required are values of contact time and COR for a higher speed impact. This need not be a particularly high speed because the normal component of velocity during an oblique impact is often not large. For example, a 90 mph serve which just clears the net has about the same vertical velocity on impact as a ball dropped from 100 inches, at around 7 ms^{-1}.

Table 3 shows that the addition of a horizontal frictional force to the model provides a very good simulation of an oblique impact. It gives the speed, angle and spin after impact to within 5% of the measured values – which is a similar figure to the experimental errors involved. Furthermore the model predicts the positional data throughout the impact period with good accuracy, as seen in Fig. 6.

The models described make one major assumption – that the impact surface is rigid. In this case, a simple friction coefficient is enough to define the properties of the court. This assumption will be valid for many tennis surfaces, but grass and clay courts may undergo significant deformation. This will affect the bounce of the ball in two ways, the surface will absorb energy as well as the ball and the deformation will provide a "ramp" for the ball to roll up, increasing the rebound angle. For these surfaces it will be necessary to include the stiffness and damping properties of the ground.

CONCLUSIONS

A model was proposed which simulates the normal impact of a tennis ball on a rigid court surface. Constant values of stiffness and damping coefficients were used throughout the impact to simplify the analysis. It was found that both these coefficients had a strong linear relationship to the impact speed up to speeds of $20\,ms^{-1}$, which is higher than the normal velocity component likely to be seen in common tennis shots.

The model was extended to allow oblique impacts. The results were extremely good for an oblique impact on an acrylic court, matching both the rebound characteristics and the centre of mass displacements during impact.

ACKNOWLEDGEMENTS

The authors would like to thank the International Tennis Federation and the University of Sheffield for their support of this project.

REFERENCES

Brody H. (1984) That's how the ball bounces. *The Physics Teacher,* **22**, 494-497

Cross R. (1999) Dynamic properties of tennis balls. *Sports Engineering,* **2**, 23-33.

Daish C. B. (1972) *The Physics of Ball Games*, The English Universities Press Ltd

Dignall R. J., Haake S. J., & Chadwick S. G. (in press, 2000) Modelling of an oblique tennis ball impact on a court surface, *3rd ISEA International Conference – The Engineering of Sport*

ITF (1997) *An Initial ITF Study on Performance Standards for Tennis Court Surfaces,* June 1997, International Tennis Federation, Bank Lane, Roehampton, London SW15 5XZ

The interaction of the tennis ball and the court surface

G.W. Pratt
Department of Electrical Engineering and Computer Science, MIT, Cambridge, MA, USA

***ABSTRACT*:** The tennis ball is modeled as a second order, damped, linear mechanical system. The details of the impact with the court surface, initially treated as non-deformable, are presented. The surface pace rating SPR is shown to be $100*(1-\mu)$ where μ is the coefficient of sliding friction between the ball and the court surface. The force on the ball is derived for a typical ground stroke and serve throughout the impact. The model is extended to explore the effects of a deformable court surface.

THE GROUND STROKE

This analysis is based on an elegant paper by Brody (Brody, 1984) in which he examined the behavior of the tennis ball during its interaction with the court surface. The analysis begins with the differential equation describing the impact of the ball falling from the maximum height of its trajectory over the net under the influence of gravity. The ball is modeled as a spring of mass m, of stiffness k, and damping constant α. Its strikes a non deformable surface with a velocity V_{on}. The deformation of the ball, or compression of the spring, is represented by the variable y whose equation of motion is,

$$d^2y/dt^2 + 2\alpha\, dy/dt + \omega^2 y = -g \quad (1)$$

where g is the acceleration of gravity. This equation has an exact solution which is,

$$y = [V_{on}/\omega]e^{-\alpha t}\, \mathrm{Sin}(\omega t) \quad (2)$$

The unknown vibrational frequency ω of the ball and the damping constant α for that vibration are determined by requiring that this solution produce the observed court contact time T for a bounce and the observed coefficient of restitution, COR, of the ball. The COR is the magnitude of the ratio of the vertical velocity of the ball just as it leaves the court surface after completing a bounce to its vertical velocity just before it hits. Hence the COR varies between 0 and 1. For a typical ground stroke, the maximum trajectory height is taken to be one metre. This makes V_{on} equal to -4.4 m/sec. Brody used values of the COR of 0.75 and a bounce contact timeT of 0.005

seconds. This determines the upward velocity V_{off} at the end of the bounce to be 3.3 m/sec.

Setting V_{off} equal to dy/dt at time T we find V_{off} as a function of the unknown frequency ω and damping constant α to be,

$$V_{off}(\alpha,\omega) = -(V_{on}/\omega)\alpha e^{-\alpha T}\sin(\omega T) + V_{on}e^{-\alpha T}\cos(\omega T) \qquad (3)$$

Equating this expression to 3.3 m/sec at T = 0.005 seconds, α was found to be 57.5 sec^{-1} and ω to be 590.5 radians sec^{-1} . In perhaps more familiar language, this means that this tennis ball has a natural vibrational frequency of 94 cycles per second and if the ball were set into vibration at t=0, that vibration would have died out 0.087 seconds later.

The acceleration of the ball during the impact is given by md^2y/dt^2 which when multiplied by the mass m of the ball produces the vertical force $F_n(t)$ on the ball. The expression for $F_n(t)$ is,

$$F_n(t)=mV_{on}\,e^{-\alpha\tau}[(\alpha^2/\omega - \omega)\sin(\omega t) - 2\alpha\cos(\omega t)] \qquad (4)$$

Taking m to be 0.057 kg., the normal force on the ball is shown in Figure 1. The average compressive force is 87.8 Newton exactly as Brody found simply by dividing the vertical momentum change, as derived from the coefficient of restitution, by the contact time. The maximum compressive force is 130.7 N.

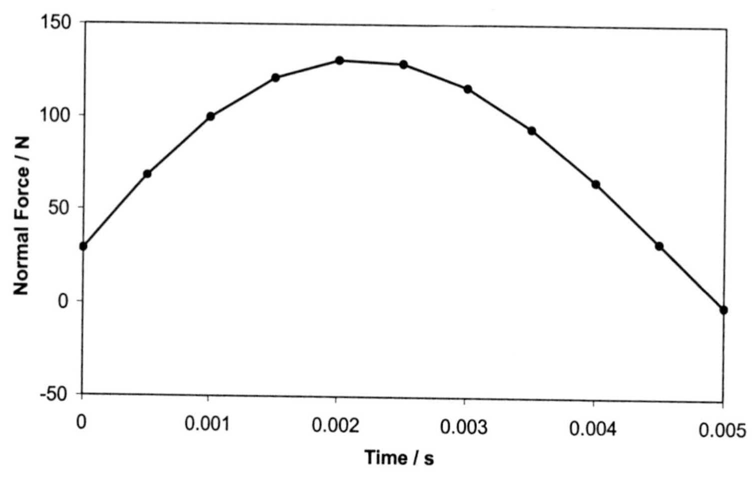

Fig. 1 Normal force on the ball during court contact for the ground stroke.

The horizontal force $F_x(t)$ is taken to be the product of coefficient of sliding friction μ and the normal force $F_n(t)$ as was assumed by Brody. Having $F_x(t)$, the horizontal velocity of the ball can be calculated during the court contact since the change in horizontal momentum from the start of the impact to a time τ later is the integral of $F_x(t)$ from t=0 to t=τ.. Dividing this quantity by the ball mass m gives the change in velocity. If the ball strikes the court with an initial horizontal velocity V_{x0}, then its velocity at time τ is,

$$V_x(\tau) = V_{x0} - 1/m\int_0^{\tau} \mu F_n(t)dt \quad (5)$$

As the ball slides forward on the court surface, the frictional force exerts a torque on the ball making it spin. Brody shows that the tangential velocity $Vt(\tau)$ of the ball is just 3/2 times the integral in the above expression,

$$Vt(t) = (3/2)\ 1/m\int_0^{\tau} \mu Fn(t)dt \quad (6)$$

The ball cannot roll without sliding until the tangential velocity of the surface of the ball equals the horizontal velocity of the ball. These two quantities are shown in Figure 2 where the initial horizontal speed of the ball was taken to be 18 m/sec representing a presumably typical ground stroke.

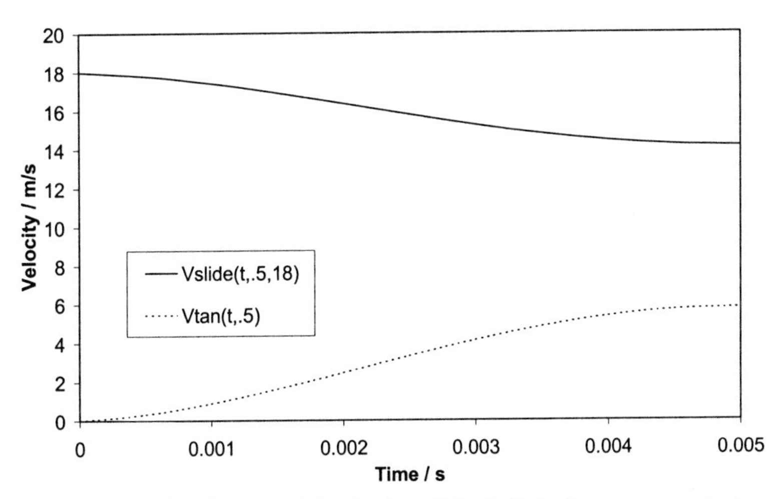

Fig. 2 Sliding speed and tangential velocity of the ball during court contact.

In Fig. 2 it is seen that the tangential velocity never comes close to sliding speed. Thus rolling without sliding cannot occur. For these two curves to touch at the end of the contact, the initial horizontal speed of the ball would have to be 9.6 m/sec which is extremely slow.

ANALYZING THE SERVE

The analysis of a serve is carried out below using parameters similar to those measured for the serve of Pete Sampras Cislunar Aerospace, Inc (1999). The initial horizontal velocity was taken to be 53.65 m/sec (120 mph) and the ball was assumed to leave the racquet at a height of 2.59 meters (8.5 feet). Air drag significantly slows the ball , decreasing its speed to 40.2 m/sec at the service line. Therefore, in order to know how much it has fallen due to the pull of gravity when it reaches the net, the horizontal velocity Vx must be found as a function of time. The equation describing the ball in flight is

$$dV_x/dt + \beta V_x^2 = 0 \quad (7)$$

The term βV_x^2 represents the effect of air drag and the coefficient β depends on the cross sectional area, the air density, and the drag due to the covering of the ball. A solution is sought that starts the ball at an initial height and initial downward velocity so that the value of the constant β selected determines that the ball should clear the net at a reasonable height and touch down inside the service line. To do this, the initial height was chosen as 2.59 meters (8.5 feet) and β was chosen as 0.016/meter. The time for the ball to travel from the baseline to the net is 0.245 seconds and 0.39 seconds to reach the service line. The height of the ball at the net turns out to be 1.096 meters, and it was assumed to have an initial downward velocity of –4.9 m/sec on leaving the service racquet. The slowing of the service ball due to air drag is shown in Fig. 3.

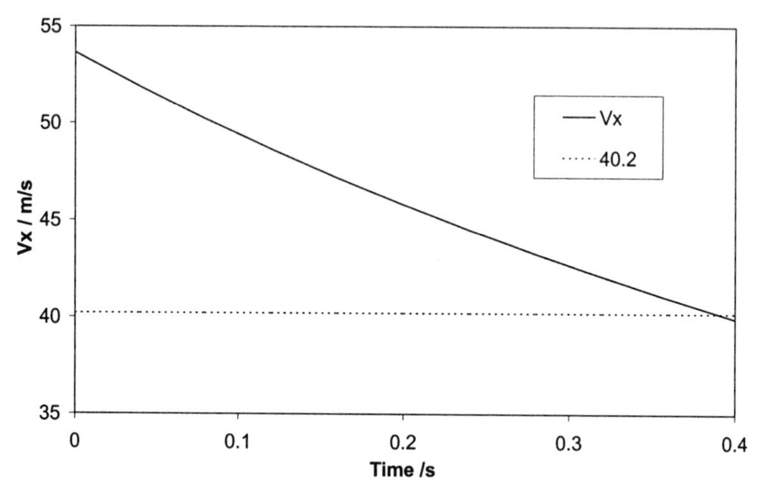

Fig. 3 Horizontal speed of the service ball vs. time.

The dotted line in Fig. 3 is at 40.2 m/sec, the horizontal speed of the ball at the service line reached after a 0.39 second flight.

The contact time on the court was taken to be 0.004 seconds for this model serve and the downward velocity V_{yo} of the ball just before impact with the court has increase from the ground stroke value of -4.4 m/sec to –8.72 m/sec. In order to satisfy the conditions that at the end of the impact the acceleration of the ball be –g and that the upward velocity be 0.75 times V_{yo}, new values of the damping constant α and frequency ω were determined. They were $\alpha = 72$ per second and $\omega = 737.2$ radians/sec. The normal force for the serve during court contact is shown in Fig. 4.

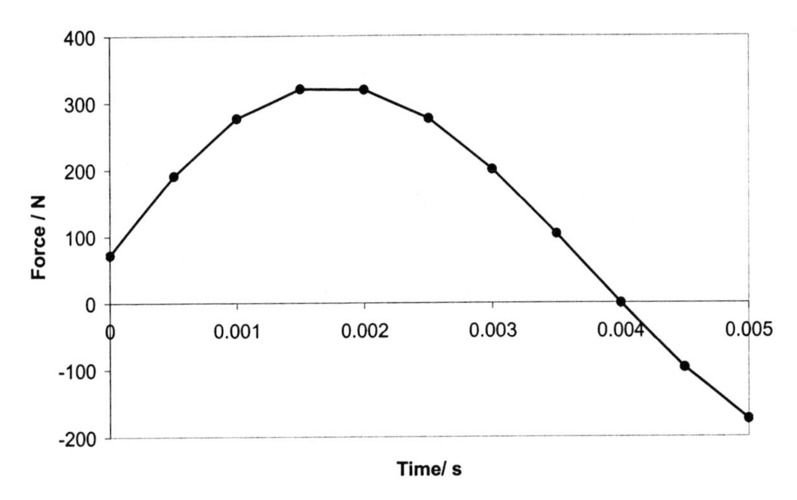

Fig. 4 Normal force on the ball during court contact for a serve.

The average normal force is 217.5 N (48.72 lb.) and the maximum is 320 N (71.7 lb.).

SURFACE PACE RATING

The SPR is defined as 1 minus the magnitude of the ratio of the change in horizontal velocity to the change in vertical velocity of the ball during court contact, this difference then multiplied by 100. Thus

$$SPR = 100(1 - \Delta V_x/\Delta V_y) \tag{8}$$

ΔV_y can be expressed in terms of COR, the coefficient of restitution, as $\Delta V_y = (1 + COR)V_{oy}$. But ΔV_x is just μ/m times the integral of the normal force over the contact time and ΔV_y is $1/m$ times that same integral. Substituting into the expression for the SPR and canceling common factors in numerator and denominator reduces the SPR to the exact result

$$SPR = 100(1 - \mu). \tag{9}$$

The mass of the ball, its speed, the height from which it falls, the coefficient of restitution never enter the final result. To find the SPR only the coefficient of friction need be measured. All of this happens because the horizontal force is taken to be μ times the vertical force in this model suggested by Brody. To first order, this seems to be a very good approximation. Because of the difference between the average normal force for the ground stroke and for the serve, and the deformable nature of the ball, the SPR may significantly differ between these two cases.

THE EFFECT OF COURT SURFACE DEFORMATION

So far the court has been treated as undeformed under the impact of the ball. This is undoubtedly a good approximation except for very soft surfaces. Following the approach of McMahon and Greene (McMahon and Greene, 1979) who examined the interaction of a runner with the running track, the effect of surface deformation can be explored by treating the ball-court impact as a first damped spring mass system, the ball, impacting on a second damped spring mass system, the court. This leads to a pair of coupled second order differential equations whose solution provides a complete dynamic description of the ball-court impact. This was done for two court stiffness values. McMahon and Greene list the stiffness of concrete or asphalt as 4376 kN/m, packed cinders as 2918 kN/m. The first case chosen here was a court stiffness of 437.5 kN/m or 15 % that of packed cinders. The maximum deflection of the court was 0.244 millimeters and 6.4 millimeters for the ball. Only 2.37 % of the initial impact energy of the ball was transferred to the court. The maximum compressive force on the ball was 129.1 N as compared to 130.7 N for the non-deformable court and the coefficient of restitution dropped from 0.75 to 0.73. The contact time for the impact was 0.005 seconds. Therefore, using this value of the court stiffness leads to an impact differing only slightly from the impact on an unyielding surface.

The second solution of the equations was carried out for a court stiffness 1/10 th of the first solution. This would correspond to a court whose stiffness was only 1.5 % of packed cinders representing a very soft court. The maximum court deflection rose to 1 millimeter while that of the ball remained the same. Now, 6.1 % of the initial impact energy was transferred to the court. The maximum compressive force on the ball fell slightly to 127.1 N and the coefficient of restitution dropped to 0.67. The contact time remained at 0.005 seconds. Therefore, for this very soft surface the biggest change was in the coefficient of restitution and in the deformation of the court. Although, the SPR would undoubtedly drop significantly for such a soft surface, this model for the impact of the ball with the court is not readily extended to describe the coefficient of sliding friction.

CONCLUSIONS

Modeling the ball as a damped linear oscillator an exact solution for its deformation vs. time upon impact with the court is given. Using measured values for the coefficient of restitution and duration time of court contact, the fundamental frequency and damping constant of the ball have been derived. The vertical force on the ball for a typical ground stroke peaks midway through the court contact time at approximately 130 newtons for the ground stroke and at 320 newtons for the serve. The differential equation describing motion of the ball including air drag is solved showing significant slowing of the service ball. The ball never rolls without slipping on the court. The surface pace rating (SPR) is shown to depend only on the coefficient of sliding friction. The impact behavior is examined for a very soft court surface and only the coefficient of restitution is significantly affected.

REFERENCES

Brody, H. (1984) *That's how the ball bounces*. The Physics Teacher pp. 494-497.

Cislunar Aerospace, Inc. (1999) http:// wings.ucdavis.edu/Tennis/Project/speed-02.html

McMahon, T.A. and Greene, P.R.(1979) The influence of track compliance on running. J. Biomechanics **12** pp 893-903.

The variation of static and dynamic tennis ball properties with temperature

P. Rose, A. Coe
International Tennis Federation, London, UK
S.J. Haake
Department of Mechanical Engineering, University of Sheffield, UK

ABSTRACT: An investigation was carried out into the effect of temperature on pressurised and pressureless tennis ball properties. Ball rebound from 100 inches, ball deformation, ball size, and ball coefficient of restitution at $25ms^{-1}$, $35ms^{-1}$ and $45ms^{-1}$ were measured for tennis balls at temperatures between 0°C and 40°C. The low velocity rebound test was found to show more difference between ball constructions than the static deformation test or the high velocity coefficient of restitution test.

INTRODUCTION

The very nature of tennis ball construction means that the properties of the ball will be temperature responsive. The aim of this project was to investigate how tennis ball properties changed within a temperature range between 0°C and 40°C, the approximate range that a tennis ball is likely to face in play. The static properties monitored were rebound height, ball deformation and ball size. These are key tests that a ball must pass to become ITF approved for tournament use. The dynamic coefficient of restitution (e) for each ball was also measured at incident velocities of $25ms^{-1}$, $35ms^{-1}$ and $45ms^{-1}$ perpendicular to a concrete target.

Both pressurised and pressureless tennis balls were tested, 4 different brands of each being used in the static testing and 2 different brands of each being used in the dynamic testing.

The balls were incubated for a minimum of 24 hours at 0°C, 10°C, 20°C, 30°C and 40°C, and tested in batches of 3 to maintain temperature stability.

METHODS

Exact test methods for each of the static tests can be found in any *ITF Approved Tennis Balls* booklet. Briefly, however, ball rebound height is measured from the bottom of a ball after it has been dropped from 100 inches (254cm) onto a flat, rigid surface. This is classed as static because of the low velocities ($<10ms^{-1}$) involved compared to the dynamic tests. Ball deformation is a displacement measurement taken when loads are applied to a ball (consequently the smaller the displacement value the stiffer the ball). Ball size is measured using ring gauges.

Coefficient of restitution is calculated from the ratio of V_{out}/V_{in} of a ball projected against a concrete block using a pneumatic cannon with the velocities calculated using light timing gates commonly used for ballistics.

RESULTS

STATIC TESTS

Eight brands of ball were tested, four pressurised brands and four pressureless brands. Six balls were tested for each brand.

Ball Rebound Height

Figure 1 shows the relationship between ball rebound height and temperature for the 8 brands of ball. It can be seen that ball rebound decreases singnificantly as the temperature reduces to 0°C. A clear distinction can be seen in the properties of the pressurised compared to the pressureless. As the properties of the pressurised balls depend on the gas in the rubber core as well as the core itself compared to the pressureless balls which rely only on the rubber core, it seems reasonable that the pressurised balls properties will change to a greater extent with temperature than the pressureless. One pressureless ball can be seen to behave as a pressurised ball. This brand has been chemically pressurised, possibly to improve its rebound height and so acts more like a pressurised than a pressureless ball.

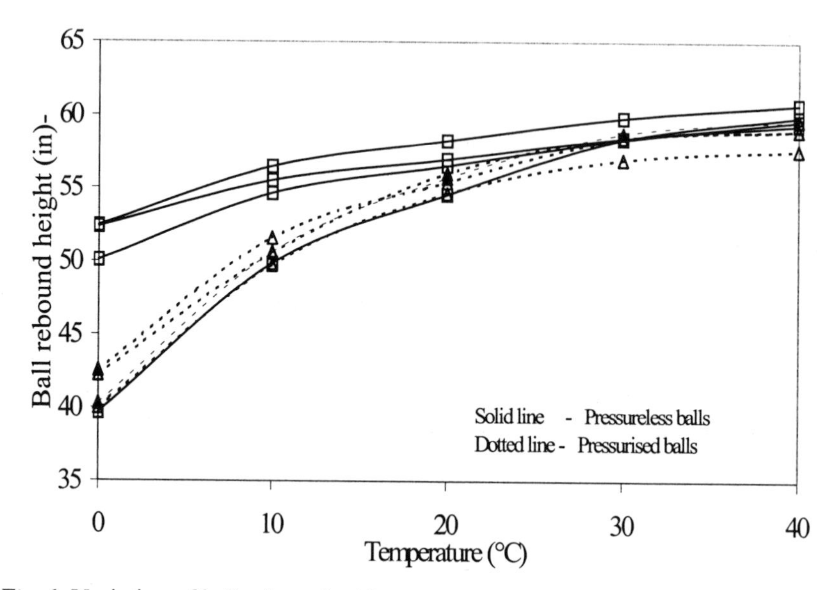

Fig. 1 Variation of ball rebound with temperature

Ball Deformation

Figure 2 shows that the deformation of both the pressurised and pressureless tennis balls does not significantly vary with temperature, although at the extremes of temperature there is some fluctuation.

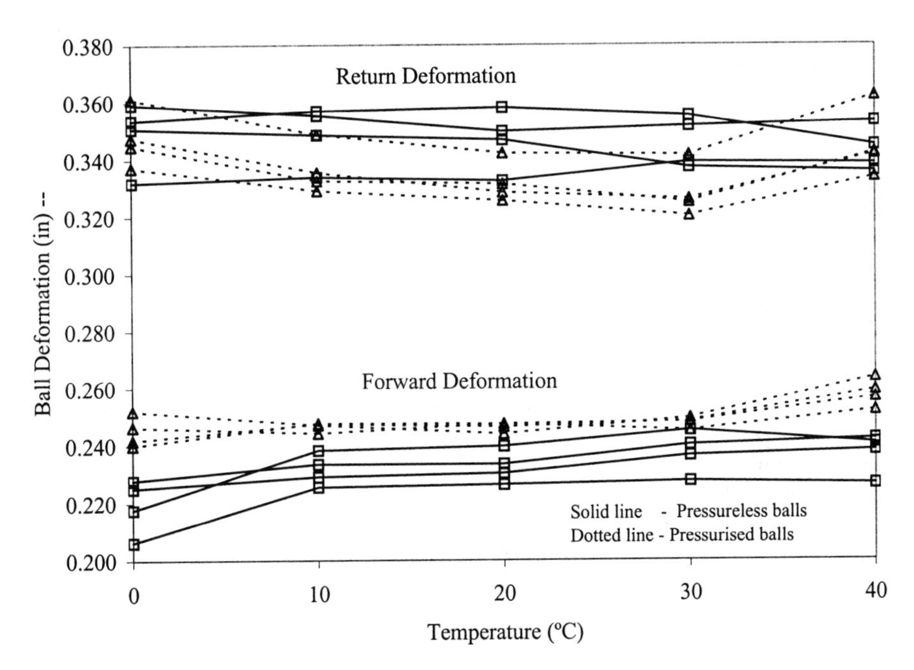

Fig. 2 Variation of Ball Forward and Return deformation with Temperature

DYNAMIC TESTS

Four brands of ball were tested dynamically, two pressurised brands and two pressureless brands.

Coefficient of Restitution

Figure 3 shows a general trend of e increasing with temperature at 25 ms^{-1}, however the results appear fairly scattered and one brand of pressureless balls results show some fluctuations.

Figure 4 shows a general increase in e with temperature at 35ms^{-1}, as in Fig. 3, although the values of e are significantly lower than at 25ms^{-1}. There again appear to be fluctuations in the results.

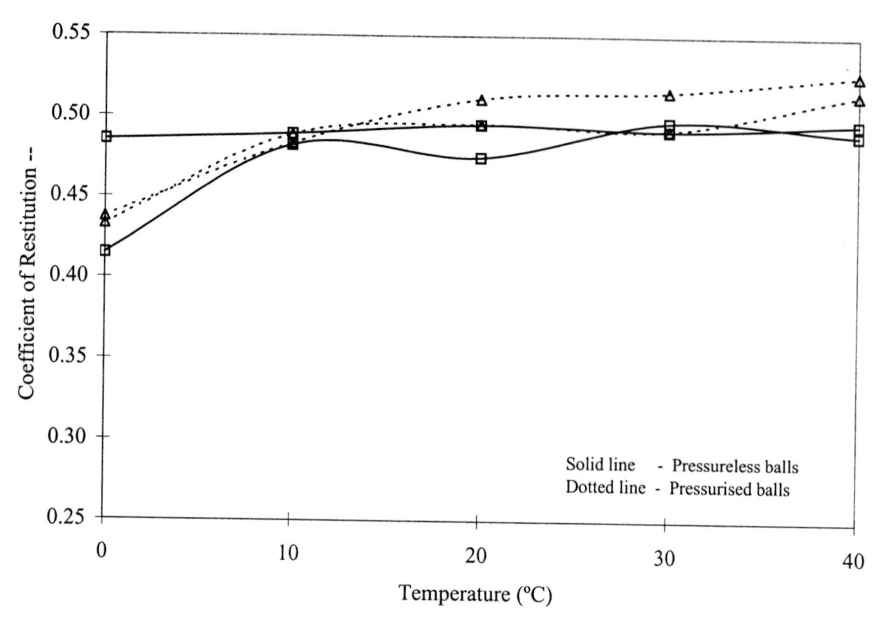

Fig. 3 Variation of e with Temperature at an impact velocity of $25ms^{-1}$

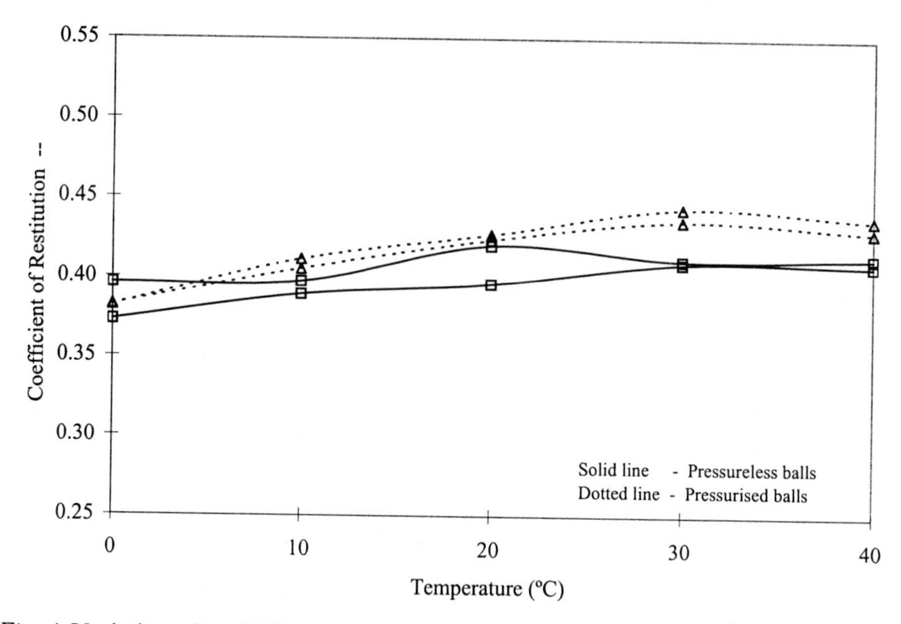

Fig. 4 Variation of e with Temperature at an impact velocity of $35ms^{-1}$

Figure 5 shows a clear and consistent increase in e with temperature at 45ms^{-1} that tends to plateau at higher temperatures. The values of *e* are lower than those at 35ms^{-1}. Brand K balls can be seen to have a noticeably lower coefficient of restitution than the other 3 brands.

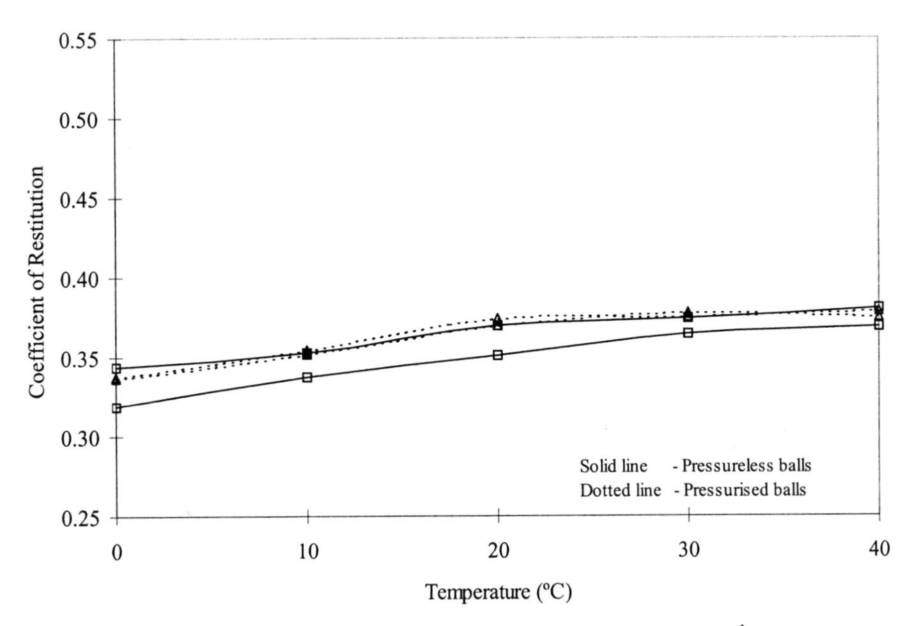

Fig. 5 Variation of *e* with Temperature at an impact velocity of 45ms^{-1}

DISCUSSION

Out of all the properties measured according to the Rules of Tennis, ball rebound is affected by temperature the most. These results indicate that, on average, a pressurised ball will bound outside ITF specification at temperatures less than about 16°C and more than about 29°C. For a pressureless ball these limits are from about 9°C to about 39°C. This may be related to the difference in 'feel' between pressurised and pressureless tennis balls.

It is quite surprising that although low and high temperatures have a profound effect on ball rebound, there is no marked change in ball deformation which is essentially a static test.

The investigation of static properties took place over a duration of 7 weeks and so a possible area of further investigation is how tennis balls change with time and if that would have a bearing on the results of this study.

There is a clear connection between the dynamic coefficient of restitution of a pressurised or pressureless tennis ball and the temperature under which it is measured. Figures 3, 4 and 5 all show *e* increasing with temperature, and a comparison of the three graphs show that the measured e value decreases with increasing ball projection velocity.

It is interesting that the static test data suggests that dynamic tests may be of more value in this type of research. Anecdotal evidence has previously suggested that differences between balls seen in static tests are magnified during play. This conflicts with the dynamic tests carried out here which shows that differences between the coefficient of restitution do not become more apparent as the testing velocity increases.

One reason for this could be that as the velocity increases, distinguishing effects caused by the ball covering become insignificant and the acquired results become a truer reflection of the core properties.

These investigations demonstrate that a careful selection of the testing technique is required when testing the coefficient of restitution of tennis balls, so that accurate, relevant data can be extracted to maximum effect.

CONCLUSION

Over a temperature range from 0°C to 40°C the average ball rebound of a pressurised ball changes from 41.3in to 58.8in respectively, and the average rebound of a pressureless ball changes from 48.6in to 59.9in.

The deformation of both pressurised and pressureless balls as measured by the standard ITF test tends to vary only at extremes of temperature.

There was no observed effect of temperature change to the size of any of the tennis balls.

A uniform trend was displayed between the coefficient of restitution of all balls tested and the test velocity and temperature. The coefficient of restitution increased with temperature and decreased with test velocity.

Different test velocities were seen to influence the coefficient of restitution and its variation with temperature. Static deformation testing did not differentiate different brands of ball clearly between test temperatures. Low velocity rebound testing showed marked differences between ball constructions at different test temperatures, which may be due to the ball covering. High velocity cannon testing (up to $45ms^{-1}$) failed to distinguish subtle differences between ball constructions.

It was concluded that the testing technique must take account of temperature fluctuation if meaningful data is to be extracted from testing.

REFERENCES

ITF Approved Tennis Ball (2000) International Tennis Federation.

Dynamic testing of tennis balls

R. Cross
Physics Department, University of Sydney, Sydney, Australia

ABSTRACT: Measurements are presented on the dynamic properties of tennis balls impacting on a force plate. The force on a tennis ball rises rapidly to about half its maximum value in the first 0.2 ms of the impact due to compression of the cloth cover and the rubber wall near the impact point. The wall then collapses inwards, resulting in a sudden decrease in ball stiffness. This explains why a bouncing ball leaves an oval mark on the court, with an undisturbed patch in the middle. The dynamic stiffness has a peak value around 90 kN/m and a time-average value about 35 kN/m. The rules of tennis require the ball to have a static stiffness of 12.6 ± 1.7 kN/m.

INTRODUCTION

The properties of tennis balls are rigidly specified by the rules of tennis. Even so, a wide variety of tennis balls with different physical properties is manufactured for the consumer. There is concern that different balls also play differently, not only because the rules allow some variation in the mass, diameter and coefficient of restitution, COR, but also because the rules are not specific regarding ball properties under actual playing conditions. The compressibility of a tennis ball is specified for static conditions, and the COR is specified for a low speed collision with concrete. When dropped from a height of 100 inches onto a concrete slab, an approved ball must bounce to a height between 53 and 58 inches. The advantage of this test is that it is easily implemented by both the manufacturers and the testing authorities. In principle, it should also be relatively easy to measure the COR at higher ball speeds, but the required apparatus has not yet been developed to a point where standards can be reliably specified or enforced.

In this paper, results are presented on the properties of several types of tennis ball, as measured by projecting the ball onto a force plate containing an array of piezo elements. This is not the same as a bounce off the strings of a racket, but it provides a valid measure of the dynamic properties of the ball and it provides data of direct relevance to the bounce of a ball off the court surface.

APPARATUS

The apparatus used in this experiment is shown in Fig. 1. The lower part shows the force plate used to measure the force on the ball. The ball was incident vertically downwards, passing through an annular, upper force plate designed to detect the time of arrival of the rebounding ball. No spin was imparted to the ball. The ball always rebounded at a small angle to the vertical, thereby striking the upper force plate. A horizontal laser beam was used, in conjunction with a photodiode, to record the time at which the ball passed the beam. The incident and rebound ball speeds were measured to within about 2%.

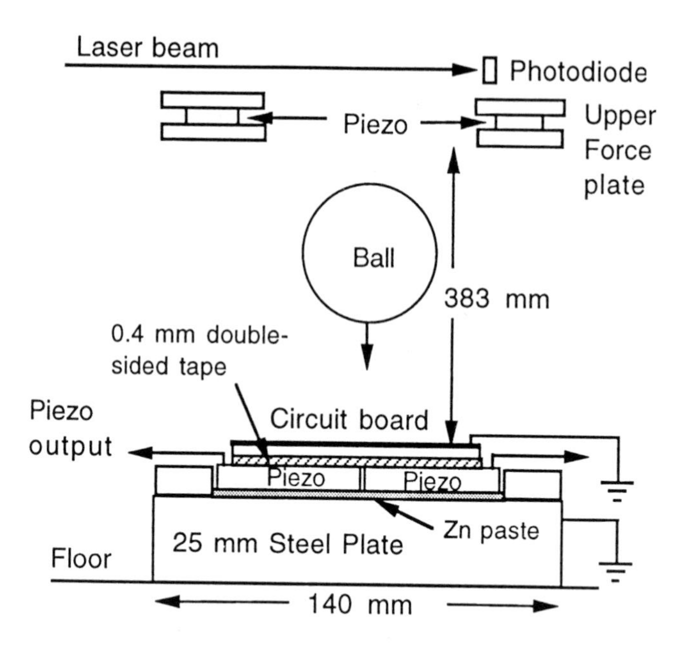

Fig. 1 Apparatus used to measure the dynamic properties of a tennis ball.

The lower force plate was constructed as an array of four square piezoelectric ceramic plates, each of dimensions 50 mm x 50 mm x 4 mm. The size of the array was chosen so that the contact area of the ball remained less than the area of the array. Each plate has two silvered electrodes bonded to and covering each of the large surfaces on opposite sides of the plate. The four piezos were arranged in a large square of dimensions 100 mm x 100 mm x 4 mm, connected electrically in parallel, and were mounted on a steel backing plate of dimensions 140 mm x 140 mm x 25 mm. Zinc paste was used to provide good electrical contact and to minimise mechanical cross--talk between the four piezos. An early version was constructed by glueing the piezos to the steel plate with a thin film of epoxy, but this arrangement resulted in strong cross-talk due to bending of the backing plate. The grounded circuit board shown in Fig. 1 was used as an electrostatic shield, since the ball charged electrically during compression. The force plate provided an accurate response, free of plate resonances, for impacts of duration between 100 μs and 100 ms, as measured by bouncing a steel ball on the plate and by walking on the plate.

THEORETICAL MODEL

The dynamics of the bounce of a ball can be predicted approximately by assuming that it obeys Hooke's law $F = -kx$, where F is the force acting on the ball, x is the ball compression and k is the effective spring constant of the ball. This leads to the result that F vs t is a half-sine waveform of duration

$$\tau = \pi\sqrt{m/k} \tag{1}$$

where m is the mass of the ball (Cross, 1999a). There is no energy loss in the ball in this case. The force on a tennis ball is in fact a strongly nonlinear function of the ball compression. Nevertheless, Eq. (1) provides a useful estimate of the impact duration, provided that k is interpreted as a time-average value of the ratio F/y where y is the displacement of the centre of mass of the ball.

When a ball of mass m impacts vertically on a rigid surface at speed v_1, it experiences an impulsive force, F, which is typically 100-1000 times larger than mg. The force is given by $F = mdv/dt$ where $v = dy/dt$ is the velocity of the centre of mass (CM) of the ball and y is the vertical displacement of the CM of the ball. A measurement of F vs t can therefore be used to obtain y vs t by numerical solution of the equation

$$d^2y/dt^2 = F/m, \tag{2}$$

assuming that at $t = 0$, $y = 0$ and $dy/dt=v_1$. A plot of F vs y represents a dynamic hysteresis curve, analogous to the static hysteresis curve obtained when one plots F vs ball compression under static conditions (Brody, 1979). The area under the hysteresis curve represents the energy dissipated in the ball. The rules of tennis currently specify bounds for such a static hysteresis curve, but do not refer directly to dynamic hysteresis measurements. Typical F vs t waveforms observed with the force plate are shown in Figs. 2 and 3. Figure 4 shows F vs t waveforms and the corresponding y vs t and F vs y curves computed from Eq. (2). A ball can rebound in either a compressed or elongated state depending on the rate at which the ball recovers from the compression and depending on the amplitude of any oscillations excited by the impact. All balls studied in this paper rebounded in a slightly compressed state, so that y remained finite when F dropped to zero at the end of the impact.

QUALITATIVE FEATURES OF THE FORCE WAVEFORMS

As shown in Figs. 2-4, the force on a tennis ball rises rapidly during the first 0.2 ms of the impact, to a value that is typically about half the maximum force at high ball speeds. This effect can be attributed to compression of the cloth and underlying rubber in a small region surrounding the initial impact point (Cross, 1999b). When the ball compression exceeds 1 or 2 mm, there is a sudden transition from a high to a low stiffness state commencing about 0.1-0.2 ms after the initial contact, depending on the ball speed. At low ball speeds, the transition occurs at a force $F \sim 50$ N. As the ball speed increases, the transition occurs at a higher force and at earlier times. The wall deforms approximately as shown in Fig. 5. The formation of an interior bubble in the ball can be attributed to a horizontal component of the applied force, transmitted through the wall to the contact area, resulting in an unstable buckling of the contact region under compression. This effect is assisted by the fact that the initial contact area compresses in a vertical direction and therefore tends to bounce off the surface while the rest of the ball is still moving towards the surface. The buckling effect was verified by visual inspection and by mounting a small piezo disk on top of the lower force plate, as described by Cross (1999b).

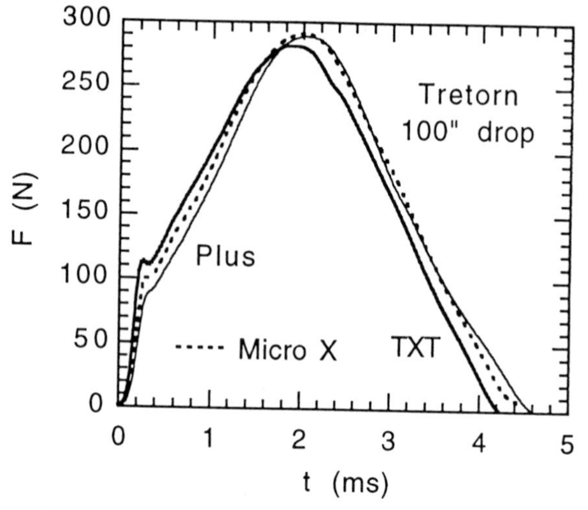

Fig. 2 Results obtained with three types of pressureless Tretorn balls dropped onto the force plate from a height of 100 inches. All three balls rebounded to a height 56 ± 0.5 inches, with COR = 0.750 ± 0.005

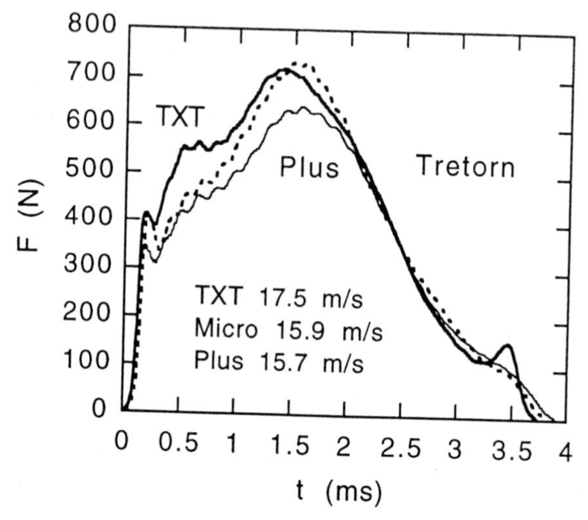

Fig. 3 Force waveforms for the three Tretorn balls shown in Fig. 2, projected at higher speeds onto the force plate. In each case, COR = 0.64 ± 0.01.

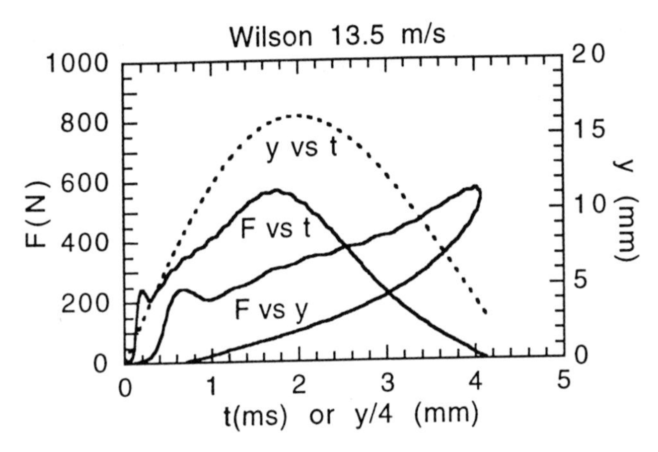

Fig. 4 Results obtained with a new, pressurised Wilson "US Open" ball.

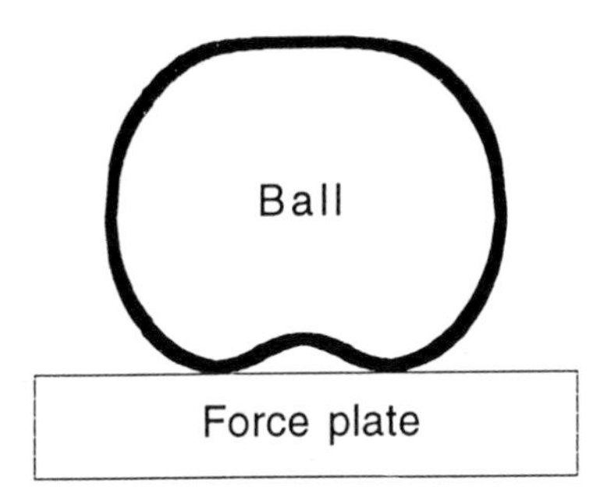

Fig. 5 Deformation of a ball about 2 ms after initial contact.

The above results help to explain why a ball leaves an oval-shaped mark when it lands on a surface such as clay where the mark is clearly visible. Furthermore, it also explains why there is an undisturbed patch in the middle of the mark. Intuitively, one might expect a diamond--shaped mark, with a pointed apex where the ball first touches and finally leaves the surface. In fact, the ball compresses quite rapidly, with the result that the mark is rounded at each end rather than being pointy.

QUANTITATIVE RESULTS

Measurements of the peak force, F_{max}, contact duration, τ, and the COR, e, were made for a variety of pressurised and unpressurised tennis balls, as a function of ball speed, v_1. For this purpose, a ball launcher was constructed using two counter-spinning wheels of variable speed. The maximum ball speed was limited to 17 ms^{-1} with this apparatus. Within this range of ball speeds, all new balls tested were very similar in performance. Consequently, results for only two balls are presented in this Section. One set of results is given for a Dunlop Airloc ball, partly because this ball can be tested over a period of several months without any observable change in properties. The ball had a mass of 58.1 gm and a diameter of 67.0 mm. The other ball was a Wilson "US Open" pressurised ball, of mass 56.5 gm and diameter 65.0 mm, tested immediately after opening a can of new balls. Results for the Dunlop ball are shown in Figs. 6 and 7, and results for the Wilson ball are shown in Figs. 8 and 9. Figs. 6 and 8 show F_{max} and τ as a function of v_1. Figures 7 and 9 show e and y_{max} as a function of v_1. y_{max} represents the maximum displacement of the CM, calculated from the measured force waveform and ball speed, as described above.

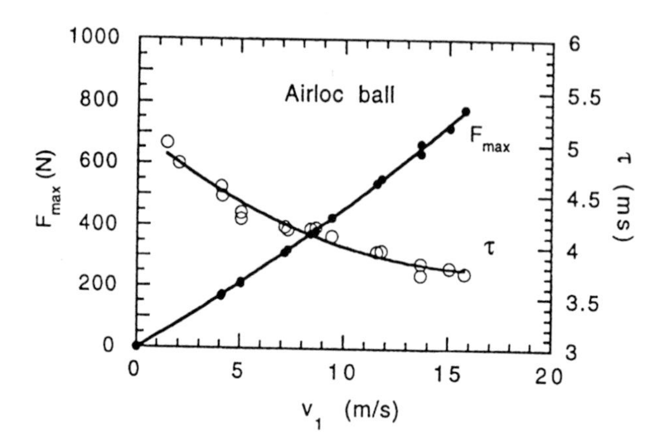

Fig. 6 F_{max} and τ vs v_1 for the Dunlop Airloc ball.

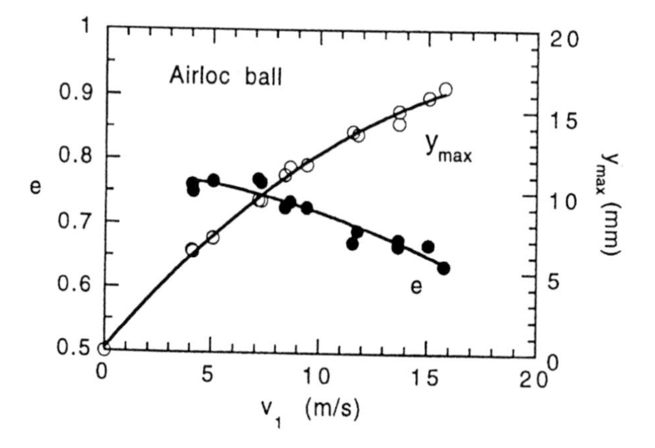

Fig. 7 y_{max}, and e vs v_1 for the Dunlop Airloc ball.

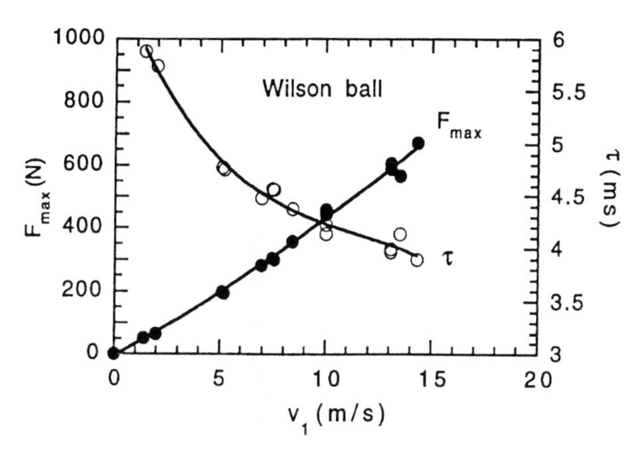

Fig. 8 F_{max} and τ vs v_1 for the Wilson ball.

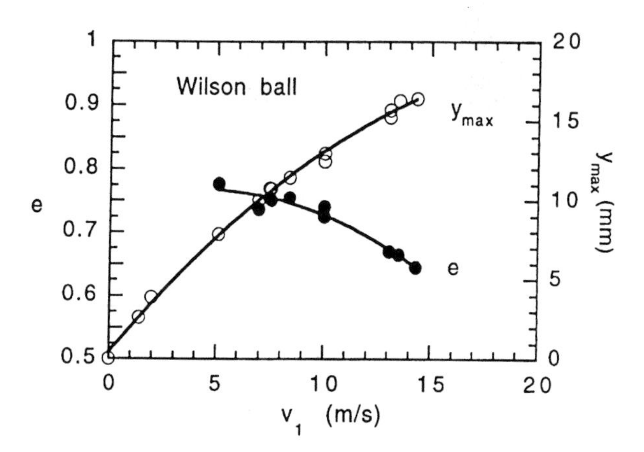

Fig. 9 y_{max} and e vs v_1 for the Wilson ball.

The standard 100 inch drop test corresponds to an incident ball speed of 7.05 ms^{-1}. The range of e at this speed, specified in the rules of tennis, is $0.728 < e < 0.762$. Both of the balls tested have very similar values of e, but the Wilson ball was slightly softer (both in terms of the qualitative feel of the ball and the measured F/y ratio) with the result that, for any given ball speed, F_{max} was smaller, y_{max} was larger and τ was longer, each by about 10%.

The actual change in ball diameter, or the dynamic compression was not measured in this experiment, but this is likely to be about 70% larger than the displacement of the CM. Qualitatively, it was found that balls with a low static stiffness also have a low dynamic stiffness. However, even if the actual dynamic compression, x, could be measured, it would be difficult to make a quantitative comparison between static and dynamic measurements of ball stiffness or hysteresis. During the initial stages of the ball compression, the dynamic stiffness, F/y, is almost an order of magnitude larger than the static stiffness, F/x. The rules of tennis specify that for a static load of 18 lb (8.165 kg), the ball shall have a forward deformation, x, of more than 0.220 of an inch (5.59 mm) and less than 0.290 of an inch (7.37 mm).

Consequently, the static stiffness, k, must be in the range $10.9 < k < 14.3$ kNm^{-1} during the compression phase. If we take the Wilson ball in Fig. 4 as an example, then the dynamic stiffness F/y = 87.5 kNm^{-1} at t = 0.2 ms and F/y = 34.1 kNm^{-1} at maximum compression (ie at t = 1.7 ms). The latter figure is a measure of the average slope of the F vs y hysteresis curve. It is therefore a reasonable measure of the time average dynamic stiffness and is also consistent with the observed duration of the impact, as indicated by Eq. (1) (τ = 4.1 ms). Similarly, the impact durations shown in Figs. 6 and 8 are also consistent with Eq. (1) if k is interpreted as F_{max}/y_{max}.

CONCLUSIONS

With the aid of a simple force plate, it is possible to obtain a large amount of information on the dynamic properties of tennis balls. The standard 100 inch bounce test, together with static compression tests, have provided sufficient information to date to regulate the game of tennis, and will continue to do so for some time into the future. Measurements of the type described in this paper provide data that may help to shape future developments in the specification and testing of ball properties. For example, the standard static compression test dating from the 1930's is somewhat operator dependent and could possibly be replaced with a simpler, more relevant, and more reliable impact duration test or a dynamic compression test as described above.

ACKNOWLEDGEMENT

Andrew Coe from the ITF kindly provided the balls used in these tests.

REFERENCES

Brody, H. (1979) Physics of the tennis racket. *American Journal of Physics*, **47**, 482-487.

Cross, R.C. (1999a) The bounce of a ball. *American Journal of Physics*, **67**, 222-227.

Cross, R.C. (1999b) Dynamic properties of tennis balls. *Sports Engineering,* **2**, 23-33.

Characterising the service bounce using a speed gun

J.I. Dunlop
University of New South Wales, Sydney, Australia

ABSTRACT: The nature of a tennis ball's first bounce following the service can be critical to the subsequent play of the game or point. The rebound characteristics are affected by the properties of a tennis court surface, which may vary substantially from one court to another, and are often referred to as "Pace" and "Bounce". A radar speed gun has been used to measure the Pace and Bounce properties of service balls on tennis court surfaces. The results indicate that Pace varies from 70 % for a "slow" to 80% for a "fast' surface. Bounce is found to be similar to the coefficient of restitution for the surface.

INTRODUCTION

The angled bounce of a sports ball off a hard surface such as a typical tennis court has been the subject of many studies, for example, publications by Thorpe and Canaway (1986), Brody (1983), conferences and in-house reports. During play the tennis ball is hit at a variety of speeds, often with spin imparted, to contact and bounce off the tennis court from a range of angles. The characteristics of these angled bounces, which depend on the physical properties of the ball and the court have a decided influence on the subsequent game. Perhaps of most importance is the nature of the ball's first bounce following service when maximum ball speeds are encountered.

The most authoritative measurements of the angled bounce are probably those of Thorpe and Canaway who made high speed cinematic studies. They identified two important aspects which affect the playing of the game. These were the Pace of the court which was the player's perception of the change in horizontal component of velocity of the ball during impact – on a "slow" surface the ball comes through slower and vice versa. The change in horizontal component of velocity is often accompanied by changes in the vertical component of velocity which they refer to as the Bounce and some players have difficulty in distinguishing between the effects of Pace and Bounce. Spin also has an influence but will be neglected in this study.

A theoretical analysis of the bounce process had been carried out earlier by Brody who derived mathematical relationships between the angles and speeds of a ball before and after the bounce. He distinguished two types of bounce – a low angle high speed bounce (such as the service) when the ball slides along the bounce contact without rolling. At higher angles and lower speeds, such as lobs, volleys etc the ball rolls during the bounce contact with the surface and the characteristics of the bounce are quite different from the sliding bounce.

There have been more recent studies of these bounces, such as Dunlop (1991), and many conference proceedings. This report describes an investigation of the kinematics of some typical tennis service bounces using a radar speed gun.

SPEED GUN

The speed gun used in this investigation was based on a Doppler radar system of nominal primary frequency 34.7 GHz. Reflections of the radar waves off a moving target are received at the transmitter and mixed with the primary frequency to produce a difference frequency related to the speed of the target towards or away from the receiver.

The Doppler equation yields the difference frequency Δf as follows

$$\Delta f = 2\, v\, f_0 / c$$

where v is the velocity of target
f_0 is the primary frequency, and
c is the velocity of microwaves (~300 Mm/s)

The calibration of the gun was first checked for these investigations. The primary frequency was measured in the University's microwave laboratory and found to be accurate (to better than 0.5 %) and stable for periods of hours.

The difference frequency is converted to a displayed speed which is checked by holding a calibration tuning fork of 4165 Hz near the receiver and noting that the correct speed of 64.8 km/hr is displayed. To ensure that the gun was in calibration over a larger range of speeds a series of tuning forks was fabricated and the calibrations checked. The results of these measurements are shown in Fig. 1 which confirms the accuracy and linearity of the system to at least 100 km/hr.

The speed measured is displayed on the gun or the data is transferred to a computer for storage and display. The sampling rate of the device was 30 Hz with a speed resolution of 1 km/hr. A higher resolution of 0.1 km/hr was also available using averaging algorithms within the system.

OPERATION OF SPEED GUN

The speed gun was capable of capturing the maximum speed of the ball after being struck by the tennis racquet and of direct tracking of ball speed (towards the gun) as a function of time. The readings obtained for a typical service are shown in Fig. 2.

For these measurements the ball was served normally into the court of play but also directly in line with the speed gun which was positioned 10 m behind the receiver's base line. These distances ensured that the direction of motion of the ball was less than 5 degrees from the straight line to the gun when it was was oriented horizontally towards the server, reducing "cosine" errors to less than 0.5 %.

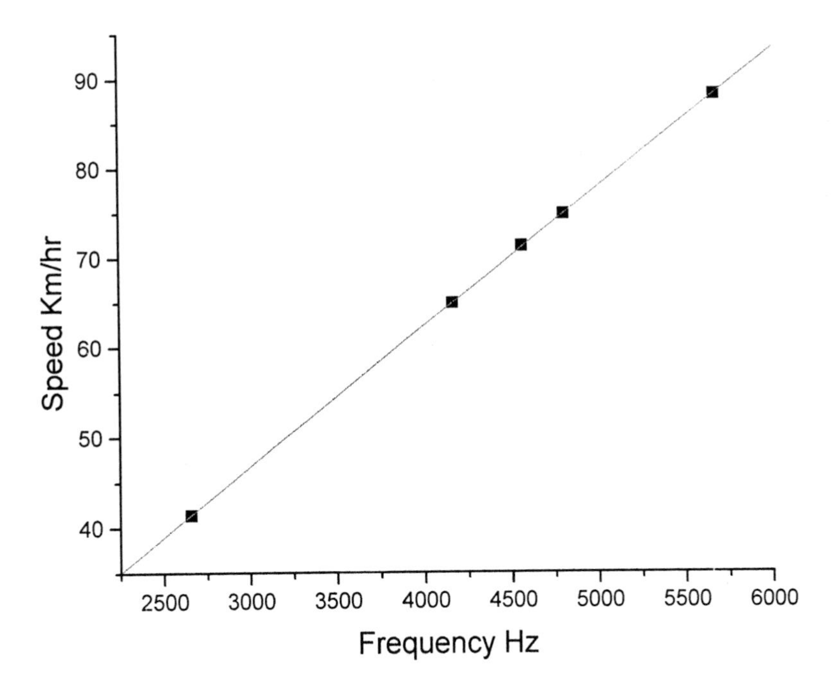

Fig. 1 Speed gun calibration

Figure 2. shows the horizontal speed of the ball soon after being struck by the racquet. The maximum speed detected was 175 km/hr. The speed at the beginning of tracking is near 170 km/hr and decreases because of aerodynamic drag to about 120 km/hr by the time it contacts the court. There is then a large discontinuity in speed during the bounce, the speed following contact being about 90 km/hr. There is a further decrease in speed due to drag during this second phase of the ball's trajectory.

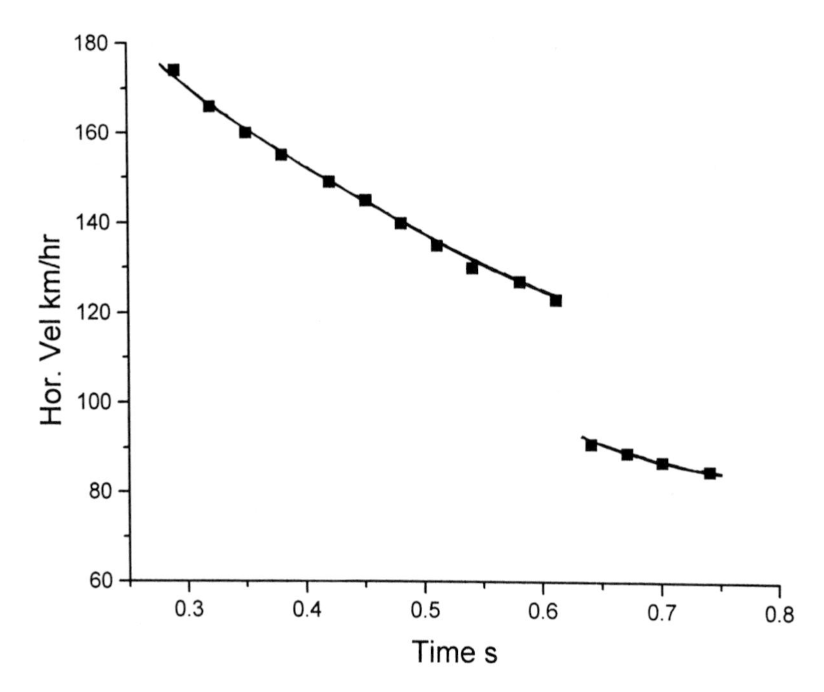

Fig. 2 Speed gun tracking of tennis ball (dots), theoretical values on solid line.

BALL TRAJECTORIES

The ballistic trajectories of a struck tennis ball are affected by three forces – the force of gravity, G, the drag force D, due to its air resistance, and the Magnus force M, caused by rotation or spin of the ball. This latter force has been neglected in this study.

The trajectory of the ball can be described by the coupled differential equations,

$$my'' = -mg - bv^2 \quad \text{and}$$
$$mx'' = \quad -bv^2$$

where $v^2 = x'^2 + y'^2$

x and y represent the two defining coordinates and the primes and double primes the first and second time derivatives.

The drag constant b is given by $b = C_D(\pi d^2/8)\rho$

C_D being the drag coefficient being 0.508 for a typical tennis ball (Stepanek, 1988) Substituting measured values of 66 mm for d, 57.5 g for m and 1.22 kg/m^3 for ρ gives a value of b/m of 0.019.

Solutions of these equations can be applied to describe a typical tennis service as measured by the speed gun in Fig. 2. This was a service from the base line contacting just inside the service line – a horizontal distance of 19.5 +/- 0.5 m. To solve the equations, initial conditions must be supplied, in this case $x(0) = 0$, $y(0) = 2.54$ m, $x'(0) = 160$ km/hr and $y'(0)$ is adjusted to –3.6 m/s giving a contact around 19.8 m. The calculated trajectory is plotted on Fig. 3. Further manipulation of the equations shows a grazing angle of contact of 11.2 degrees. The equations can be solved to give horizontal speed against time and this yields a good fit to the measured values as in Fig. 2, but often shows slightly less drag than measured. The drag is affected by slight variations in ball dimensions, particularly diameter and non sphericity, and imparted spin.

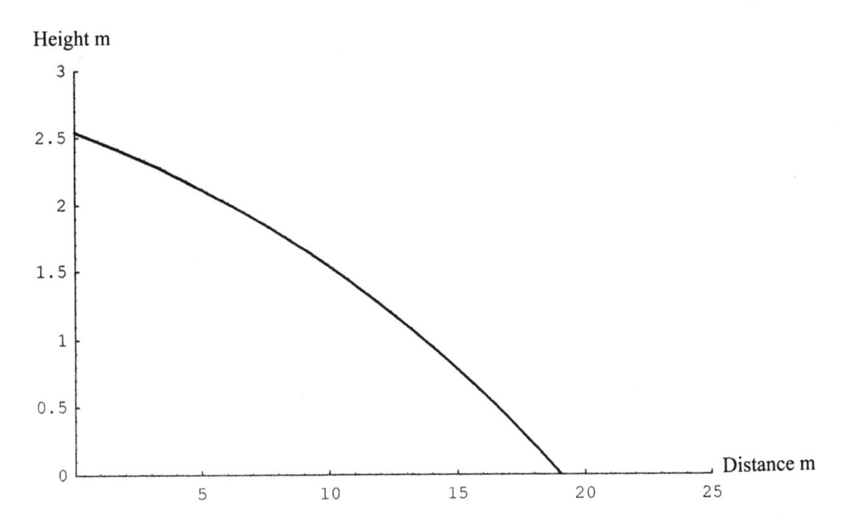

Fig. 3 Trajectory of typical tennis serve

PACE DETERMINATION

A direct determination of Pace can be made from Fig. 2. The figure indicates that the ball contacts the surface with a horizontal velocity v_{xi} of 113 km/hr and leaves contact with a velocity v_{xf} of 85 km/hr. This gives a Pace of the surface for this delivery of 85/113 = 0.75 or 75 %. The precision of the measurement depends on the limit of resolution of the speed gun - to the nearest integer km/hr - but with extrapolation and curve fitting this precision can be significantly better than the +/- 0.5 km/hr used here.

Brody's equations indicate that the Pace is dependent on the angle of contact, so it should be noted that this Pace measurement relates to the angle encountered here. There is also an assumption of zero spin, whereas the players often impart top spin on their services. The value stated above therefore relates to the actual Pace perceived by the player for a particular service.

BOUNCE DETERMINATION

For determination of Bounce, or the vertical coefficient of restitution, it is necessary to track the trajectory of the ball to its second bounce. The time interval between the two contacts can then be measured and from this and trajectory theory the vertical velocity after the first bounce, v_{yf} can be determined. The vertical velocity before the first bounce, v_{yi} can be also determined from solutions of the trajectory equations. A typical trajectory from a ball projector, with the speed gun positioned behind the projector is shown in Fig. 4. The precision of the time of flight measurement can be better than the limit of resolution of 33 ms, but the precision of the final determinations will be degraded by the trajectory calculations required.

RESULTS

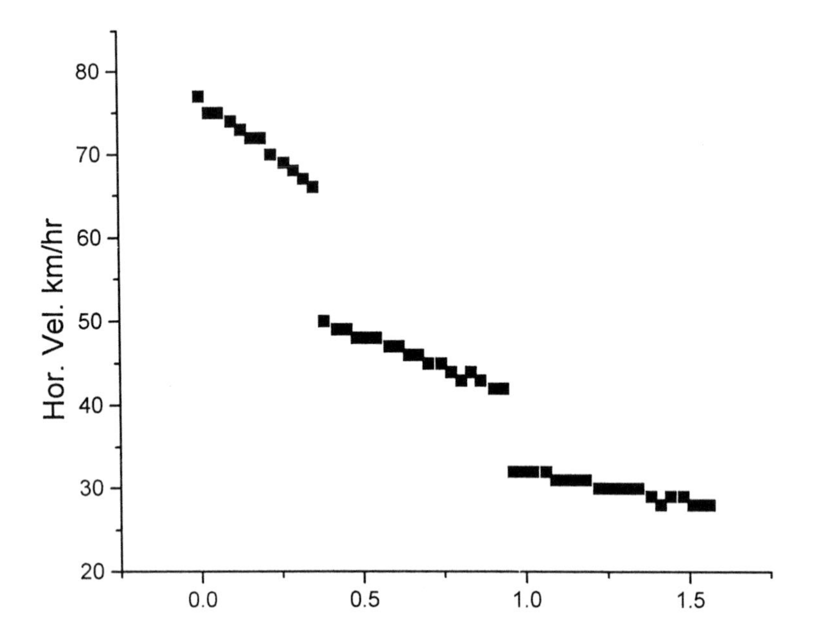

Fig. 4 Speed gun tracking of tennis ball showing second bounce

Two professional tennis players were asked to strike normal first services for measurement on three different tennis surfaces – Rebound Ace, grass (White City centre court) and a sand filled synthetic tennis surface. The serving actions of both players were recorded on video camera to determine the heights of service for trajectory determinations.

Similar measurements were carried out using a pneumatically operated ball projector mounted horizontally at a height of 1.00 m. The projector ejected balls without appreciable spin at a muzzle speed of 80 km/hr and produced contact speeds of 66 km/hr and contact angles of 12.7 degrees.

Fresh Slazenger "Hard Court " tennis balls which had first been tested to meet the rebound specifications and temperature conditioned to the temperature of the courts (22 degrees) were used. The service balls were tracked using the speed gun and various parameters as described above determined.

Typical results are shown in Table 1.

Table 1 Measurement of Pace parameters

Surface: Rebound Ace

Source	V_{xi} km/hr	V_{xf} km/hr	Pace %	Average
Dan	128	92	72	
	135	98	73	
	123	92	74	
	135	101	74	
	138	98	71	73 +/- 2
Ivan	113	78	69	
	108	74	69	
	93	66	71	
	113	81	73	
	111	81	73	
	113	85	75	72 +/- 2
Projector	66	48	73	
	66	48	73	
	66	48	73	
	66	47	71	73 +/- 1

Results of tests on other surfaces are summarized in table 2.

Table 2 Measurement of Pace on several courts

Surface	Source	Pace
Grass	Dan	76 +/- 2
	Ivan	78 +/- 2
	Projector	79 +/- 1
Sanded turf	Dan	71 +/- 2
	Ivan	70 +/- 2
	Projector	69 +/- 1

The ball projector was also used to determine the Bounce characteristics of various surfaces. It was set to project balls horizontally from a height of 1.00 m. as before, striking the surface at 12.7 degrees with a vertical velocity v_{yi} of 4.22 m/s.
Results of measurements relating to the determination of Bounce are listed in Table 3.

Table 3 Measurement of Bounce parameters

Surface	Δt ms	V_{yf} m/s	Bounce %	Coeff. Rest. from 1 m
Rebound Ace	590	3.00	0.71 =/-.03	0.75 +/- .01
Grass	520	2.62	0.62 +/- .03	0.68 +/- .01

DISCUSSION

The measurements show considerable variation in the Pace of the tennis service off any particular type of surface as perceived by the receiver. This is due mainly to the variations in speed (and hence contact angle) and perhaps spin of each service from a particular player. The angle of contact was noted to range from 10.4 to 11.6 degrees for various typical serving parameters – racquet speeds 180 to 205 km/hr and heights 2.54 to 2.9 m. The Pace values measured range from 70 % for a slow surface to 80 % for a fast one. The Pace measurements using the ball projector, albeit at lower speeds, were much more consistent, and considering their precision might be used to distinguish one type of surface from another.

Bounce, which is essentially the coefficient of restitution for the angled rebound, can also be determined to a lesser precision than Pace, but sufficient to suggest that it is very close to that measured from the vertical rebound, supporting Brody's analysis.

CONCLUSION

The results of the measurements indicate that the radar speed gun offers a simple and effective method for determining the Pace and Bounce characteristics of a tennis court surface.

ACKNOWLEDGEMENTS

The author wishes to thank the Australian Institute of Sport for its support.

REFERENCES

Brody H.F. (1984) That's how the ball bounces, *The Physics Teacher*, **November** p494-497.

Dunlop J.I., Milner C.J. & White K.G. (1991) Sports surface/Ball interactions, *J. Sports Turf Res. Inst.* **68** 114-123.

Stepanek A. (1988) The aerodynamics of tennis balls – The topside lob, *Am. J. Phys.* **56**, 138-142

Thorpe J.D. and Canaway P.M. (1986) Performance of Tennis Court Surfaces II, *J. Sports Turf Res. Inst.* **62**, 101-117

3 The court and facilities

Facilities equals growth

P.A. Sandilands
The Lawn Tennis Association, UK

ABSTRACT: This paper focuses on the role of facilities in the drive for growth in tennis. It puts the case that appropriate and modern facilities for the sport make a real difference and without which any sustainable sports development objectives cannot be met. It shows an excellent illustration of one particular scheme developed in partnership with a local authority that has transformed the perception and the attitudes towards tennis in its locality. In conclusion ways are suggested in which the ITF should involve itself to encourage and enable facility enhancements.

THE NEED FOR TENNIS FACILITIES

Do facilities make a real difference? There is no question that facilities on their own cannot make the game more accessible and encourage a significant amount of new participants. A winning combination must also have at its core a compliment of energetic coaches, an exciting competitive structure, comprehensive development programmes, a social dimension, and financial sustainability as a result of good management. But without the essential ingredient of a suitable facility, any short-term success cannot hope to be sustained.

A new or refurbished facility can assist in attracting new participants to the game, and in the case of covered or floodlit courts can increase the opportunities for play by 100% and 33% respectively – given the vagaries of British weather! This growth in the number of court hours available will also help enable the improvement of standards of play – more training and competitive opportunities created during the year. A new facility in particular can raise increased awareness of tennis in a locality, and if it is in an area which suffers from a lack of sports provision can act as a regenerating agent. Better facilities provide the opportunity for partnerships – with industry, sponsors, central and local government, and the media, which in turn increases the profile of the sport to those not traditionally orientated towards tennis. Good and modern tennis facilities and environments will have a better chance to compete with other leisure pursuits, particularly in attracting the youth of today.

I don't intend to concentrate on the high profile tournament venue developments we have seen recently – the new Number 1 court at Wimbledon,

Key Biscayne, Homebush Bay, Qatar, Indian Wells. These and many other major developments are excellent expressions of the modernisation of the sport and take account of the growing pressures to provide better facilities for the competitor, spectator, sponsor, and media, and to enable tennis to maintain its high profile as a truly international sport. Whilst these investments are applauded they do not in reality contribute significantly to increasing participation.

The recent ITF research programme into tennis participation and attitudes provides us with an interesting analysis of the role facilities may play in growing the game compared to other key factors believed to have an influence on the development of the sport. The survey suggests that nations with easier tennis court access may be seeing higher growth in participation. The difficulty in confirming this however is due to the fact that few nations have accurate data on their infrastructure, that is, numbers of courts and their utilisation.

The correlation between participation and provision is however mirrored by a number of other sports that have concentrated on the development of facilities as a means to grow their sport and to improve playing standards. Tennis itself provides good examples where higher investment in facilities (usually public) results in significant increases in participation or improved standards of play. Sweden's international success in the 80s was linked to the investment in covered courts. France's depth of tennis talent has as its foundation, the development of tennis centres in small villages and communities in the 70s. The Federal Republic of Germany exceeded the targets of its first Sports Golden Plan by the early 1980's and promptly re-wrote the Plan to embrace higher targets. Spain's enviable successes in Barcelona were founded, at least in part, on planned investment in promising performers and in the sports services made available to them. Other such stories abound throughout sport.

Tennis, as for many sports, relies on giving participants enjoyment, health, ambition, and fun. Tennis has more going for it than other sports in that it is truly sexless, ageless and global. However, without the tools at our disposal, and in particular the environments in which to play, we cannot expect to make an impact in the world of sport and leisure.

Tennis, through the work of the ITF and its National Associations, has a great opportunity to make an impact. This impact will be short lived unless there is a commitment to enable the development of facilities. Facility strategies should be produced which respond to clear sports development needs rather than merely identifying and filling gaps in provision. In each nation, it will be different. It will depend upon existing provision, existing participation levels, and projected increases. It will be based upon the growth or reduction in the delivery agents – clubs, education, and public sector. It will depend upon the nature of the sports development need – is it to increase the base number of participants or improve the environments for the performance player? It will concentrate the mind on what priority should be given to the provision in education, in the community, and for sporting excellence. Strategies should ask the questions: What is needed? Why is it needed? Where should it be built? How can it be achieved? When can it be achieved? Strategies should not be long term but should reflect the reality of building programmes and the availability of funding. Facility strategies should also reflect any particular policies adopted by the agency in relation to this area, for example, embracing green policies in building programmes, accessibility for players with disabilities etc.

Once a strategy is in place, individual projects can then be justified and selected by reference to the adopted strategy. Where key locations have been identified within the strategy, these can be dealt with as opportunities arise. Where there are no specific locations identified, the strategy will act as a balance and a filter to ensure that any opportunist schemes can be justified and followed through. Each building project should be carefully assessed in the context of the sports development need. By doing so, the successful projects at a local, regional or national level should contribute significantly towards the growth of the sport.

Of course each individual project requires considerable thought and planning. Failure to plan is planning to fail, and the result of inadequate planning for building projects can be catastrophic. Planning, design, development and management are the key stages in delivering a successful facility project and each of these stages would warrant a session in this Congress to give delegates a better understanding of some of the principles and methodologies involved.

THE INDOOR TENNIS INITIATIVE

Can facilities make a real difference? The experience in Great Britain suggests it can and has. The Indoor Tennis Initiative (ITI) is a programme run by the British Tennis Foundation (BTF), and supported by the Lawn Tennis Association (LTA). In partnership with the public sector, the programme was introduced in 1986 and now boasts 50 centres, providing 254 covered tennis courts, and 233 outdoor tennis courts. Most of the centres have been built at new sites, with many located in areas not traditionally associated with the sport. From the outset, this was a facility-led programme as there was a belief in 1986 that the traditional tennis club in GB was unable to play a full developmental role in the growth of the sport because of the size of the units. To this day, over 50% of tennis clubs have 3 courts or less, and the majority have little chance of increasing the size and scope of their facilities.

The original objectives of the ITI were: to introduce a healthy lifetime recreational activity to a new range of players from the broadest possible social spectrum; to maximise the usage of public park and school courts by introducing a year round tennis programme; to increase opportunities to play all the year round and encourage those regular players to play more; to improve national playing standards resulting from increased facilities for coaching, practice and competition; to expand the game of tennis in Great Britain.

The success of the programme was to harness the funding and support of other agencies that wanted to invest in the game. The interests of the LTA, BTF and the All England Lawn Tennis Club (AELTC) were obvious. The Sports Councils in England, Scotland and Wales had a 'Sport for All' philosophy and in particular wanted to facilitate the development of tennis in urban and inner city areas. Local authorities saw the potential opportunities the programme gave to their communities and were keen to help regenerate areas and make better use of their existing stock of tennis facilities. The building and court construction industry also saw great opportunities for work and a number of national and local sponsors found that the programme and built facilities gave them an excellent opportunity to reach the local community. Central government agencies were also interested in providing some significant financial assistance to strategic areas, and invested in ITI centres where they were located within their own re-generation programmes and projects.

The outputs from the 43 ITI centres in operation are impressive. Annual usage of covered courts is averaging at 65% and usage of outdoor courts at 40%. It is estimated that the centres have created some 200 tennis specific jobs – coaches, development officers, etc, and many other new job opportunities. Some 65% of the indoor tournaments and events run by the LTA take place at these centres. But more importantly, it is estimated that each centre has introduced the game to 2500 new participants. Yes, some of these could have started tennis by another route, but the club sector due to the quality and scope of their facilities generally, could never hope to emulate such an increase in participation. Interestingly, whilst the ITI concept was designed to make it easier for the casual infrequent player to have an occasional game or two, over 80% of users now play at least once per week at the centres.

PUMA SUNDERLAND TENNIS CENTRE

There have been many good examples of ITI partnership in growing the game. The partnership forged with Sunderland City Council in the north of England is an excellent example of the success of the ITI concept and is a benchmark for others to emulate in relation to tennis development. It was the third centre to be built in 1988. It has changed the local community's perception of tennis, from an elitist middle class game, to a sport which is accessible, exciting and modern, and which enables players to reach their full potential. This has been achieved through a dynamic partnership with a local authority dedicated to sports development principles whilst embracing the commercial opportunities such a venture presents.

The Sunderland centre's mission statement is ' At the Puma Sunderland Tennis Centre, we are dedicated to becoming the best ITI centre through knowing our customers and providing a service to the standard they desire'. The centre was opened in July 1988 and cost £788,000 funded by Sunderland City Council, the ITI, and a sponsorship package from Puma Ltd. A successful Sports Lottery bid together with further BTF and County LTA funding expanded the centre from 4 to 8 indoor courts and was until recently the largest purpose-built tennis facility in the north of England.

The aim of the centre is 'To act as the focal point for the development and promotion of the game of tennis to all sections of the community and to make maximum use of existing facilities throughout the city'. The objectives include the provision of a structured and co-ordinated approach to increase participation in the game of tennis by all sections of the community; to provide a co-ordinated coaching development plan for the acquisition of tennis skills; to encourage individuals showing natural ability to progress to the highest level; and to provide a varied programme of tennis activities which cater for the needs of the community. The facilities now include 8 covered courts, 4 outdoor acrylic and 6 artificial floodlit grass courts, a bar/function room, gym, conservatory and play area, changing rooms and toilets.

The success of the development of the ITI partnership in Sunderland is reflected in the huge growth in the number of people of all ages and abilities playing tennis throughout the year. Prior to the Centre being built, the vast majority of tennis played by the wider community, was on outdoor shale or tarmac courts provided in parks or at the many Colliery Welfare Sports Grounds that were in existence at that time, and was to all intent and purposes a summer only activity.

Opportunities to play all year round, to play on courts that provided consistent playing qualities, to receive coaching, to compete regularly, and (for those with the desire and ability) to progress through a structured development programme were extremely limited and most certainly did not reflect the City Council's aspirations of providing 'sport for all'.

As well as providing a focal point and impetus for the development of tennis as a recreational sport, that could be enjoyed by all members of the community, the Centre has been the much needed stimulus to kickstart the development of other tennis facilities in the City, in particular, the 'Community Tennis Partnership' (CTP) which was developed at Biddick School and Specialist Sports College. This facility services the needs of young people from the Washington area of the city and acts as a feeder centre for the City and County squads based at the ITI Centre. A further measure of the success of the ITI Centre, supported more recently by the CTP, is the number of young players, both boys and girls, who first started playing tennis at the Centre and who have gone on to achieve national and international honours (a total of 14 and 7 respectively) and the success achieved in establishing a flourishing and successful wheelchair tennis club/squad.

Without the vision and support of the ITI, tennis in Sunderland would not be where it is today, providing quality opportunities for people to play tennis for their own intrinsic reason and, for those with the talent and desire, to reach the highest possible standard they can achieve.

Some of the success of Sunderland has been replicated in many of the other centres in the programme, whatever their scope of facilities and locations. Statistically for all the centres, overall 54% of users are male and 46% are female, with the number of females increasing where there are additional facilities, such as crèches. In all cases the ITI centres are attracting a higher percentage of ethnic minority users than the corresponding proportion in the catchment of the population. Indeed the average proportion for all centres on a year round basis is precisely double the proportion of the centre's catchment population. Although pricing policies for tennis are inevitably higher than for team or multi-sports activities, the proportion of users who are unemployed is only slightly less than the corresponding unemployment rate amongst the catchment population. Centres are also frequently used by players registered as disabled, and it is perhaps no coincidence that the great strides made in GB in wheelchair tennis is reflected in the use of the new facilities which are specifically designed for wheelchair tennis activities.

An ITI centre is truly a local sports provision with over 90% of users travelling from within a 30-minute drive time. The clearest single reason for using an ITI centre emerging from a recent study is the quality of the facilities. On average more than 4 out of 5 users participate at their chosen centre at least once per week, which illustrates a committed clientele for whom the indoor facilities afforded by their centre form the core venue for their tennis play. Almost half of the users of ITI centres were beginners when they first used the centres which emphasizes the valuable role which they have fulfilled in attracting new players into the game. Almost 50% of users have improved their standard of play as a result of using the centers. This may be seen as a relatively disappointing figure and suggests that work still needs to be done on improving the competitive opportunities within the centres. However, it is also accepted that there will always be a considerable number of players who are merely content to maintain their standard of play for recreational purposes only. Providing they remain in the sport, there is nothing wrong in this. Even so, and perhaps surprisingly however, over 90% of users rate the standard of coaches as good.

In summary therefore, the centres attract a remarkably representative cross-section of users from the catchments they serve. A high proportion (81.9%) plays at the centres at least once a week. Just under half (48.4%) have improved their standard of tennis, but opinions on the standard of coaching are very positive. Perceptions of design quality are generally positive, and is the top reason why people play at the centres. In addition, through the proactive way in which the centre sports development policies are delivered, many more players, young people in particular, are targeted through outreach work in schools, clubs, and parks. These external venues act as feeders to the main covered court centre.

The ITI centres are playing a major part in the promotion of tennis in each locality. The programme and its success is probably unique to GB, but a similar concept could be reproduced in other countries. This particular programme has concentrated its efforts on providing localised modern facilities accessible for all at a cost that is generally affordable by all. Where it is not, mechanisms are found to subsidise targeted groups to enable new players to take up the sport.

The ITI has been a key part of the LTA's drive to improve facilities in GB. Tennis funding support totalling £20million has levered an additional £70million of investment into these new centres. Over the past 10 years investment in tennis building and associated facilities in GB has been estimated to exceed £135 million with the LTA committing some one third of this sum. This does not include investment by the private health and racket club sector that has resulted in the building of 50 clubs with 8-12 covered courts in each. And yet in GB we have a long way to go to even compete with other established and mature tennis nations. The sobering statistic is that there is only 1 covered court for every 56,000 of the population.

ROLE OF THE ITF

So what role should the ITF have in this area? Providing capital funding is not a realistic option given its global remit and even a significant increase in financial help, particularly to developing nations, would merely result in token contributions towards improving their tennis infrastructure. But in its excellent Drive for Growth, the development of good and modern facilities for the sport must be a vital part of its objectives. In my opinion, the ITF should help enable the improvement of facilities to grow the game in many countries. It should advise on appropriate facility strategies. It should provide design standards. It should identify good practice in planning, design, development and management. It should help to convince those that can potentially fund facility developments the value of tennis to its communities and the improvements it needs to make to its infrastructure. It should help co-ordinate the establishment of true partnerships. It should use its influence on government agencies to enable facility development to be made easier and to encourage industry to work with the sport to develop products that make the tennis experience valued.

CONCLUDING REMARKS

I believe the growth of tennis must go hand in hand with the growth and improvement in tennis facilities. This is a major and long term challenge which will need the commitment and energies of all those who wish to see the sport flourish.

REFERENCES

ITF Participation and Attitude Trends 1997-1999, (2000) *Sports Marketing Surveys Ltd*

ITI Centres Research Study 1995-1998 (1998) *Whiteley International*

Sandilands, P.A. (1998) LTA Facility Strategy 1998-2002

Sandilands, P.A. (1992) LTA 5 Year Facility Plan 1993-1997

Construction and maintenance of championship grass courts

E. Seaward
The All England Lawn Tennis & Croquet Club, Wimbledon, UK

INTRODUCTION

The construction, subsequent annual maintenance and preparation of a grass tennis court has but one main requirement, i.e. to provide a surface on which the players can confidently use their complete range of skills.

The object of the programme is to produce a court which has the characteristics of:

(1) A firm surface that provides good traction for the players.
(2) An even bounce of a good height throughout the court.
(3) Even density and colour of a weed free grass sward.
(4) Good presentation with well defined lines.

What is required from a grass court can be defined by considering the players' requirements. How to achieve the desired playing surface requires a mixture of groundsmanship skills, management and scientific knowledge.

For ease of explanation, the production of the grass court can be split into three distinct sections:

(1) Construction.
(2) Annual maintenance programme.
(3) Preparation for Championship play.

CONSTRUCTION

There is no particularly defined way of constructing a court. It is, however, essential that, at the tournament venue, the courts are constructed in a manner which will allow for consistent playing characteristics from one court to another. Having said this, it takes years for a newly constructed court to mature, even though it will gve a good surface from the start.

The profile at Wimbledon, from the base upwards, is (Fig.1):

(1) 150mm depth of angular stone.
(2) 50mm depth of pea gravel.
(3) 238mm depth of root zone.
(4) 12mm depth of turf.

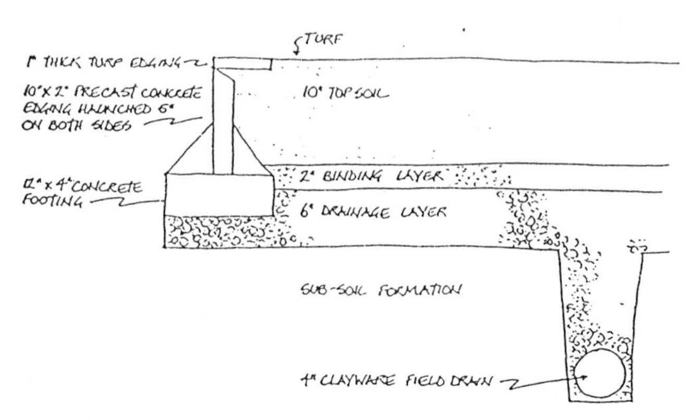

Fig.1 Section Through a Court

The function of the stone and pea gravel is to provide adequate drainage and a barrier between the natural soils and the root zone.

Along with the grass plant, the other vital ingredient for success is in choosing the correct soils to provide the root zone. These two ingredients, if correctly produced and managed, will provide the desired surface. Ref. Grass Tennis Courts, STRI.

THE SOIL

The ideal soil will not only provide the anchorage for the grass plant, but its soil particles must also have the property to enable it to bind together to create a mass which, when dry, will have the strength to withstand the pressure of the players' movement without breaking. Also, it must be possible for it to be rolled to form a hard surface, so allowing the ball to bounce.

As it is necessary to have a ready supply of the same soil over many years, it cannot just be excavated from a source. It has to be produced by separating the various elements and then mixing them to the correct proportions. The analysis of the soil used at Wimbledon is:

(1) Fine gravel - 1%.
(2) Very coarse sand - 2%.
(3) Coarse sand - 5%.
(4) Medium sand - 17%.
(5) Fine sand - 20%.
(6) Very fine sand - 9%.
(7) Coarse silt - 10%.
(8) Silt - 15%.

(9) Clay - 21%.

When introducing the soil to a new construction, it will be installed in four layers with each layer being consolidated before the next one is introduced. This will greatly reduce the amount of sinkage that will naturally occur as the soils settle.

THE TURF

In order to get the court into play as quickly as possible, the grass will be in the form of turf rather than seed. The turf is purpose grown in the same type of soil as used in the construction in order to ensure the consistency of material used. The seed mixture of the turf is:

(1) 70% perennial ryegrass, cultivar - Lorina.
(2) 30% creeping red fescue, cultivar - Barcrown.
Ref. Newell Jones 1995 and Newell Crossley Jones 1996

This current mixture is the result of research started in 1992 in conjunction with Dr Andy Newell at the Sports Turf Research Institute. This work is ongoing as new grasses are brought into the market each year. It is vital that these are tested for the demands of tennis to ensure that we are taking advantage of modern developments. Ref. Newell, Hart'Woods, Wood, Lister, Richards 1999.

Prior to the turf being laid, the final levelling of the root zone is carried out and checked with a laser level.

The turf is laid using the modern technique of the big roll system. Each individual turf will be some $8m^2$ and will weigh approximately 1000 kilos. Once the turf laying is completed, the court is left for approximately two weeks before work starts on producing the final surface.

If the timing of the construction is correct, the turf laying will be completed in the Autumn so that play can commence the following season. Normally, from start to finish, the work will take in the region of 18 months.

Fig.2- Laying the Turf, Big Roll System

ANNUAL MAINTENANCE PROGRAMME

Once the season is finished, the work in the Autumn is vital. Failure to carry out the operations correctly will certainly have a detrimental effect on court performance the following season.

The programme is designed to produce the desired grass cover, a level surface and to ensure that the soil remains healthy. It has to be remembered that we are dealing with a living item and therefore such things as bacterial activity have to be encouraged. The programme will consist of (Ref. Grass Tennis Courts, STRI):

SCARIFICATION

This process will remove any weak plants and lateral growth. Failure to do this adequately will lead to a spongy surface prone to fungal diseases and, from the players' point of view, will give poor ball bounce the following season.

AERATION

During the season the court will have been rolled a great deal to produce the required surface. As a result of this, the soil would have become compacted and must therefore be relieved by introducing holes into the soil. This process allows passage of air into the soil and encourages bacterial activity, which in turn enables the food chain to function so encouraging grass plant development.

OVERSOWING

Throughout the season the number of grass plants in the court has been reduced and scarification has reduced this still further. These plants have to be replaced, so new seed is sown with the same mixture as that used in the production of the turf. The seeds are sown at 35 grams per square metre over the entire area. A little soil is mixed with the seed as this helps the new plants to grow after the seeds have germinated. Germination will take place in a short period and the court will be ready for mowing within three weeks.

TOP DRESSING

Once the new grasses have started to establish themselves, top dressing is applied to give the final levels for the following season. The top dressing material is soil with the same analysis as that used in the construction.

Throughout the winter, whenever necessary, the courts will be mown at a height of 13mm. Additionally, regular checks will be made to ensure that the grass remains disease free and healthy.

PREPARATION FOR PLAY

This will commence March/April, depending on weather conditions. The aim is to provide courts for play by mid May and to get the playing surface to peak approximately 7 days prior to The Championships (i.e. in time for the players arriving for practice).

Included in the management is the reduction of the height of cut to the playing height of 8mm, fertilisation and rolling programmes, plus the drying out of the soils by means of managing the covers. Finally, it must be ensured that the presentation is correct.

Fig.3 Courts Prepared for Play

MOWING

As the growth of the grass increases, the frequency of mowing should also be increased. To help the grass remain dense and healthy the aim should be to remove no more than 3mm at any one time. Towards the end of March the height of cut should be slowly reduced from the winter height of 13mm to the playing height of 8mm.

FERTILISING

Ideally an application of fertiliser should be made four weeks prior to The Championships. as this will allow the grass to benefit from the nutrients. It will also ensure that the initial flush of growth caused by the nitrogen will have had time to slow down before the start of play. This is vital, because during this initial flush the grass becomes very green which might be pleasing to the eye, but can make the grass slippery for the players. It is very important to remember that we are not merely growing grass, but producing a surface for tennis.

ROLLING

This will start in March and continue throughout the Spring and Summer until The Championships. It is difficult to quantify just how much rolling is needed; the ideal is to produce a hard surface to provide that all important ball bounce.

MARKING OUT

There are many methods of marking out, all of which cannot be faulted providing the end product produces an accurate measurement with bright, clearly visible lines.

PREPARATION FOR CHAMPIONSHIP PLAY

With the court rolled firm, marked out correctly and the mowing shades in place, the court is almost ready for Championship play.

Clay soils can only be rolled to a certain degree of firmness. To obtain that final hard surface much of the moisture held in the soil is removed from the top 50mm by covering the courts each time it rains during the day and also every night for approximately 10 days prior to The Championships. The exact timing of when the covers are brought into use depends on weather conditions. The aim is to have the soil moisture content at about 13% by Day One.

Once the court is hard, the daily maintenance throughout the two weeks is to mow and roll. In hot dry weather a little water may be applied to the grass in the evening to relieve the stress within the plant, but the court must again be dry and ready for play by the following morning. To this end, the Meteorological Office is consulted daily regarding the weather forecast for the following day before any application of water. The details requested will be the amount of cloud (if any), temperature, wind and humidity. If there is any doubt about the court being dry then water will not be applied; as previously stated, the players' requirements are the first consideration.

Fig.4 Preparing Courts - Mowing Courts Under the Covers

Along with regular maintenance each morning, measurements are made on the courts to include hardness, moisture content, ball bounce and wear. This information helps us to ensure that we are achieving consistency with the courts.

Once The Championships are completed, we can again turn our thoughts to the all important Autumn Renovation and preparation for the next season.

CONCLUSION

To ensure that the courts meet the requirements of the players, satisfactory playing characteristics and uniformity from one court to another can only be achieved by constructing with the correct materials. This, followed by a considered maintenance programme, will allow the courts to peak in time for the event and can, with a little help from nature, provide the desired court surface for players to be able to use their full range of skills.

REFERENCES

Grass Tennis Courts, How to Construct and Maintain Them (2000). Published by *The Sports Turf Institute* , Bingley, West Yorkshire, BD16 1AU - 192 pages.

Newell A.J. & Jones A.C. (1995) Composition of Grass Species and Cultivars for the Use on Lawn Tennis Courts. *Journal of the Sports Turf Institute*, Volume 71.

Newell A.J., Crossley F.E.M. & Jones A.C. (1996) Selection of Grass Species and Mixtures for Lawn Tennis Courts. *Journal of the Sports Turf Institute*, Volume 72.

Newell A.J., Hart'Woods J.C., Wood A.D., Lister R.A. & Richards J. (1999) Selection of Grass Species and Mixtures for Lawn Tennis Courts. The Sports Turf Institute.

Ground conditions in tennis court construction

J. Bull
Weeks Technical Services plc, Tonbridge, Kent, UK

ABSTRACT: If poor ground conditions are not identified before construction, there is the risk of avoidable delay, cost over-run and contractual claims. This paper covers the need for quality site investigation for tennis court construction. It describes some common ground-related problems affecting tennis courts and techniques for their investigation. Contractual issues are also briefly covered.

INTRODUCTION

It is widely appreciated that ground conditions present the main technical and financial risk in construction projects. Tennis courts are structurally light and impose minimal loading on the ground except during their construction. Courts do have to remain to a tight tolerance on surface regularity throughout their design life. Ground movement after construction is therefore of great concern.

In spite of their lightness, tennis courts are still engineering structures in contact with the ground. Soil and rock are natural materials for which variation and movement is perfectly normal. Ground investigation is therefore an essential part of court design if the risk of ground-related problems is to be minimised.

This paper covers court design from an engineer's perspective involving a 'bottom-up' rather than 'top-down' approach. This paper will enable you to:

1) Understand how the base layers are designed.
2) Appreciate the main ground-related problems encountered with tennis courts.
3) Understand basic ground investigation techniques.
4) Be aware of the contractual implications of ground conditions.

BASE LAYER REQUIREMENTS

Figure 1 shows typical court construction layers. The ground surface on which the court is constructed is called the formation. The lowest layer consists of graded crushed rock called sub-base and provides a firm platform to compact subsequent macadam layers. A geotextile mat is sometimes specified to prevent loss of sub-base from punching into the formation although it provides no structural strength. As in-service loading from playing tennis is very low, the sub-base's primary task is to support macadam-laying plant. The finished sub-base must not be used as a haul road for earthmoving vehicles.

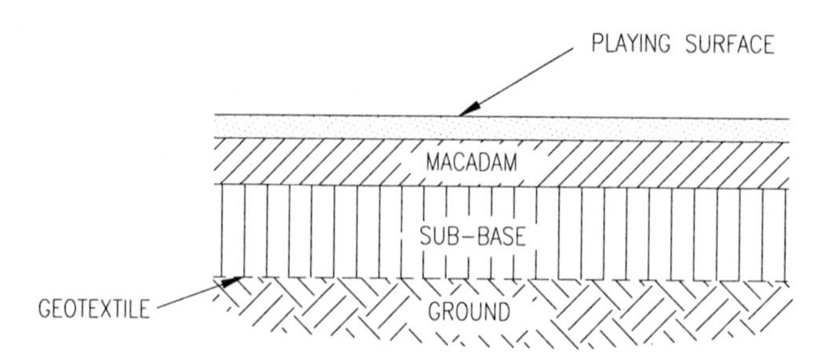

Fig 1. Typical tennis court construction layers

The principal test for sub-base adequacy is that it must not move under plant loading. If not, then it will not be possible to properly lay and compact the macadam layer above. Similarly, moving sub-base is unlikely to be sufficiently compacted. On porous courts, any ruts in the formation may retain water, which then risks localised softening and movement.

GROUND STRENGTH AND SUB-BASE THICKNESS

Sub-base thickness varies according to the ground strength. In highway engineering, mainstay of pavement design is the CBR test. CBR stands for California Bearing Ratio as the California Division of Highways first developed the test in the 1930s.

Briefly, the test consists of a plunger driven into the ground. The CBR is expressed as a percentage based on the resulting load/deflection curve. Tests can be completed using a vehicle-mounted rig or in the laboratory. CBR is affected by soil water content and must be taken into account in laboratory testing or if winter construction is being considered. Further details can be found in Croney (1977).

Table 2 shows a classification chart for ground strength in terms of CBR. It is arbitrary and based on our previous experience on sports pitches.

Table 2 Ground Strength Classification

CBR	Formation Strength	Implications for Courts
>20%	Excellent	Hard digging possible
10-20%	Very good	Minimum sub-base thickness in respect to ground strength
5-10%	Competent	No special requirements in respect to ground strength
2-5%	Soft	High sub-base thickness Care needed during construction
<2%	Very soft	Sub-base thickness excessive Special measures required

Sub-base thickness depends on ground strength but should be 150mm minimum. This is the minimum thickness that can be placed to a reasonably consistent depth using normal construction plant. However, this will only be appropriate on stronger formations. On very soft ground, sub-base thickness becomes disproportionately high and special measures are needed as discussed below in the section on soft ground.

SOFT GROUND

Soft ground can occur for a variety of reasons including:

1) Poor drainage or high groundwater level.
2) Geologically recent soils that are relatively loose.
3) Soils with high organic content such as peat.
4) Former ponds or watercourses that have silted up.
5) Loosely tipped fill.

In these situations, your court will be considerably more expensive compared with more competent formations because the base design needs to be much more robust. Never build on peat because it can be highly compressible leading to a high risk of differential settlement.

Unlike firmer ground, soft sites often have a top crust overlying poor ground as shown in figure 2 below:

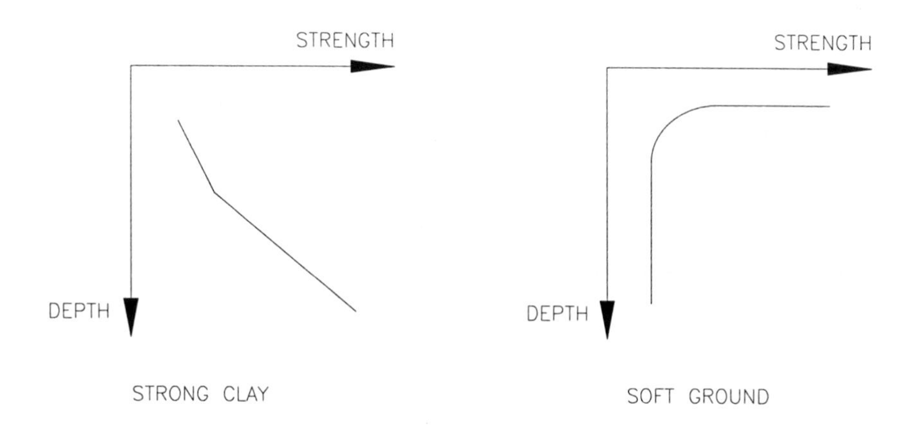

Fig. 2 Ground strength with depth profiles

On soft sites, understanding ground strength variation with depth can be useful in optimising the base design. The court designer will be interested in trying to make use of the crust for the sub-base design. Digging through the crust is clearly not a sensible idea. Methods of treating soft ground include a capping layer, excavation and replacement or increased drainage. A capping layer consists of crushed rock between formation and sub-base but it has a less tightly controlled specification compared with sub-base.

FROST HEAVE

Frost heave can disrupt playing surfaces out of tolerance as shown in figure 3 below:

Fig. 3 Suspected frost heave.

For frost heave to occur, two things need to happen:

1) Frost penetration into the formation.
2) Moisture entering the frost zone.

Background information on frost heave mechanism is given in O'Flaherty (1998). Sieve analysis will quickly and cheaply confirm if a soil is at risk from frost heave. Frost cabinet testing will confirm the risk beyond doubt if required.

Standard highway engineering practice is to provide 450mm minimum construction thickness on frost susceptible soils. Frost penetration clearly depends on climate and geographical location. Some authors have suggested that the 450mm requirement is too conservative in most areas of the UK. However, we are unaware of any definitive guidance on relaxing the minimum depth.

SHRINKABLE CLAYS

Clay soils can shrink or swell if their moisture content changes significantly. As clays have very low permeability, movements take time to occur and can take many years. Movements often take place when clays dry out either due to evaporation or through desiccation that occurs from tree root activity. This is a frequent cause of insurance claims for building subsidence and we have seen cases of movement under sports pitches and also floodlight column bases. Factors affecting clay shrinkage include:

1) Soil clay content.
2) Seasonal and climatic factors.
3) Tree species because this determines water demand.
4) Tree size; the tree root zone is assumed to be 1.5 times the tree height.

The risk of shrinkage is a property of the site. The size, extent and timing of movement are not predictable, as climate is a factor. Some pitch owners accept the risk and have to do localised base reconstruction during resurfacing. This approach may be inappropriate in high risk areas. It is best to build courts outside tree root zones, allowing for future tree growth. BRE Digest 240 (1993) gives a useful introduction to this subject.

SLOPING SITES

Sloping sites need much more care to investigate and develop than flat sites. To engineer a plateau on a slope for tennis courts, construction may involve:

1) Using excavated soil as fill
2) Retaining structures
3) Risk of landslip

Steeply sloping sites carry the risk of major failures and you should not consider developing sloping sites without professional advice.

USING EXCAVATED SOIL AS FILL

Court levels can be built up using excavated soil, as shown in fig. 4:

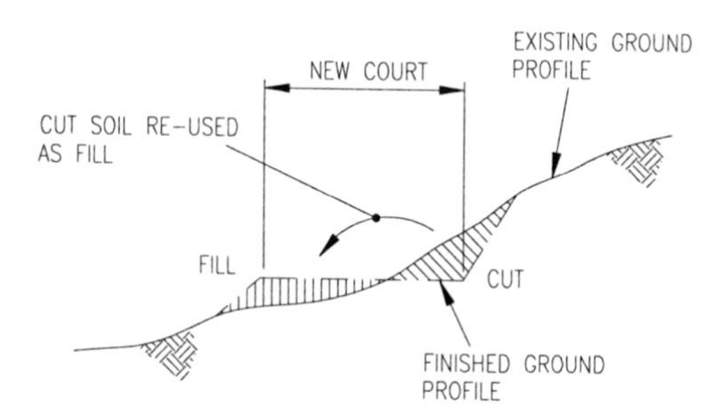

Fig. 4 Balanced cut and fill

Importing engineered fill, such as crushed rock, is expensive because of haulage and material cost. Likewise, removing surplus soil to tip has a similar unit cost due to tipping charges arising from landfill tax. The cheapest and most environmentally sustainable option is always to re-use excavated soil from site. However, this does require careful design and specification. To avoid long-term settlement or rutting by construction plant, the soil must be properly compacted. If the soil is too wet or too dry, it can not be compacted to a high enough density and therefore strength. Compaction has to be designed to suit site conditions and relies on laboratory testing. It is not sufficient to insist on well-compacted soil; the required density and permitted moisture content must be specified. Therefore, to obtain construction cost savings, you need to allow for increased design costs. For small volumes of soil, imported fill will the safest option, especially if the ground is very wet. Refer to Lambe and Whitman (1979) for an introduction to the theory of soil compaction.

RETAINING WALLS

If abrupt changes in level are required, retaining walls may be needed although they increase design and construction costs and can only be used in stable ground. Additional ground investigation will be needed to establish wall design parameters. There are many different methods of constructing retaining walls, depending on site conditions and geometry.

LANDSLIP

Unstable ground brings the risk of landslip, which consists of large-scale ground movements. Landslip risk depends on geology, groundwater conditions and slope geometry. Making cuttings too high or too steep can trigger landslips. In addition, slopes can include pre-existing failure planes as a result of past geological conditions. Excavating in such slopes can cause the unstable material to slip off, as shown in figure 5. These slopes are particularly hazardous because only a shallow cut may be needed to cause slippage.

Fig. 5 Translational landslip in an existing slope.

Clay slopes may not reach their worst case design condition until several decades because of the slow rate of change in groundwater conditions. Ground investigation for landslips therefore requires careful ground logging, sampling and testing.

DERELICT SITES

If your site is not 'greenfield', former site uses must be investigated. So-called 'brownfield' sites have inherent engineering problems including:

1) Buried obstructions such as foundations or tanks.
2) Landfilling of waste materials which may not have been properly compacted. Refer to Charles (1993) for detailed advice.
3) Underground voids, such as abandoned mineshafts.
4) Chemical contamination. This is a continually evolving subject as legislation develops. An introduction is provided by Harris and Herbert (1994).

Specialist engineering advice must be obtained at the earliest possible stage as these problems can make tennis court construction uneconomic.

GROUND INVESTIGATION

Ground investigation is always site specific and the basic approach described here may not be adequate for particularly unusual or difficult ground. A desk study before starting site work can often give important clues on ground conditions. This would involve consulting information sources such as geological maps and memoirs together with archive maps to investigate former site uses. A site walkover by a civil or geotechnical engineer can also highlight potential difficulties.

The ground investigation techniques we use are chosen as follows:

1) Using hand-held equipment in case there is no vehicle access.
2) Provide an indication of the ground strength with depth profile.
3) There is seldom a need to log or sample the ground at depth.
4) Cover the widest possible area in the shortest possible time.

Instead of measuring CBR directly, we prefer to measure ground strength using a dymanic penetrometer from which CBR can be deduced. We prefer to use a driven tube sampler for logging and sampling for the reasons given above. This can penetrate up to 6m, which is ample for most tennis projects.

Fig. 6 Dynamic penetrometer and driven tube sampler.

This equipment provides a large number of readings will enable the designer to be aware of any variation across the site.

A ground investigation must at least confirm the following:

1) The nature of the ground and groundwater level.
2) The presence of soft ground.
3) Any risk of frost heave.
4) Any risk of clay shrinkage.

It is important that the ground profile is accurately recorded in accordance with BS 5930 (1981). Ground investigation costs vary according to the amount of laboratory testing and the number of courts. Economies of scale are possible on larger projects. Ground investigation costs rise for steeply sloping and derelict sites.

CONTRACTUAL IMPLICATIONS

Our experience of industry practice is that the ground as seldom examined before construction with the risk being transferred to the contractor. For larger projects, this risk transfer is frequently done formally using 'design and build' forms of contract. In a competitive tendering, contractors are under pressure to minimise their costs and therefore there is the temptation to avoid ground investigation. A primary purpose of a construction contract is to clarify where responsibilities lay and how each party must perform. Ground conditions must not be left to chance and therefore the contract documents have to leave the contractor in no doubt his duties. The minimum risk option is for the Employer to commission a designer to do this on his behalf. However, if the contractor has to design the base and earthworks, this must be stated explicitly. The contractor should be given ground investigation information to complete the design. Transferring the risk without detailed information of what is expected puts the customer at an unacceptably high risk of construction problems or in-service failure.

CONCLUSIONS

This paper has been written in response a sizeable number of ground-related problems reported to the author by tennis players and clubs. Many of these failures could have been avoided if the ground conditions had been examined in advance. Investigating flat sites can be done cheaply but sloping or derelict sites need special care and may only be cost-effective to design and develop for multiple courts. There is a saying amongst civil engineers that you pay for a ground investigation whether you have one or not.

REFERENCES

BRE Digest 240, (1993), Low rise buildings on shrinkable clay soils: Part 1, Building Research Establishment.

BS 5930, (1981), Code of Practice for Site Investigations, British Standards Institution.

Charles, J. A. (1993), Building on fill: geotechnical aspects, Building Research Establishment.
Croney, D. (1977), The design and performance of road pavements, HMSO.
Harris, M. and Herbert, S. (1994), Contaminated Land, *ICE Design and Practice Guide*, Thomas Telford.
Lambe, T. W. and Whitman, R. V. (1979), Soil Mechanics, SI Version, Wiley.
O'Flaherty, C. A. (1988), Highway Engineering, Vol. 2, 3rd Edition, Arnold.

The use of quality control procedures during the construction of tennis courts

V. Watson
Centre for Sports Technology, London, England

ABSTRACT: All construction contracts, no matter how large or small, benefit from the application of a quality control system. Such a system, if correctly drawn up and implemented, offers protection to all parties to the contract, including the suppliers of raw materials and components, the installer, the consultant and the facility owner who has the ultimate reassurance that all is as it should be when they accept handover.

RATIONALE

We live in an age when any purchaser considers it quite natural to expect that any goods or services that they buy should be of 'reasonable quality' or 'fit for purpose', and they have become increasingly vociferous in demanding that these basic requirements are met. The concept of quality and consistency should therefore be at the forefront of any plan to design, supply, build or buy a tennis facility. However, it is very easy to simply pay lip-service to quality control by tacking on a last minute requirement once work starts on site and hoping that everything will be in order. Proper quality control must run right through a building contract from conception to completion.

BACKGROUND

There are many standardised schemes which attempt to ensure consistency of output from an organisation, whether they are a manufacturer, installer, consultant, test laboratory or major purchaser. The exact format of such schemes depends on the type of organisation (for instance in the UK, test laboratories have their own quality assurance scheme) and they vary from one country to another, although International Standards exist for quality monitoring. However, many schemes control only consistency and not necessarily absolute quality. It is perfectly possible to conform to some so-called quality control schemes and consistently produce rubbish!

The scheme which is outlined in this paper has been developed and used by us over a period of more than 15 years to quality control the synthetic surface aspects of sports facility provision (reference 1). It was originally developed for use with large facilities such as athletics tracks and hockey/soccer pitches, but the principles are valid for a project of any size. It does not claim to be a perfect method of control, since certain practical difficulties remain, particularly over the sampling of in-situ cast systems. However, the scheme is subject to regular review and its on-going development over that period has been in response to problems that have arisen on specific projects and with particular systems. We are always happy to receive suggestions for improvement!

OPERATION

There are various stages to the operation of this quality control scheme and each subsequent stage depends upon the satisfactory completion of the one preceding it.

STAGE 1

When writing the specification for the job, it is important that it contains clauses that define performance requirements for the surfacing system, in conjunction with its base as appropriate, which are measurable both in the laboratory and on the completed installation. In the case of tennis facilities, these should include all relevant ITF performance requirements. Consistency of certain constructional aspects such as density, thickness or pile height etc, should also be stipulated in the specification.

STAGE 2

A sample of the surfacing material and all individual components comprising the total system, should be requested at the time of tendering. These reference samples should be identical in every respect to the materials that will be used to form the completed installation and should be retained and referred to in cases of dispute or litigation. These reference samples should be tested for relevant performance requirements and certain characterisation measurements should also be undertaken such as thickness or pile length, weight per unit area, tensile properties etc. The results of these tests and measurements form part of the basis for comparison with samples prepared or taken on site during the construction phase, and also with the performance of the final installed surface upon completion. It is important that these preliminary tests are completed and the results obtained are confirmed as being acceptable before work starts on site. Any notable discrepancy between the results obtained and what was expected from that product or system, should be taken up with the supplier/contractor immediately.

STAGE 3

The third stage requires that samples of all materials and components of the surfacing system are taken or prepared on site throughout the construction phase, with a detailed log book kept by the site supervisor. This log would include information about the areas of the facility which are represented by each sample, batch numbers etc. In the case of prefabricated surfacing systems, these samples should be removed from an agreed proportion of the rolls of material delivered to site, and should also include a number of samples of adhesive-bonded or stitched seams. In the case of systems prepared by casting, spreading or spraying in-situ, the samples should either be cut from the previous days laying (i.e. after cure), or they should be prepared as sample 'trays' trying to simulate as best as possible the techniques used over the main area of the facility, and left to cure alongside the area they represent for an appropriate time.

It is acknowledged that trying to obtain properly representative quality control samples of surfacing prepared 'in-situ' is difficult. One arrangement is to allow an area of surface to fully cure and then cut a sample from that area, making good with fresh material. Not surprisingly, many contractors are unhappy about deliberately introducing patch repairs! Trying to make small sample 'trays' of in-situ surfacing can also be difficult because the technique used cannot be the same as that used to prepare larger areas. This is especially the case with resin-bound rubber crumb normally laid by paving machine, or spray-applied finishes where the operator needs to develop a particular 'action' in order to apply a consistent amount of paint.

There can also be problems with obtaining representative samples of adhesive-bonded seams in prefabricated systems. Sometimes, when installing rolls of such a product, excess material is rolled out beyond the perimeter kerbs of the synthetic area and the seams also continued out beyond the perimeter. Finally all the excess material is trimmed off and these off-cuts (including the seams) can be used as quality control samples. Even this is less than ideal because the application of adhesive to the seam backing tape beyond the perimeter kerb line can be made difficult by changes in level etc., and the application of pressure to that part of the seam during cure of the adhesive may be difficult or virtually impossible. If the perimeter fence to the facility, or any boundary wall, abuts the synthetic area, such an overlap of sheet systems during installation may not be possible. If this is the case, specially laid lengths of seam can be prepared for quality control testing, but again these could not be considered as properly representative of the seams made over the entire area.

All site samples are then submitted for testing to determine their consistency of construction and performance, by comparison with each other and with the reference samples tested before the contract was awarded.

STAGE 4

The final stage of a good quality control scheme should always involve a comprehensive inspection and testing survey of the completed facility. This survey should include aspects of workmanship such as surface smoothness, as well as all specified performance requirements. If the results obtained on the site samples prepared during construction indicate a potential problem with a certain area of the surfacing, this survey should be used to confirm the problem (or to prove its insignificance) and to clearly identify the exact area affected. If any repairs prove to be necessary, a repeat survey should be undertaken afterwards, before the facility is formally accepted by the owner. There may be occasions when, by mutual agreement between the contractor and the client, the facility is put into use and a careful check is made on the performance of the surface over time to ensure that any problem identified initially does not spread to other areas or deteriorate to an extent where the use of the facility is being compromised in some way.

CONCLUSION

It is important to remember that an effective quality control scheme cannot guarantee the trouble-free installation of the surfacing. Some of the materials involved are too crucially dependent on weather and workmanship for this to be possible. The great advantage of such a scheme, however, lies in the fact that when problems do occur, they are more easily detected and the precise area and extent of any problem can be identified. All reputable installers should welcome such quality control schemes when they are operated by suitably experienced and fully equipped test laboratories. In this respect, the laboratory should be a member of an appropriate organisation such as the International Association for Sports Surface Sciences (ISSS). The owner of the facility has the most to gain, in the reassurance they receive about the consistency and performance of their completed job. For this reason, they should look upon an investment in a quality control scheme as a vital and integral part of their contractual requirements and obligations.

REFERENCE

Tipp G. & Watson V. J. (1982) *Polymeric Surfaces for Sports and Recreation*, Applied Science Publishers, London & New Jersey.

Alternative structures for covered tennis

J. Smith
The Lawn Tennis Association, London, England

ABSTRACT: British Tennis has had considerable success in obtaining pound for pound funding over the past five years, but with increasing pressure on Sports Lottery resources and ever extending waiting periods before projects can move forward, twelve months ago the LTA decided that it was time to review the alternatives! Some interesting concepts, technology and products that had been previously overlooked in favour of "permanent" buildings were uncovered.

A SERIOUS OPTION

Beginning with the idea of reducing capital cost to about half that of a traditional building, the "non-traditional structures" market, the realm of airhalls and framed tents, was reviewed for project cost. It was found that not only were these systems competitive on cost but they showed a number of advantageous characteristics that make them a serious option alongside the more common "traditional" structures.

MID-MARKET PRICE

There are three methods of creating covered tennis. Air supported structures, framed fabric structures (both substantially "off the peg") and traditional built structures (commonly "one-offs"). These types conveniently allow us to split the market into three price sectors.

The cheapest way of covering tennis is with an air-supported structure. Air supported structures rely on mechanical ventilation systems to provide the air that inflates a plastic membrane that forms the hall.

Mid-cost are products like framed fabric structures with lightweight cladding materials and airhalls of advanced design. It is this area of the market that the LTA decided it needed to stimulate to help provide wider covered tennis in the UK.

At the top end of the market are the traditional built structures that are generally designed and constructed from scratch and in detail. In this sector you can spend as much as you like!

ALTERNATIVE ALTERNATIVES

"Alternatives", can be grouped into three sub categories:

(1) Airhalls - Lightweight air supported structures or bubbles are the simplest types. They are single skinned visqueen sheets held down by a concrete ringbeam. They have simple mechanical equipment to keep them inflated, often with no backup systems. Sometimes a net is used to hold down the skin. More advanced airhalls are high quality systems which have heavier membranes and often are double skinned. They have more complex mechanical systems with full backup incase of failure.

(2) Framed fabric Structures – They use aluminium or timber portal frames in a similar way to a traditional tennis structure but the sections used are far lighter due to the use of membrane technology for the outer skin. They may be held down to a ringbeam or ground anchors but need no mechanical to support them.

(3) The third category covers tensile structures that are commonly designed to a specific site and are often rather bold and expressive in their form.

Fig. 1 Alternative types of structures for covered tennis; top to bottom – an airhall, a framed fabric structure and a tensile structure.

MATERIALS

The materials that are used for these structures are generally more modern and noticeably lighter in weight than those used in traditional construction. In the same vein, construction techniques are orientated towards system building and are more akin to the factory production line than the more common handmade approach. Less material, controlled manufacturer and finer tolerance construction help reduce capital cost and construction time.

For the roof (and the walls) technically advanced membrane (fabric) systems are the norm. These are strong, have a long lifespan (20-25 years), have good light transmission performance and are special coated to shed dirt. As these materials reduce the weight of the roof, then the size and weight of the members needed to support is consequently decreased. Aluminium or timber frames are the materials commonly used for the structural frame and have a lifespan in excess of the 60-year design life of a traditional structure. They also have the potential to be taken down and stored or quickly re-built elsewhere which explains their popularity in the marquee hire business. The timber used is generally softwood-laminated beams and although aluminium is not a renewable resource, it is reusable.

The main factor governing foundation design becomes uplift rather than the need to support an imposed load.

SPORTS ENVIRONMENT

The purpose of a covered tennis court is to extend the potential availability of the tennis court whilst providing an environment suitable for playing the game. One of the characteristics that is significant about most of the alternative structures, is the even distribution of natural light (transmitted through the roof) across the court surface. This is particularly good with a white translucent membrane which gives a milky white overcast sky effect. This means that diffused light is evenly distributed onto the tennis court, very important in a high-speed projectile game like tennis where the ball can be travelling at speeds in excess of 110mph. Artificial light will be necessary on days with poor external daylight conditions and at night. The light transmission through a typical fabric is approximately 22% of the external lighting level. The wider use of membrane roofs would mean that glare problems from rooflights cut into opaque metal deck roofs could be avoided and cost savings made through the reduced periods of artificial lighting that would be required.

Membrane technology does have a downside; it has negligible thermal resistance, therefore some condensation will inevitably occur, even on double skinned roofs. Double skinned roofs also further reduced light transmission by another 20%.

Noise from mechanical ventilation systems is a constant problem in tennis halls making it impossible for player and coach to communicate. Echo's caused by the shape of air structures returning and focusing sound energy also lead to difficult conditions. Natural ventilation in framed structures avoids excess noise and also provides energy savings. Although formed from similar fabric materials, the angular shape of the traditional portal shape breaks down the focus of energy and gives a reverberation time similar to a traditional structure.

SHORTER BUILD TIME

Because the construction process can be broken down into packages with many components manufactured off-site in controlled conditions, the time required to develop a scheme is reduced by as much as half. The packages are court construction (although it may be possible to install on existing courts), structure, services and external works. The shorter build process gives potential savings by reducing the cost of inflation to the project, by making the finished product available for use earlier and reducing the downtime of a site that would normally be in use. The kit of parts nature of some elements means less design time is required. The simpler build process means that less time and consequently, fewer fees are required to hire a professional team.

PLANNING PERMISSION

Planning permission is one of the most significant factors in the failure of building projects in the UK, aesthetics therefore become an important consideration. Although it is possible to create an air-supported roof on traditional style walls, the most potential in this area lies with the portal-framed fabric structures. On the continent it is quite common for these buildings to have a diversity of cladding materials from crinkly tin sheet to insulated sandwich panel systems, from timber boards to blockwork, render and paint. There is also no reason to exclude windows or just have open sides for ventilation and external views!

A constant issue with planning authorities is colour and mass. There are some simple designs available to cover single courts that can be placed next to each other and linked to allow views between courts. Materials can be flexible to suit local needs and membrane roofs can be coloured or patterned to blend into the environment, although darker colours reduce light transmission qualities.

NOT JUST STAND-ALONE

Attaching air-supported structures to more traditionally built clubhouses is not uncommon and there are examples where the series of portals of a framed fabric structure has been extended and the full range of clubhouse facilities built within the single space. Flexibility is good with these types of modular building, but costs will rapidly increase where areas of traditional construction are incorporated into a scheme.

CONCLUSION: THE WAY FORWARD

In order to stimulate the UK market, the LTA are developing through 1999/2000, a series of alternative structure pilot projects around the country. They will test materials, massing, project management and construction techniques and cost.

Impact-absorbing characteristics of tennis playing surfaces

S.J. Dixon
Department of Exercise and Sport Science, University of Exeter, UK
A.C. Collop
Department of Civil Engineering, University of Nottingham, UK
M.E. Batt
Centre for Sports Medicine, Queen's Medical Centre, University of Nottingham, UK

***ABSTRACT*:** The influence on running biomechanics of three artificial playing surfaces providing differing amounts of mechanical shock absorption was investigated. Measurement of ground reaction forces during running demonstrated that although peak impact force was maintained at similar values for the different surfaces, there was a significantly later occurrence of peak impact force relative to initial ground contact as mechanical shock absorption was increased ($p<0.1$). It was also found that subjects demonstrated marked changes in joint angles when running on the different surfaces. The later occurrence of peak impact force suggests that increased mechanical shock absorption increases the cushioning of impact forces in running. Although this finding appears to support the use of mechanical tests to identify surfaces with adequate shock absorbing properties, the identification of joint angle changes with surface variation suggests that biomechanical testing is required to detect undesirable movements that may influence either performance or the likelihood of injury occurrence.

INTRODUCTION

Recent guidelines published by the International Tennis Federation (I.T.F) include a description of a mechanical test to assess 'shock absorption ability' of surface materials (I.T.F., London, 1997). This test involves the release of a mass to impact the test surface and the measurement of peak deceleration during impact. Although mechanical tests provide a measure of the relative hardness of surface materials, the resulting influence on the player performing on the surface has not been found to be predictable from these values (Nigg, 1990).

Tennis is played on a large range of surfaces, including synthetic materials, clay, grass and asphalt. Information on the relative hardness of these materials and the influence on human performance would be of benefit when selecting a playing surface. In the present study the impact-absorbing ability of three playing surfaces is assessed using both mechanical and biomechanical test methods, the objective being to compare the relative impact-absorbing ability of the three playing surfaces during running.

METHODS

Three test surfaces were assessed: an acrylic surface used for tennis courts (ETC Holdings Ltd., Melton Mowbray, UK); a conventional asphalt surface; and a commercially available rubber-modified asphalt surface (Sureflex®). For testing, the acrylic surface was attached to a conventional asphalt material, as typically occurring in commercial application of this surface. For each surface condition, a slab of test material of thickness 25 mm and dimensions 280 mm x 400 mm was placed over a force plate. In addition, an approach runway of approximately eight metres length was constructed using slabs of the same thickness material and dimensions 800 mm x 700 mm. The approach runway allowed the subjects to adjust to the surface condition prior to force plate contact, and provided a realistic setting.

Biomechanical tests were performed using six heel-toe runners as subjects. For each surface condition, the subjects each performed 10 running trials at a running velocity of 3.3 $m.s^{-1}$ (±5%). GRF data were collected using a force plate (Kistler 9261 A) located beneath the test surface. Synchronised kinematic data were collected using an automatic opto-electronic tracking unit (CODA, Charnwood Dynamics, UK). All data were recorded at a sampling rate of 800 Hz. For each running trial, the GRF data were collected for a single left-footed step. Kinematic data were collected to include the step corresponding to force plate contact, and a period of approximately 0.5 s either side of force plate contact. A trial was accepted if the required running speed was attained, and the subject made left foot contact with the slab covering the force plate without obvious alterations in running stride.

The vertical GRF data were used to calculate peak impact force magnitude and time of occurrence relative to initial ground contact. Two-dimensional (sagittal plane) kinematic data were collected using active CODA markers placed on the left side of the body. The marker placement and joint angles were defined as illustrated in Figure 1. Initial ankle and knee angles were determined for the movement field immediately prior to ground contact. Peak joint angles during the ground contact phase were identified. Heel impact velocity was calculated by numerical differentiation of the heel marker displacement data. For each variable, the mean of the 10 trials was used to represent the value for each subject-surface combination. For the resulting group mean values, an analysis of variance (ANOVA) with repeated measures was used to compare the selected biomechanical variables for the three surface conditions (significance level of 0.1).

An impact rig adhering to British Standards for sports surface testing (BS 7044) was used to determine the mechanical shock-absorption properties of the test slabs of material. The impact test procedure involved the release of a mass of 6.8 kg from a specified height to impact the test surface. The magnitude and time of occurrence of the peak deceleration were measured for each surface for a drop height of 10 cm, providing an impact velocity of 1.4 $m.s^{-1}$. Three measurements were made for each surface, and mean values calculated.

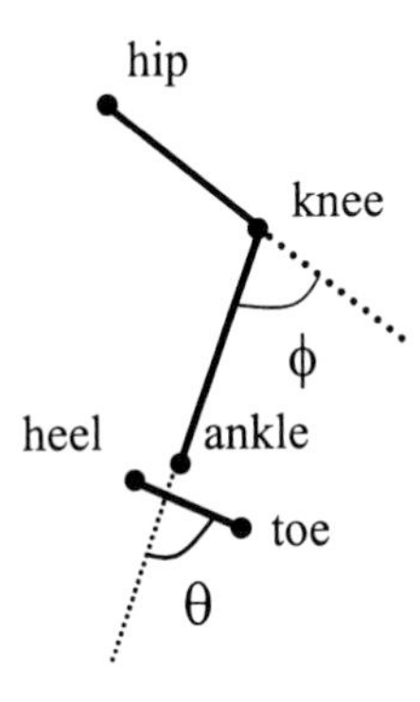

Fig. 1. Marker locations and joint angle definitions.
(θ = ankle angle; ϕ = knee angle)

RESULTS

The biomechanical impact forces and times of occurrence are presented in Table 1. To allow for differences in subject mass, impact force values are provided in multiples of subject bodyweight (BW). No significant differences were identified in the magnitude of peak impact force across the three surfaces. Compared with the conventional asphalt surface, the rubber-modified surface resulted in a significantly later occurrence of peak impact force ($p<0.1$). Average rates of loading of impact force, calculated by division of peak impact force by time of occurrence, are presented for each individual subject in Figure 2. Observation of these individual subject results indicates that all but Subject 5 experienced a reduction in average rate of loading for the rubber-modified surface compared with the conventional asphalt.

Table 1 Mean peak impact forces in bodyweight (BW) units and times of peak force occurrence, for each subject for the three test surfaces. Standard deviations are provided in parenthesis.

Peak Force (BW)			**Time of Occurrence (ms)**		
Asphalt	**Acrylic**	**Modified**	**Asphalt**	**Acrylic**	**Modified**
1.60 (0.16)	1.62 (0.13)	1.58 (0.09)	31.1 (2.2)	32.6 (2.0)	33.1* (1.3)

*significant difference from asphalt ($p<0.1$)

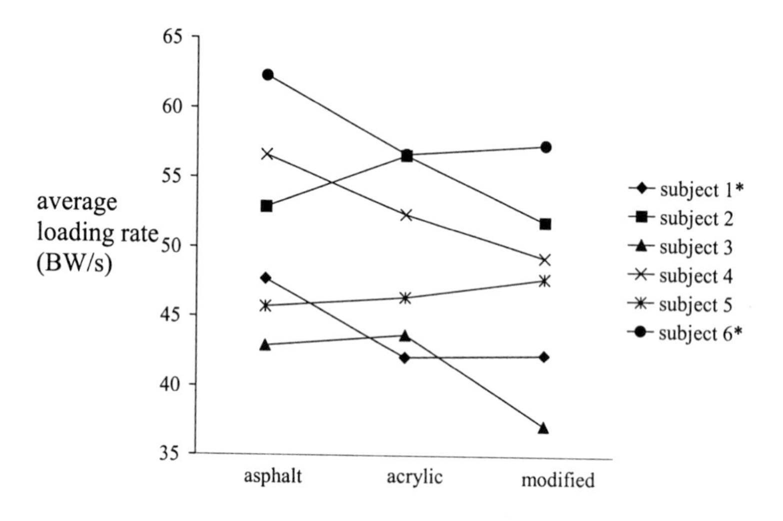

Fig. 2 Average loading rate for each subject when running on the conventional asphalt, the acrylic surface and the rubber-modified surface

Ankle and knee joint angle results are provided in Table 2. There were no significant differences in the group angles and heel impact velocity when running on the acrylic surface or the rubber-modified surface compared with the conventional asphalt surface ($p<0.1$). However, individual subject analysis revealed marked changes in peak joint angles (Figure 3; Figure 4). For five of the six subjects, a trend was demonstrated for peak ankle and knee joint angles to be increased as surface mechanical shock-absorption was increased, indicating an increase in joint flexion. In contrast, Subject 5 showed a reduction in peak joint flexion as mechanical shock absorption was increased.

Table 2 Ankle and knee joint angles for the three test surfaces (degrees), with standard deviations provided in parenthesis.

	Asphalt	**Acrylic**	**Modified**
initial ankle angle	101.6 (3.1)	101.9 (4.1)	101.8 (4.8)
initial knee angle	17.9 (5.9)	18.2 (5.0)	17.9 (5.9)
peak ankle angle	118.4 (1.8)	118.6 (1.6)	119.4 (1.9)
peak knee angle	46.5 (3.2)	47.2 (3.7)	47.4 (3.9)

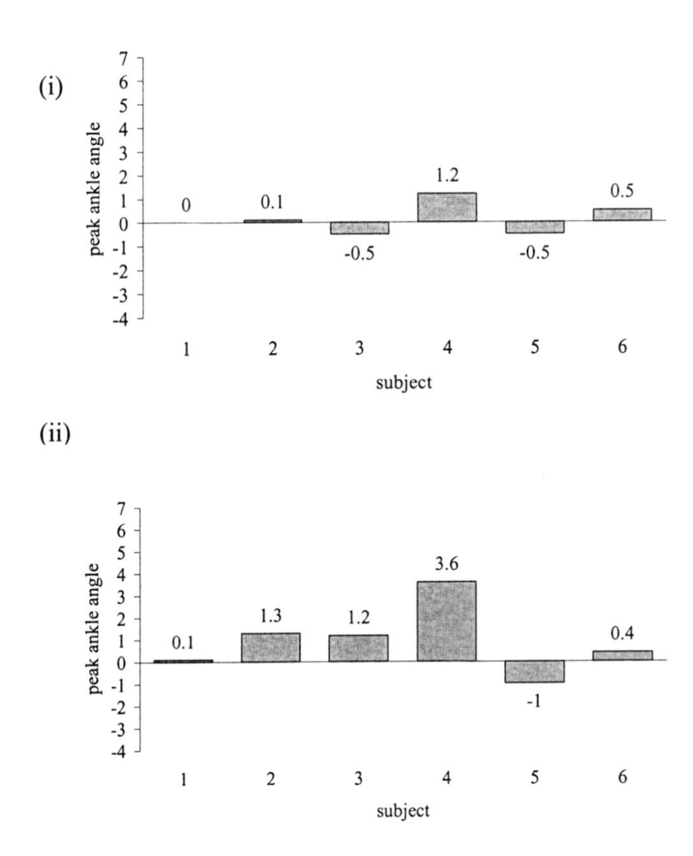

Fig. 3 Changes in peak ankle angle for the acrylic surface (i) and the rubber-modified surface (ii) compared with the conventional asphalt surface (degrees).

The results obtained using the impact rig are presented in Table 3. It can be seen from the table that marked differences were detected between the two materials, with the peak deceleration being reduced by 82% by the use of the rubber-modified material compared with the conventional asphalt material and by 65% by the use of the acrylic material. It can also be seen from these data that the peak impact deceleration occurs later for both the rubber-modified material and the acrylic material compared with the conventional asphalt material.

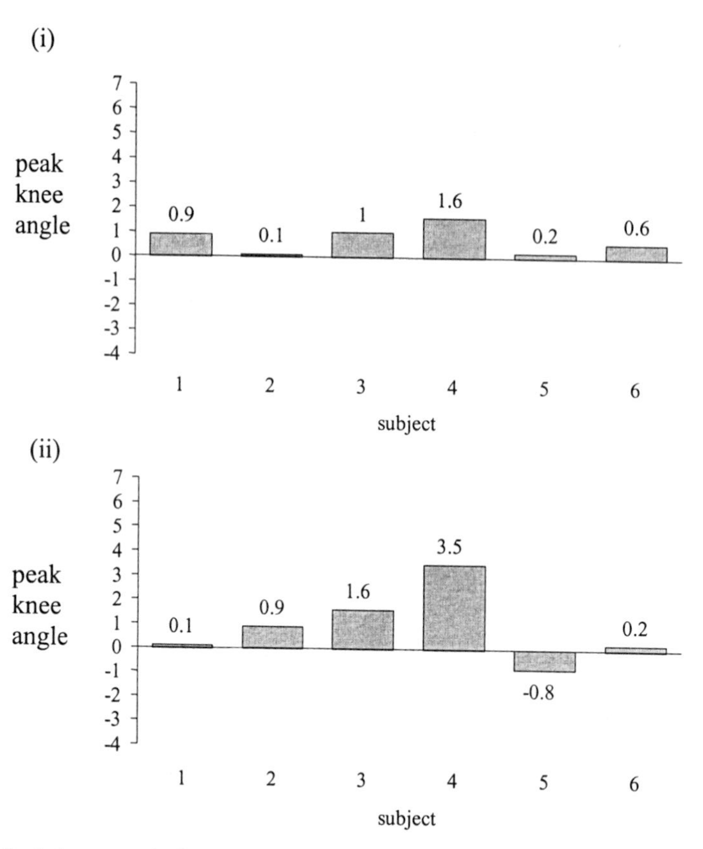

Fig. 4 Peak knee angle for the acrylic surface (i) and the rubber-modified surface (ii) compared with the conventional asphalt surface (degrees).

Table 3 Mean peak impact deceleration and time of occurrence for the test surfaces for a drop height of 10 cm

	Asphalt	**Acrylic**	**Modified**
peak deceleration (g)	300	105	55
time of occurrence (ms)	1	3	4

DISCUSSION

It has been demonstrated in the present study that an acrylic surface and a rubber-modified asphalt material both provide increased mechanical shock absorption compared with a conventional asphalt surface. However, the study of six heel-toe runners has indicated that this increased mechanical shock absorption does not result in a reduction in the peak impact force. This result is in agreement with many previous studies in which similar peak impact force values has been obtained when changes in shoes or surfaces have been used to manipulate shock absorption (Clarke et al., 1983; Nigg and Yeadon, 1987).

The finding of the present study that the average loading rate of peak impact force is reduced for a surface with increased mechanical shock-absorption is consistent with the findings of several previous shoe and surface studies (de Wit and de Clercq, 1995). In the present study, the group finding of a significantly later occurrence of peak impact force for the rubber-modified surface has been supported by individual subject analysis, with all but Subject 5 showing this response. These findings support suggestions that loading rate is more sensitive to changes in shock absorption during running than the magnitude of force.

Similar impact forces observed for different materials have previously been attributed to kinematic adjustments made by runners, including changes in ankle and knee joint angles (Clarke et al., 1983). In the present sports surface study, a general trend has been demonstrated for an increased peak ankle dorsi-flexion and peak knee flexion for the acrylic and the rubber-modified surfaces compared with the conventional asphalt surface, indicating increased joint flexion with increased mechanical shock absorption. This is contrary to suggestions that the provision of increased mechanical shock absorption results in a reduction in joint flexion (Clarke et al., 1983).

The trend observed in the present study for less ankle and knee joint flexion for the asphalt surface compared with the surfaces providing increased shock absorption suggests that subjects have not used joint flexion to provide cushioning of impact forces when running on the stiffer surface. This is supported by the finding that all of the subjects showing this behaviour exhibited a greater loading rate of impact force on the stiffer surface. In contrast, Subject 5 has shown no increase in impact force variables but appears to have adjusted to the stiffer surface by exhibiting an increased joint flexion. A requirement for individual subject analysis is therefore indicated.

Although the magnitudes of force clearly differ, the general trend of peak impact force results in running and using the mechanical tests has resulted in the same ranking of the materials used in the present study. This may indicate that the use of a mechanical test is sufficient for the identification of desirable playing surfaces. However, the demonstration of clear differences in joint angles for the different surfaces indicates that individuals adjust their movement patterns in response to surface changes. Since adjustments in movement have been associated with injury occurrence (Nigg and Segesser, 1988) and will influence performance, the importance of biomechanical testing of sports surfaces is highlighted.

CONCLUSIONS

In conclusion, it is suggested that the routine use of mechanical testing to identify differences in playing surfaces should be supplemented by biomechanical testing of subject responses to the surfaces. In addition to the running tests described in the present study, testing of tennis surfaces should ideally include monitoring of players performing typical tennis movements on the surfaces. It is proposed that only by using a combination of mechanical and biomechanical tests will the most suitable playing surfaces be developed for reducing the likelihood of injury occurrence whilst not inhibiting player performance.

REFERENCES

British Standard BS 7044 (1990). Artificial Sports Surfaces. Section 2.2 Methods for determination of person/surface interaction.

Clarke, T.E., Frederick, E.C. and Cooper, L.B. (1983). Biomechanical measurement of running shoe cushioning properties. In: *Biomechanical Aspects of Sport Shoes and Playing Surfaces*. (Ed. B.M. Nigg and B.A, Kerr). Calgary, AB: University of Calgary.

Nigg, B.M. (1986). Biomechanical aspects of running. In: *Biomechanics of Running Shoes* (Ed. by B.M. Nigg), pp. 1-26. Human Kinetics Publishers, IL.

Nigg, B.M. and Yeadon, M.R. (1987). Biomechanical aspects of playing surfaces. *Journal of Sports Sciences,* **5**, 117-145.

Nigg, B.M., Bahlsen, H.A., Luethi, S.M. and Stokes, S. (1987). The influence of running velocity and midsole hardness on external impact forces in heel-toe running. *Journal of Biomechanics*. **10**, 951-959.

Nigg, B.M. (1987). The validity and relevance of tests used for the assessment of sports surfaces. *Medicine and Science in Sports and Exercise* **22**, 131-139.

Nigg, B.M. and Segesser, B. (1988). The influence of playing surfaces on the load on the locomotor system and on football and tennis injuries. *Sports Medicine* **5**, 375-385.

Sureflex® (1995). Patent PCT/GB96/01240. SARCO Ltd. Nottingham, UK.

de Wit, B., de Clercq, D. and Lenoir, M. (1995). The effect of varying midsole hardness on impact forces and foot motion during foot contact in running. *Journal of Applied Biomechanics* **11**, 395-405.

Investigation into the sports characteristics of short and medium pile synthetic turfs and textile tennis court surfaces

A.L. Cox
Centre for Sports Technology, Cromford, Derbyshire, England

ABSTRACT: The development of short and medium pile synthetic turfs and the introduction of needle punch textile surfaces has stimulated great interest amongst British tennis clubs. In order that they may form a view on the desirability of these surfaces and advise their affiliated clubs, the Lawn Tennis Association commissioned a series of sports performance tests. Surfaces were tested for surface pace, slip resistance, traction and shock absorbency. A procedure was developed to simulate the effects of wear so the longer-term performance of the surfaces could be assessed.

BACKGROUND

Over the last fifteen years synthetic turf has proved to be an extremely popular form of tennis court surfacing with many British recreational and club players. The majority of synthetic turf courts have been surfaced with 19 to 23mm piled sand filled tufted carpets. It is the presence of sand in the surface, however, that has been attributed as the cause of the unsatisfactory playing characteristics described by some players, generally competing at the higher levels of the game. This has led manufacturers to develop carpets with lower pile heights and greater pile densities allowing the ball interaction with the surface to be influenced primarily by the fibres resulting in a higher and more consistent ball bounce.

The development of these surfaces has been viewed with interest by the Lawn Tennis Association (LTA) the governing body of tennis in Britain. Recognising the need to gain a greater understanding of how these new forms of surfacing perform, so that they may provide basic information to their affiliated clubs, they appointed the Centre for Sports Technology (CST) to undertake a programme of testing and evaluation.

SAMPLE RANGE

Tests were carried out on the most common forms of short and medium pile synthetic turfs and needle punch textiles offered on the U.K. market in 1997/98. Samples of the products and details of recent installations were supplied by U.K. contractors.

The basic physical properties of the surfaces are shown in Table 1. All of the short and medium piled carpets were manufactured using a tufting process. The textile surfaces were manufactured using a needle punch process. All of the synthetic turfs where manufactured with a split pile fibre except sample SP2 which was manufactured from a monofilament yarn. All of the carpets had piles manufactured from polypropylene or modified polypropylene yarn. Sample MP1 was manufactured from a Low Sliding Resistance (LSR) yarn.

Table 1 Descriptions of surfaces

CST code	Pile Height	Tuft density	Pile Profile	Infill
MP1	15mm	52,500 m^2	Straight	Sand
MP2	15mm	42,000 m^2	Straight	Sand
MP3	15mm	42,000 m^2	Straight	Sand
MP4	15mm	42,000 m^2	Straight	Sand
SP1	10mm	84,000 m^2	Curly	Sand
SP2	9mm	65,000 m^2	Straight	Sand
SP3	9mm	59,400 m^2	Straight	Sand
SP4	9mm	70,200 m^2	Straight	Sand
SP5	9mm	72,500 m^2	Straight	Sand
SP6	9mm	46,200 m^2	Straight	Sand
SP7	10mm	58,800 m^2	Curly	Sand, rubber crumb top dressing
TX1	9mm	N/A	Smooth	Sand
TX2	12mm	N/A	Dimpled	Sand
SP7	10mm	N/A	Diamond profile	Rubber granules

TESTING CRITERIA

The playing qualities of the surfaces were assessed using the testing procedures adopted by the International Tennis Federation (ITF) and detailed in the *Initial ITF Study on Performance Standards for Tennis Court Surfaces.*

Tests were undertaken on site, on laboratory samples and on artificially conditioned samples. Site tests were made, wherever possible, on courts that had been installed in the preceding three to twelve months. Tests were made on areas of high use (near the base line) and areas of low use (near the net on the side of the court) so that any changes resulting from the initial use of the court could be identified.

In order that the variations in performance caused by moisture could be determined tests were carried out under dry and wet surface conditions where weather conditions permitted.

To allow predictions of how the performance of the surfaces may change as they are used a conditioning procedure was developed. This was designed to reproduce the pile flattening and sand consolidation that may be expected to occur after four years typical club use. The procedure was based on a modified Lission Carpet Tester fitted with a profiled drum that was allowed to traverse across the test specimen for a prescribed number of cycles. The effects of the conditioning procedure were verified by making Surface Pace and Force Reduction tests on long pile synthetic turfs that had known characteristics when new and after four years actual use.

RESULTS AND FINDINGS

SURFACE PACE

Site tests

The behaviour of the tennis ball as it impacts the court is of primary importance to a player. The 'speed' or 'pace' of a surface will have a direct bearing on whether players find the court suitable for their styles of play and whether a surface is suitable for an organisation's particular needs. The pace of a surface is generally considered to be a player's assessment of how quickly a ball is travelling towards them and the height of the ball at the point the return stroke is played.

To measure the Surface Pace standard test balls were delivered by a ball launcher onto the surface at an angle of 16° ± 2° and at a velocity of 30 ± 2 m/s (108kph). The velocities and angles of the ball before and after it impacts the surface were measured and the *ITF Surface Rating* calculated from the resulting data. The Surface Pace values recorded on the various surfaces is detailed in Table 2.

Table 2 Surface Pace values

CST code	ITF SURFACE PACE RATING			
	Area of high use		Area of low use	
	Dry	Wet	Dry	Wet
MP1	44	50	36	46
MP2	41	49	36	40
MP3	No data	63	No data	53
MP4	43	43	44	43
SP1	No data	59	No data	58
SP2	34	48	36	42
SP3	No data	46	No data	41
SP4	35	44	34	41
SP5	No data	42	No data	42
SP6	43	38	39	38
SP7	41	56	37	52
TX1	No data	41	No data	44
TX2	No data	34	No data	27
TX3	50	59	48	53

Note : No data indicates weather was wet at time of test

The results of the test programme show:

- New medium and short pile synthetic turfs have surface pace values that fall into the medium or medium fast bands described in the *Initial ITF Study on Performance Standards for Tennis Court Surfaces.* Long pile synthetic turfs would generally be classified as medium fast or fast (CST has recorded surface pace values in the range 40 to 50 on these types of carpet).

- As with most forms of synthetic outdoor surface, the presence of moisture increases the surface pace due to a reduction in ball/surface friction.

- The surface pace values recorded on straight yarn, short pile synthetic turfs are generally similar to the range of values recorded previously by CST on impervious acrylic surfaces.

- Greater variations between high and low use areas are noted on medium pile surfaces than short pile carpets. Previous test carried out by CST suggests the variations on short pile surfaces are generally similar to those found on impervious acrylic courts.

- The use of curly piles in synthetic turfs seems to increase the surface pace when compared to carpets of a similar pile height and density but formed from straight yarns. It is thought this is due to the ball impacting the turned over sides of the tufts which offer less resistance to the ball than vertical tuft ends.

- The performance of surfaces incorporating rubber granular top dressings is very dependent on the top dressing. the presence of moisture appears to effect the ability of the rubber granules to move, causing relatively large changes between dry and wet conditions.

- Needle punch textile surfaces cover a range of surface pace values that encompasses the performance of short and medium pile carpets.

Effects of use on surface pace

As a surface is played on the pile of the carpet tends to flatten and the sand infill consolidates. These changes normally mean that the tennis ball comes off the surface at a greater speed and bounces higher. The degree to which these changes occur are a function of a carpet's design. As the usage of a court's surface is uneven it is important that these changes are as small as possible if court consistency is to be maintained.

Surface pace tests were made on the samples conditioned to simulate four years typical club use. The results are shown in Table 3.

Table 3 Surface pace values after simulated use

CST code	ITF SURFACE PACE RATING			
	As new		After 4 years simulated use	
	Dry	Wet	Dry	Wet
MP1	36	44	39 (+3)	52 (+12)
MP2	33	36	47 (+14)	58 (+12)
MP3	37	38	42 (+5)	55 (+17)
MP4	37	48	43 (+6)	60 (+12)
SP1	37	42	43 (+6)	53 (+11)
SP2	27	36	36 (+9)	43 (+7)
SP3	33	37	38 (+5)	43 (+6)
SP4	32	33	36 (+4)	48 (+15)
SP5	34	44	40 (+6)	49 (+5)
SP6	35	42	37 (+2)	45 (+3)
SP7	36	52	34 (-2)	47 (-5)
TX1	23	34	32 (+9)	48 (+14)
TX2	46	57	38 (-8)	46 (-11)
TX3	50	59	34 (-16)	53 (-6)

The results show:

- The synthetic turf surfaces will increase in surface pace as a result of use (as will most other forms of tennis court surfacing). These changes will become more noticeable when a court is wet.

- A significant increase in surface pace, particularly if the surface is wet, is likely to occur on medium pile synthetic turf carpets. Some forms of short pile carpets will also show large increases, whilst others will remain fairly consistent and only change by amounts similar to surfaces such as impervious acrylic.

- Textile surfaces show no general trends. In certain cases they were found to actually become slower, although it is thought that this is primarily due to the loose of the particulate dressings, which may be detrimental to other aspects of the surfaces' performance.

SLIP RESISTANCE AND TRACTION

Site tests

The ability of a court to provide players with sufficient grip to allow them to stand, move and turn with confidence is one of the most important aspects of a tennis surface's design. It has also been one of the largest areas of concern to players using long pile synthetic turfs.

Slip Resistance was measured using a rubber foot mounted onto a pendulum that was allowed to slide across the surface and the resistance to movement recorded. The test gives a numerical value, the higher the number the greater the slip resistance.

Traction characteristics were determined by measuring the rotational force required to turn a weighted foot resting on the surface.

The results obtained on high use court areas are detailed in Table 4. The high use areas are considered to be most representative of how the surface performs because the particulate infill materials have "settled in".

Table 4 Slip resistance and traction results – high use areas

CST code	SLIP RESISTANCE		TRACTION COEFFICIENT	
	Dry	**Wet**	**Dry**	**Wet**
MP1	78	64	0.85	0.68
MP2	78	70	0.86	0.99
MP3	No data	74	No data	1.14
MP4	No data	81	No data	1.24
SP1	No data	74	No data	1.19
SP2	96	73	0.85	0.69
SP3	No data	86	No data	1.24
SP4	97	85	1.12	1.00
SP5	No data	90	No data	0.97
SP6	74	80	0.67	0.62
SP7	99	74	0.86	0.91
TX1	No data	80	No data	0.89
TX2	No data	62	No data	0.57
TX3	92	65	0.59	0.75

The results show that:

- The slip resistance results on all of the surfaces fell within the preferred range detailed in the *Initial ITF Study on Performance Standards for Tennis Court Surfaces.*

- The traction results indicate that short and medium pile synthetic turfs can be expected to fall within the preferred range detailed in the ITF document. Some textiles may, however, have traction values that are below the ITF's preferred range.

- The presence of moisture, as would be expected, reduces the slip resistance and traction properties of the surfaces, but not to unacceptable levels. The variations may, however, be a problem on courts that dry unevenly.

Effects of use on slip resistance

One common complaint about long pile carpets is that they become slippery when they age. To assess whether the new forms of surfacing will also suffer from this problem slip resistance tests were made on test samples conditioned to simulate four years use. The results are shown in Figure 5.

Table 5 Slip resistance values after simulated use

CST code	As new		After 4 years simulated use	
	Dry	Wet	Dry	Wet
MP1	78	65	76 (-2)	56 (-9)
MP2	64	64	62 (-2)	59 (-5)
MP3	74	72	67 (-7)	65 (-7)
MP4	66	70	69 (+3)	73 (+3)
SP1	78	60	70 (-8)	65 (+5)
SP2	92	88	80 (-8)	69 (-19)
SP4	94	88	89 (-5)	73 (-15)
SP5	76	85	74 (-2)	83 (-2)
SP6	66	65	70 (+4)	72 (+7)
SP7	81	65	72 (-9)	67 (+2)
TX1	88	56	79 (-9)	65 (+9)
TX2	70	59	65 (-5)	67 (+8)
TX3	50	61	66 (+16)	50 (-11)

The results suggest that short pile synthetic turfs will retain acceptable levels of foot friction. Some medium pile synthetic turfs and textiles have relatively low slip resistance values when new and after a period of use their values may drop below the ITF's preferred range.

SHOCK ABSORPTION

One of the reasons that synthetic turfs have proved popular to many tennis players is due to the surface providing a level of player comfort not available from various forms of hard court.

The ability of a surface to absorb the impact energy of a player moving across it can be measured in a number of ways. The ITF have selected a test that simulates a person running on the surface and measures the dynamic response. As the hardest surface a player may use is concrete the test compares the result obtained on the synthetic surfaces to the value recorded on concrete and expresses it as a percentage reduction. The higher the value of Force Reduction the greater the level of shock absorbency. The results are detailed in Table 5.

Table 5 Force Reduction values

CST code	Area of high use		Area of low use	
	Dry	Wet	Dry	Wet
MP1	13%	18%	12%	16%
MP2	20%	20%	17%	18%
MP3	No data	16%	No data	16%
MP4	No data	15%	No data	14%
SP1	No data	17%	No data	18%
SP2	13%	11%	12%	13%
SP3	No data	8%	No data	7%
SP4	8%	9%	10%	10%
SP5	No data	12%	No data	10%
SP6	7%	9%	10%	9%
SP7	20%	29%	22%	30%

Table 5 Force Reduction values (continued)

CST code	Area of high use		Area of low use	
	Dry	Wet	Dry	Wet
TX1	No data	18%	No data	21%
TX2	No data	6%	No data	11%
TX3	26%	26%	31%	30%

All the synthetic turfs and textile surfaces give values that are significantly better than most forms of cushioned acrylic commonly used in the United Kingdom. The high values recorded on some of the surfaces are similar to values found on clay courts. The use of a court tends to reduce the level of Force Reduction.

CONCLUSIONS

The new forms of synthetic turf and textile tennis court surfaces offer a wide and interesting range of sports characteristics. The development, in particular, of short pile synthetic turfs has created a group of surfaces that have characteristics that seemed to be much slower and more consistent than earlier forms of carpet. In many cases the short pile carpets appear to have similar pace characteristics to impervious acrylics although the consistency or trueness of the bounce is not as good. It is also noted that the effects of slice have not been considered as part of this investigation and player feedback indicates that the responses of synthetic turfs and textiles are different to that of acrylic surfaces.

REFERENCES

Cox A.L. (1998) *Investigation into Synthetic Turf & Textile Surfaces.* (Technical Report prepared for Lawn Tennis Association). Centre for Sports Technology, Derbyshire, England

International Tennis Federation (1997) *Initial ITF Study on Performance Standards for Tennis Court Surfaces*. International Tennis Federation, London

Smith J, Thomas N, Cox A. L. (1999) *Tennis Facilities Review, Autumn 1999*, 8-10. Lawn Tennis Association London

Comparison testing of let-cord sensors – the Trinity system by GTE and the Sharpline system by Signal Processing Systems, Inc.

J.R. Fisher
Signal Processing Systems, Inc., Sudbury, MA, USA

ABSTRACT: The ATP Tour and three of the four Grand Slam tournaments have used for a number of years an electronic device replacing the net judge. The device comprises a pair of accelerometers attached to the headband at either end of the net and a handset, held by the chair umpire, which emits a beep if the ball strikes the net.

Although any decision-making device inevitably makes some errors even when working perfectly, chair umpires experienced with this device sensed an inconsistency in response. Signal Processing Systems, Inc. (SPS) was asked to investigate the situation and potentially to develop an alternative design which would be more consistent and at the same time be more reliable and less expensive to maintain.

It was found that the existing net device exhibited a marked variation in sensitivity to ball strikes depending on where the ball hit the net. A striking force which was detected with roughly 90% probability when the point of contact was above the singles line was never detected at the center of the net, rather the sensitivity control had to be increased to achieve the same level of detection. This variation has been attributed to attenuation of the vibration as it propagates along the net (primarily through the net cable) from the point of impact of the ball to the sensors near the ends of the net.

To overcome this problem, SPS has designed a sensor cable extending over the full width of the service boxes and having very uniform sensitivity, independent of the point of ball contact. This new device is the final stage of evaluation by the ATP Tour.

SUMMARY

Signal Processing Systems, Inc. (SPS) has performed extensive side-by-side testing of the Trinity let-cord sensing system, now used by the ATP Tour as well as by other tennis organizations, and several configurations of the Sharpline let-cord system developed by SPS. The systems tested can be divided into two groups: those (including the Trinity system) which use localized sensors mounted at one or two points along the net and which rely on transmission through the net support cable of the vibration due to ball impact, and those (including two configurations of the

Sharpline sensor) in which the sensor itself is distributed along the net. All systems provide an audible indication (a beep) when a ball hits the net and all incorporate a sensitivity adjustment so that the force on the net necessary to cause a beep can be adjusted over a wide range. In general terms, all systems tested proved reasonably effective in detecting let-cords. In more specific terms, the systems based on localized sensors (including the Trinity and one version of the SPS sensor), showed a marked variation in sensitivity along the length of the net. Thus, if the sensitivity control is set at the desired force level for balls impacting near the sideline, then some let-cords with equal force at the center of the net will be missed. Conversely, if it is set to detect impacts of the desired force level at the center on the net, then it may be vulnerable to false indications due to wind and other spurious disturbances. This variation is the source of the inconsistent operation noted by chair umpires who sometimes were unable to find a sensitivity control setting which would achieve the desired level of sensitivity.

In contrast, the systems based on distributed sensing (including the two preferred versions of the Sharpline sensor) respond with equal sensitivity along the entire active length of the net so that at any control setting (for lighter or heavier ball touches) a given level of ball impact force will cause the same response regardless of position along the net.

BACKGROUND

The ATP Tour and three of the four Grand Slam tournaments have used for several years the Trinity let-cord sensing system developed by GTE gmbH of Kunzelsau Germany. The device comprises a pair of accelerometers attached to the headband at either end of the net just outside the singles sidelines and a handset held by the chair umpire which emits a beep equivalent to the "Net!" call by a net judge if the ball strikes the net.

Each accelerometer has a sensitivity of approximately 0.0005 V/g along the preferred axis which is marked by a red dot. From each accelerometer a 4-conductor, 2 mm diameter connecting cable is threaded through the weave of the net to the netpost and then across the court apron to the control handset at the umpire's chair.

The handset has an adjusting control by which the chair umpire can set the device to beep in response to greater or lesser striking force. The scale ranges from 1 to 10 with 10 being the most sensitive. Chair umpires are advised to set the control initially to 5 for singles and 8 for doubles and then to adjust it downward if wind or other sources of disturbance cause incorrect beeps (referred to as "phantom lets").

This system has proven to be generally satisfactory and, along with the Cyclops service line system, has helped to gain acceptance for the use of electronic devices in support of tennis officiating.

Signal Processing Systems, Inc. (SPS) of Sudbury, MA has developed the Sharpline electronic system for tennis line calling based on a very sensitive triboelectric contact sensor which is embedded in the boundary lines of the tennis court. This sensor, which is manufactured by SPS, is in the form of a thin ribbon, approximately 2/3 inch wide and 0.0075 inch thick -- slightly thinner than standard adhesive tape. It comprises sensing wire enclosed within a protective aluminum coated shield. The shield is sealed to protect the sensing elements both from the physical environment and from electromagnetic interference. In most applications the sensor is provided with a second rugged outer cover for further protection. The sensor is produced in continuous lengths and then is cut to the length required for a specific installation (here 35 feet). The sensing function is similar to that of a common microphone -- albeit a very long, thin microphone -- except that it senses only direct contact and does not couple to (i.e., does not respond to) sounds in the air.

SPS NET SENSOR CONFIGURATIONS

The SPS Sharpline let-cord system consists of three pieces: the sensor itself, a control handset (similar to the Trinity control handset) held by the chair umpire and a connecting cable between the sensor and the control handset. ATP Tour experience with the Trinity system has shown that the cable between the sensors and the umpire's control handset are subject to both heavy wear and abuse, so SPS has made this cable (only one is needed rather than two) both heavier, to extend its life, and physically separate to simplify replacement. A transparent vinyl provides further protection.

Three versions of the basic SPS Sharpline sensor were tested. In one version (the center strap version) the basic Sharpline sensor ribbon is attached under the center strap of the net and the connecting cable is threaded along the bottom of the net to the netpost and across the court apron to the control handset at the umpire's chair.

In the second version (the net-weave version) the basic Sharpline sensor ribbon is folded inside a protective black polyester braid cover and is threaded through the weave of the net just below the headband for a length of approximately 35 feet (from the far singles sideline, where one Trinity sensor is normally mounted, to the near netpost). The connecting cable is then run down to the bottom of the netpost and across the court apron to the control handset. The black braid cover is very similar in appearance to the twine used for the weave of tennis nets but is slightly larger in diameter.

The third version of the Sharpline sensor (the headband version) consists of a very thin white vinyl band 5.5 inches wide and 41.5 feet long, similar to the headband of the net, with the Sharpline sensor ribbon attached along the center 30 feet. With this configuration, the white vinyl band is placed over the existing net headband. It extends 2.5 inches down either side of the net, making a new headband incorporating the Sharpline sensor ribbon. This headband is held in place by a narrow black fastening tape along the bottom edge which is fastened through the net just under the original headband. (During different test sessions, several headband variations were used having different weights of the vinyl material, as well as cloth, and attached with either Velcro or snap fastener tape.) The connecting cable runs from the netpost to the control handset at the umpire's chair.

TEST CONDITIONS

In each testing session, the Trinity system and the SPS system being tested were set up on the same net so that the response of different systems could be compared on each net hit. The Trinity sensors were installed according to the standard ATP Tour installation procedure with each of the two sensors placed in mounting clips on the net headband about one foot outside the singles sideline. The two Trinity sensors were therefore approximately 29 feet apart and each was 14.5 feet from the center of the net.

The Trinity sensors were connected to the Trinity control handset which was adjusted as necessary to achieve the desired sensitivity. When used by the ATP Tour a sensitivity setting of 5 or 6 (10 is most sensitive) is normally recommended for singles play. A more sensitive setting of 8 is recommended for doubles to reduce the number of times that players (two of whom are close to the net during the serve) detect lets which are missed by Trinity.

In all tests the SPS Sharpline sensor being tested was connected to the same SPS control handset which is generally similar to the Trinity control handset. The most notable difference is that in addition to a beeper with adjustable detection sensitivity, the Sharpline handset has a connector into which optional earphones can be plugged. When earphones are used, the user can hear and easily distinguish the sound of a ball hitting the net, a player touching the net or any other contact with the net. The earphones can be used either in addition to or instead of the beeper.

SPS set up a test facility with the capability to cause a tennis ball to strike the net with repeatable speed and force adjustable in two ranges. Low speed "brush by" tests were conducted with a tennis ball suspended as a pendulum. More forceful tests were conducted using a ball machine. For the "brush-by" tests a tripod was placed over (but not touching) the net and a ball on a light string was suspended from the tripod. The length of the pendulum was adjusted and the amplitude of the swing was controlled to vary the impact force with reasonable repeatability. For the higher speed tests, a Playmate ball machine (model BP-F by Metaltek, Inc.) was positioned very close to the net (one foot away) to minimize the effect of the slight ball-to-ball variation in firing elevation angle. The machine was set to its maximum firing speed, which, according to Metaltek is 95 mph. Taking into account the decrease of ball speed due to drag through the air, this impact speed is roughly equivalent to a radar gun reading of 115 mph at the normal point of serving.

TESTS PERFORMED

Four types of tests were performed:

(1) Range of sensitivity tests to determine the range of ball striking force necessary to trigger the detector with varying settings of the sensitivity control.
(2) Consistency tests in which the net was hit many times with approximately the same force to determine if the sensors would respond consistently.
(3) Uniformity tests in which the net was hit with the same force at different places along the length of the net to determine if the sensors would respond equally well at all points along the net.
(4) High-speed tests in which balls were fired from a ball machine adjusted so the balls (ideally) would just clear the net to determine if sudden air disturbance would cause the sensors to give false detections.

A fifth test was performed only with the SPS headband sensor to determine if the sensor could withstand repeated ball hits at high velocity.

TEST RESULTS

RANGE OF ADJUSTMENT

In preliminary test sessions, the range of sensitivity adjustment of the SPS handset was found to cover a much wider range of sensitivity settings than the Trinity handset -- more than necessary -- so the SPS handset was modified to achieve roughly the same range of sensitivity as the Trinity.

CONSISTENCY OF DETECTION

Consistency testing was performed at the singles line (one Trinity sensor was about one foot away). With the Trinity handset initially set at 5 (10 is most sensitive) the ball impact force was adjusted so that the Trinity sensor achieved nearly 100% detection. This hit was just loud enough for the sound of the touch to be heard close-up (but not from the umpire's chair). With the SPS sensor, the handset was adjusted to the least sensitive setting which would achieve essentially 100% detection. This setting was 5 for the headband sensor (10 is most sensitive) and 6 for the net-weave sensor. This "brush-by" test was repeated many dozens of times with varying forces of ball impact. The two systems beeped together when the ball brushed the net with a force equal to or greater than the set threshold level. All sensors responded consistently over the achievable range of ball striking force.

UNIFORMITY OF DETECTION

In these tests, the sensitivity of the Trinity and of the several SPS sensors was tested at different points along the length of the net. Specifically 9 locations were used: at each singles sideline (designated as the near and far sidelines), one foot, three feet and seven feet in from each singles sideline, and at the center. When tested at these points along the length of the net, the SPS sensors were uniformly responsive, but the Trinity sensitivity was found to drop off markedly toward the center of the net.

When the impact force was held constant along the net, the Trinity control had to be increased to achieve the same level of detection as one or the other of the SPS sensors (see Figure 1) reaching a setting of 8 at the center of the net.

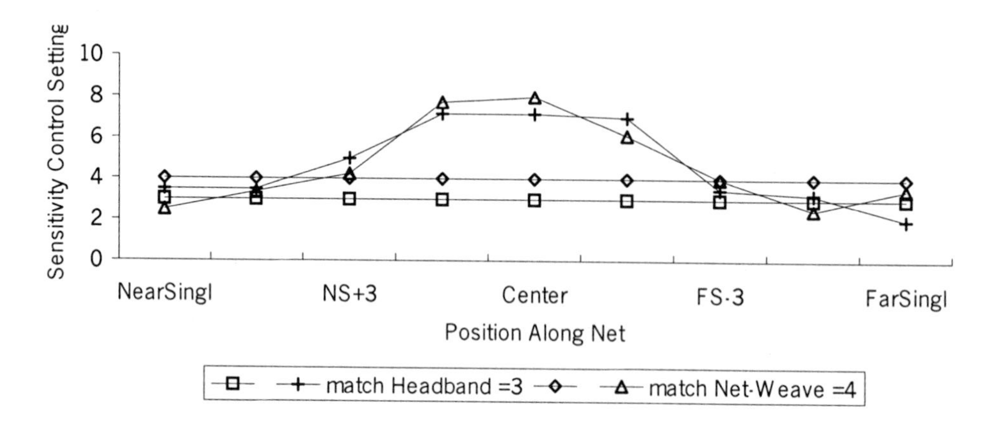

Fig. 1 Comparison of Trinity & SPS Sensors as a function of position along the net: The SPS headband sensor is slightly more sensitive than the SPS net weave sensor. With the force of the ball impact held constant at different locations along the net, the two SPS sensors achieved 90% probability of detection at settings of 3 and 4 respectively. The sensitivity setting on the Trinity sensors was adjusted to achieve the same response to ball hits as the SPS sensors. Higher sensitivity settings were required for the Trinity near the center of the net and lower settings near the ends.

Conversely when the Trinity setting was held constant at 5 and the impact force varied, equal performance could be achieved with reduced SPS control settings (see Figure 2). At the center of the net with the Trinity at the original setting of 5, it was difficult to cause the test ball on the pendulum to hit the net hard enough to cause a beep, so for some tests the ball was held in the observer's hand.

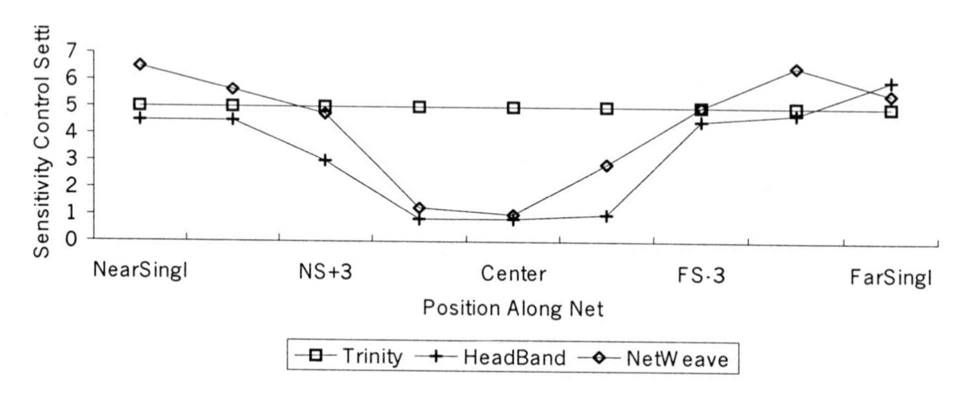

Fig. 2 Comparison of Trinity & SPS Sensors as a function of position along the net: With the sensitivity of the Trinity set to the standard value of 5, the force of the ball impact was adjusted at each location along the net so that the Trinity achieved a 90% probability of detection. The sensitivity setting was then adjusted on each of two SPS sensors to achieve the same response to ball hits as the Trinity. Near the center of the net the SPS sensors matched the Trinity at significantly lower sensitivity settings.

The SPS net-weave sensor required a sensitivity setting of 6 to respond at the same level as the SPS headband sensor setting of 5 but, like the SPS headband sensor, response was also uniform across the net.

HIGH SPEED SENSITIVITY TESTS

The sensors were set up as with the lower speed tests described above. The ball machine was adjusted so that the ball should just graze the top of the net; however, even at one foot range, the ball-to-ball variation in firing was such that many balls missed the net by a margin of 1/4 inch or more and others hit the headband with great force. Fewer than 25% of the balls fired could be characterized as "grazing" the net. With the ball crossing the net approximately one foot from the Trinity sensor (i.e. at the singles sideline location) either both the SPS Sharpline sensor and the Trinity responded (when the ball actually touched the net) or neither responded (when the ball missed the net). When the test was repeated at the center of the net (with the Trinity in its normal position just outside the singles sideline) neither sensor responded when the ball clearly missed the net and both responded when the ball clearly hit the net. But in a percentage of instances when the ball was in the "grazing" category, the Sharpline sensor generated a beep when the Trinity did not. This result was expected based on the previous low speed tests, but it was not possible to determine if the ball actually touched the net in the instances where Sharpline detected the ball and Trinity did not. This was because the ball machine made significantly more noise than the net hits in these instances and because for safety the observers were 6 to 8 feet from the ball machine and from the hit point on the net.

HEADBAND SENSOR ENDURANCE TESTS

The purpose of this test was to determine the endurance of the SPS headband sensor under high-speed, direct ball hits. This test was conducted after the tests described above were completed. During the course of the high-speed sensitivity test describe previously, roughly half of the roughly 300 balls fired struck the headband with considerable force. Because of these previous hits, this final test should really be considered as a <u>continuation</u> of endurance testing. For this test, the ball machine position was retained at approximately one foot from the net but the elevation was lowered so that balls were fired directly at the headband. The ball hopper was filled with balls and the ball machine was set at the maximum firing rate. At the beginning of this test the headband was slightly dirty (this test was done on a clay court) at the spot where the balls fired in the high-speed sensitivity test hit the headband. After each hopper (about 100 balls) was used, the headband was inspected for signs of wear or damage. After three hoppers of balls (in addition to the roughly 150 hits during the high-speed sensitivity tests) careful inspection showed that the vinyl headband cover had a small cut (about 1/2 inch long) along the line where the vinyl was joined to the Velcro which holds the headband to the net;, nonetheless, the Sharpline sensor continued to function properly with normal sensitivity. Testing was stopped at this point.

CONCLUSIONS

Three SPS Sharpline let-cord sensor designs were tested. Two of these sensors (the headband sensor and the net-weave sensor) were of the distributed sensing type which extend over the entire 27+ foot span of the net where let-cord ball hits can occur. The third sensor (the center strap configuration) was of the localized sensing type which, like the existing Trinity sensors, detect net vibration at only one or two points and therefore depend on the net support cable to transmit the vibration of the ball impact from the point of the hit to the location of the sensor.

As would be expected, the distributed sensors exhibited uniform sensitivity over the entire active span of the net whereas the localized sensors (both the Trinity sensor and the SPS center strap sensor) exhibited marked variation in sensitivity along the length of the net. Since the SPS center-strap sensor was located in the middle of the span over which let-cords occur, it showed somewhat better sensitivity than the Trinity sensors which were positioned several feet from the closest plausible impact point.

When the sensors were tested for high-speed "breeze-by" sensitivity, the results were consistent with the "brush-by" lower speed tests: at points close to the Trinity sensors the SPS distributed sensors responded in the same way as the Trinity: either both responded or neither responded. Near the center of the net the Trinity responded less often, presumably due to the same reduction in sensitivity it showed in the lower speed tests.

Among the SPS Sharpline sensors tested, the center strap sensor configuration is least desirable because is suffers from the same variation in sensitivity shown by the Trinity system. Either of the two SPS distributed sensors is preferable.

The net-weave sensor is as sensitive as the currently used Trinity sensor but has the advantage of uniformity of response. The headband sensor is slightly more sensitive than the net-weave sensor and also uniform in response.

Any of the SPS sensors can be used either with the beeper or with earphones. The beeper may be most suitable for use by the chair umpire, but tournaments which prefer to use a let-cord judge might find that using earphones not only is more sensitive than the net judge's hand on the net but also would allow the judge to retreat to a safer location.

Comparing the two SPS distributed sensors, the net-weave sensor, although somewhat less sensitive than the headband sensor, is more compact for transporting between tournaments (the sensors are carried as luggage by ATP Tour chair umpires). The headband sensor is more sensitive but somewhat larger and, even though it looks much like the normal net headband and is therefore inconspicuous, it is nonetheless very visible. Selection between the two distributed designs requires judgement concerning the relative importance of these diverse aspects. This judgement can be made only by prospective users.

CFD modeling of full-size tennis courts

Z.B. Gradinscak
Department of Mechanical and Manufacturing Engineering, RMIT, Australia

ABSTRACT: The harsh effects of hot summer winds on the court surface and while playing are well known to the thousands of Australian tennis enthusiasts. Red porous surfaced tennis courts for example, due to the materials used and construction, are highly susceptible to the damage in extreme weather conditions. Current evaluation of the computational fluid dynamics (CFD) tennis court models developed at RMIT aims to elaborate the wind effect on tennis court surface and to determine the parameters for CFD models as to simulate the microenvironment of a full-size tennis court. The author initiated a study to assess potential problems associated with different windbreak constructions inquiring on the effect that different mesh positioning would have on wind conditions over the tennis courts. This paper briefly presents the testing procedure and developed CFD models and discusses potential directions of research offered at the RMIT wind tunnel and computational laboratories.

INTRODUCTION

In practice the tennis court windbreaks are developed to suit individual courts and specific situations. Windbreaks, coming in a variety of configurations, provide a number of features and characteristics, with main function being to reduce the velocity of wind across the tennis court. The density of mesh used and particularly the positioning of the mesh varies in each situation, attempting to provide the same overall effect of reducing wind flow over the court.

The permeable windbreaks around the tennis courts perform two major functions:

(1) They create a physical barrier to airborne debris, and
(2) They create their own microenvironment of moderated air movement.

These two functions are very important for the tennis courts where there is no, or is very little, natural windbreaks or protection from changeable weather conditions.

It is important to note that tennis court windscreens are designed to reduce court wind fluctuations but not to completely eliminate the wind. The research in this area carried out by the Australia's Commonwealth Scientific and Industrial Research Organization (CSIRO) in conjunction with Tennis Australia (TA) and Victorian Tennis Association (VTA) resulted by recommendation for the use of woven green mesh as the most effective wind barrier for tennis courts [VTA 1992].

Although no direct data on testing of windbreaks around the tennis courts can be found, the research conducted on erosion in desserts and areas of unstable topsoil exist [Richardson 1989, Richardson and Richards 1993] and to some extend can be applied for the tennis courts general description and as a guide for more specific studies.

The research on various aspects of windbreak effects over large areas is a useful reference for the initial setup of tennis court investigations. The detail information and recommendations related to the construction of red porous surface courts and on construction of wind screens around the courts can be found in the VTA instruction manuals. However, the direct information relating to the testing setup or testing results on the windbreaks around the tennis courts for an accurate prediction of wind conditions for a full-size tennis court can not be found in the existing reports [Austin and Bower, 1998].

An experimental wind tunnel rig was developed at RMIT [Gradinscak 2000]. The reported study aimed to reproduce the airflow across the court surface in the wind tunnel replicating the conditions encountered on a full-size tennis court and to develop a CFD model simulating these conditions. The results from a number of set experiments were used as input parameters for CFD model. The small RMIT wind tunnel was used for preliminary assessment of potential problems associated with the setup of experiments. The apparatus and experimental procedure is briefly explained in the next section giving an overview of relationships between the experimental outcomes and CFD model.

INITIAL APPARATUS AND EXPERIMENTAL PROCEDURE

To investigate the wind velocities within the microenvironment of an experimental tennis court model, the small wind tunnel in the RMIT Laboratories was used. This wind tunnel operates a flow through configuration, with a standard working section of 340mm in height and 300mm in width.

To facilitate for the velocity profile measurements and visualization a perspex working section was constructed and installed in the working section of the wind tunnel, as shown in Figure 1. The perspex working section consisted of several features, which allowed pressure tapings to be fitted along the bottom surface of the working tunnel section, while slides integrated into the top face allowed a pitot static tube to be utilized for velocity profile measurements, as illustrated in Figure 2.

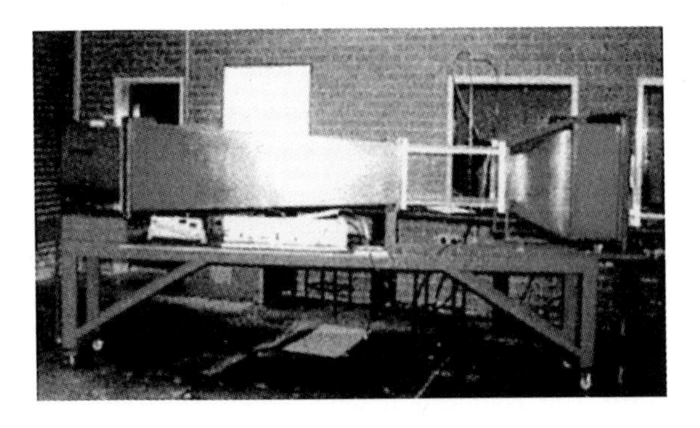

Fig. 1 Used Wind Tunnel with Installed Perspex Working Section.

The size of wind tunnel working section limited the construction of new perspex working section to 1:40th scale, with the dimensions being scaled off the outer dimensions of a full-size tennis court.

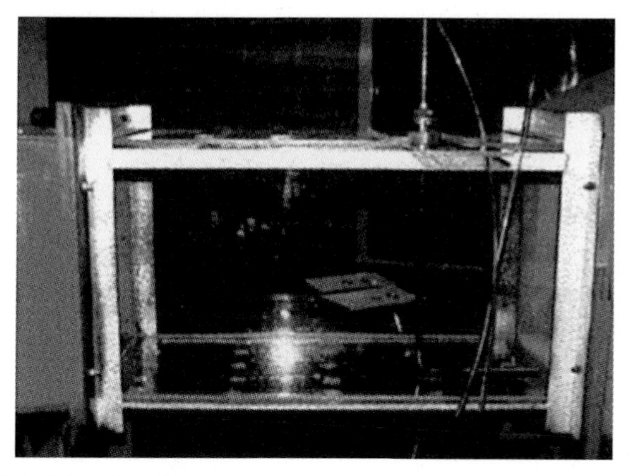

Fig. 2 Perspex Working Section.

The constructed perspex working section allowed several mesh models to be tested. The windbreak wind tunnel models consisted of geometries for which the mesh was positioned at:

(1) Full fence height of 115mm.
(2) Top two-thirds fence height of 75mm.
(3) Bottom two-thirds fence height of 75mm.

The above geometries were tested with the mesh density of 90%, 80% and 60%. Pressures were measured across the surface of the tennis court model using multi-tube inclined manometer at the average freestream wind speed of 7.72 m/s. In the experiments the static pressure was sensed through the tapping connected to a manometer. The reference point for pressure readings was the central hole in the base of the perspex working section.

Several measurements were also taken to establish a velocity profile across the tennis court model. The profile was generated by using a directional pitot tube placed at twenty-six selected positions across the perspex working section.

A series of flow visualization recordings were also undertaken using the fog machine to identify any turbulent areas within the perspex working section and to allow comparison of flow patterns with the computer model. To record what was observed from the flow visualization the photographs were taken for all mesh placements and densities.

To develop inputs for the CFD models, the data was required on characteristics of the model itself and the type of flow that is occurring around it. The required data were calculated from the data obtained from the wind tunnel experiments. The wind tunnel results were used to develop a profile of pressure drops at various displacements behind the mesh defining the properties of the porous jump used in the computer model.

The CFD models were developed using user-defined boundaries and defining the court surrounding. Two models, a 2D and a 3D model, were created to gain an overall picture of what was occurring at the court ground level and above the ground. The models were scaled to represent the wind tunnel working model with one unit on the computer model being equal to that of one centimeter on the wind tunnel working model.

The 2D model was representative of cutting a plane through the longitudinal centerline of the tennis court model to obtain a picture of the wind flow and velocities above the ground. The mesh fence around the court was modeled by defining it as a porous jump.

The results from a number of 3D computer model runnings showed that the curves created from the pressure calculations, some of which are illustrated in Figures 3, 4 and 5, are similar in shape to that of the curves generated from the wind tunnel readings.

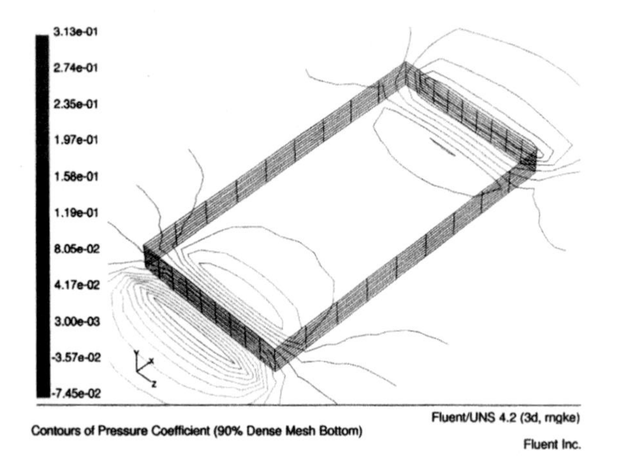

Fig. 3 Contours of Pressure Coefficients at Ground Level of the 3D Computer Model for 90% Mesh Density for Ground Level Geometry.

The plots illustrated in Figures 3, 4 and 5 show the contours of pressure coefficient (C_p) at the ground level around and over the court surface resulted from the computer runnings for the 90% dense mesh models of all three geometries considered.

It has to be noted that aim set in preliminary research to develop a validated computer model that replicates wind conditions of a full-size tennis court was not achieved with the described procedure, because of the time and resource constraints. However, the comparative pressure coefficients and velocity profiles were established for the average Melbourne wind-speeds and were integrated in developed CFD model. Satisfactory comparison of CFD model runnings and wind tunnel model testing were obtained, indicating that a broader research would result in a thorough and validated computer simulation model. Such a CFD model has potential to provide reliable tool for not only the planar investigations at the ground level, but also for a full 3D inquiry of a tennis court's microenvironment.

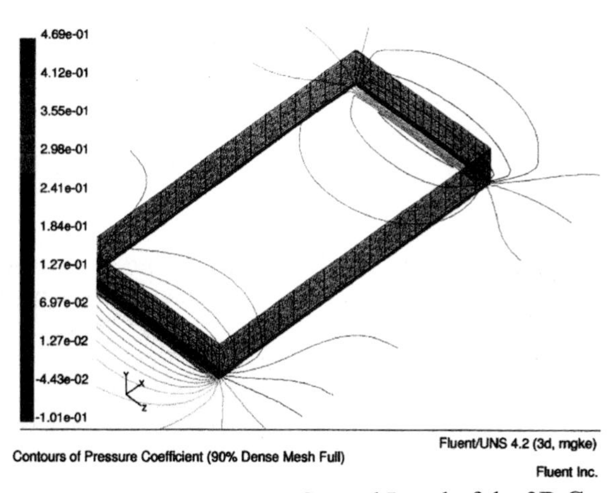

Fig. 4 Contours of Pressure Coefficients at Ground Level of the 3D Computer Model for 90% Mesh Density for Full Fence Geometry.

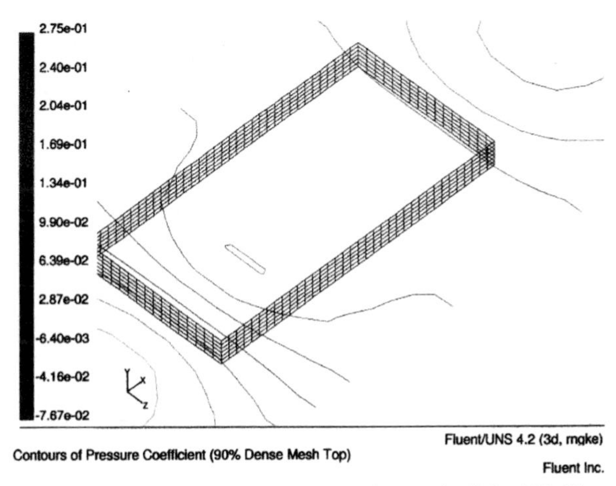

Fig. 5 Contours of Pressure Coefficients at Ground Level of the 3D Computer Model for 90% Mesh Density for Above the Ground Geometry.

It is important to know that for experiments in the small wind tunnel, because of the blockage phenomenon, the air velocity readings around the model were showing unreal wind conditions in comparison to those encountered in a full-scaled tennis court. Therefore the industrial wind tunnel with reduced blockage effect would produce more realistic results and its configuration and experimental strategy is described in the next section.

INDUSTRIAL WIND TUNNEL AND EXPERIMENTAL STRATEGY

The industrial wind tunnel at the RMIT Mechanical Engineering Laboratories can be used to assess the wind velocities encountered in the microenvironment of an experimental tennis court model. The wind tunnel operates a flow through configuration with a standard working section of 2000mm in height and 3000mm in width.

A standard testing set for an automotive experiment conducted in the tunnel is shown in Fig. 6, illustrating the size of tunnel's working section.

Fig. 6 Standard Testing Setup in Working Section of Industrial Wind Tunnel.

Size of the industrial wind tunnel working section limits the construction of the working tennis court model to 1:20th scale, with dimensions scaled off the outer dimensions of a full-size tennis court. The working section size allows several tennis court models to be tested with the inclusion of winds from diagonal directions that may induce pressure variations, and hence scouring of the court not seen in the testing conducted so far.

Pressures measured across the surface of the tennis court and measurements of velocity profile across the tennis court would be possible for wide range of wind velocities including the average and extreme wind speeds allowing a database required for validation of the CFD models to be build.

CONCLUDING COMMENTS

Described preliminary research at RMIT aimed to determine the experimental setting and required parameters for a CFD model development that would simulate microenvironment conditions encountered over a full-size tennis court. Being a preliminary research the study has had a limited budget and as a consequence no publishable results from the actual wind tunnel experiments have been produced. However, satisfactory comparison of CFD and experimental results has been recorded from the computational and experimental models producing useful data for further research.

The aim for this year is to extend the work by setting new experiments using the large industrial wind tunnel. The plan is to utilize small wind tunnel for a quick assessment of potential problems and then to utilize RMIT 3m x 2m industrial wind tunnel to assess and identify parameters of a full-sized tennis court microenvironment. A number of identified advantages in such an approach are:

- The industrial wind tunnel covers speed ranges commonly encountered – having max speed of 45 m/s.
- Using industrial wind tunnel will provide a reduction of blockage effect possibly to a negligible level.
- The industrial wind tunnel has acoustically treated vanes, providing low background noise.
- The multi-component built in force balances in the industrial wind tunnel will allow vary quick setting of new experiments.

The CFD and experimental models, once fully validated, would allow further testing into the layout of a multiple court setup as to the effect of meshing between the courts and the effect of meshes of varying densities between courts within the multiple court area.

The testing can be drawn also for positioning the windscreens relative to both, the height of the fence and court surface. Some indications in favor of positioning the mesh all the way down to the court surface were drawn from the comparison of the results from computer runnings executed in preliminary study. The reduction of wind velocities for the full length of the court can also be investigated in relation to the height of the windbreaks and fence.

With further development and evaluation of the CFD and experimental models a broader study of the effects that the different microenvironments of tennis courts would produce in relation to the playing conditions can be undertaken.

REFERENCES

Austin R. and Bower C. (1998) Wind Scouring of Tennis Courts. *Final Year Project*, RMIT, Melbourne.

Gradinscak Z. B. (2000) 3D Modeling of Wind Conditions for a Full-size Tennis Court. 3rd International Conference – *The Engineering of Sport*, June 10-12, Sydney.

Richardson G. M. (1989) A Permeable Windbreak: its Effect on the Structure of the Natural Wind. *Journal of Wind Engineering and Industrial Aerodynamics*, Vol. 32, pp101-110.

Richardson G. M. and Richards P. J. (1993) Full Scale Measurements of the Effect of a Porous Windbreak on Wind Spectra. *3rd Asia Pacific Symposium on Wind Engineering,* Dec 13-15, Hong Kong.

Victorian Tennis Association (1992) Windscreens for Tennis Courts. *Technical Instructions No. 4*, 3rd Issue.

4 The game

Proposals to slow the serve in tennis

H. Brody
Physics Department, University of Pennsylvania, Philadelphia, USA
R. Cross
Physics Department, University of Sydney, Sydney, Australia

ABSTRACT: Various methods of slowing the serve in tennis are evaluated, in terms of calculated ball trajectories, serve windows and transit times. A larger tennis ball appears to be the best option, but other methods are also worthy of trial evaluation.

INTRODUCTION

There has been a lot of discussion about slowing down the game of tennis and many suggestions about how to do this have been put forth. It is generally accepted that the problem that needs to be addressed is the speed of the serve on fast courts such as the grass courts at Wimbledon. There it has become a first serve contest, comparable to the quick draw gun fights in the old west. Some players win 40% of their good 1st serves as aces. The serve has become so dominant at Wimbledon and service breaks so rare, that some players have over 30% of their sets end in tie-breaks. The proposed solutions, described below, are based on changing some of the tennis parameters to reduce the serve speed so that the receiver has more time to react to the serve. Most of the proposed solutions will work, but many of them have other consequences that are undesirable. It is recognised that the proposals need to be evaluated by experimental trials, due to the difficulty of predicting possible outcomes. It is also recognised that many people feel that there is nothing wrong with the game of tennis as it stands.

TYPICAL SERVE PARAMETERS

The speed and trajectory of a ball can be calculated in terms of measured drag and lift coefficients (Haake et al., 2000). A 57 gm, 65 mm diameter standard ball served at 200 km/hr takes 0.594 s to cross the baseline if it bounces off a clay surface, and 0.568 s if it bounces off grass. These figures are shown in Table 1, together with data for a serve speed of 180 km/hr, plus data for larger diameter balls. In Table 1, θ is the vertical angle of the ball, served down from the horizontal, so that the ball lands on the service line. It was assumed that the ball is served by a tall player from a height of 2.9 m above the centre line and is served down the centre of the court to land on the service line. It was also assumed that clay has a coefficient of friction $\mu = 0.7$, and that $\mu = 0.4$ for grass.

Table 1 Transit times

D (mm)	V (km/hr)	θ	Surface	T (sec)
65	200	6.92°	Clay	0.594
65	200	6.92°	Grass	0.568
65	180	6.43°	Grass	0.631
69	200	6.85°	Grass	0.586
72	200	6.80°	Grass	0.600

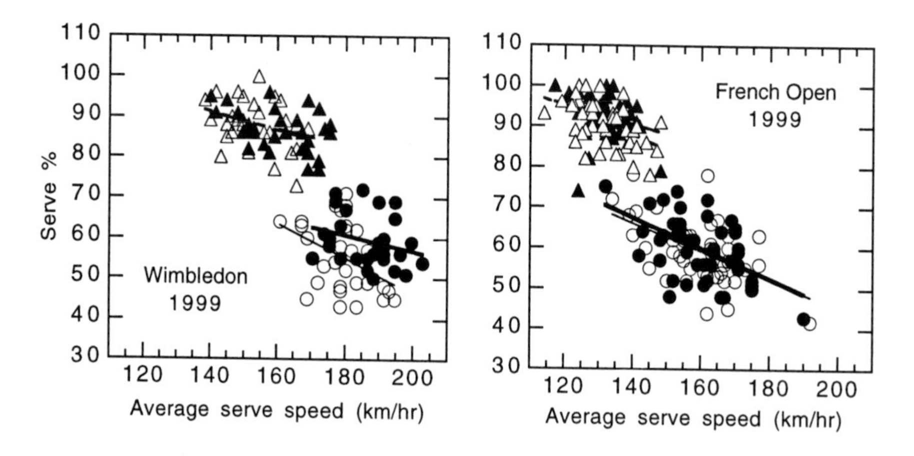

Fig. 1. Wimbledon and French Open 1999 men's singles serve % statistics. The serve % is the % of serves that are good (ie not a fault). Solid circles = 1st serve (winners), Solid triangles = 2nd serve (winners), Open circles = 1st serve (losers), Open triangles = 2nd serve (losers).

The transit time on grass is 0.063 s longer at the lower serve speed, giving the receiver a significantly better chance of returning the ball. If the receiver is running at 6 m/s to reach the ball, he can cover an extra 38 cm in 0.063 s. The effect of this extra 0.063 s is clearly illustrated by the statistical data shown in Figs. 1--3 for the 1999 Wimbledon Open men's singles matches. Figures 1-3 also show data for the 1999 French Open, where matches are played on clay, at a much lower pace. Part of the difference can be attributed to the speed of the surface, which is determined by μ. A 200 km/hr serve on clay takes 0.026 s longer to cross the receiver's baseline. In effect, a 200 km/hr serve on clay is equivalent to a serve speed of about 190 km/hr on grass, in terms of the transit time. However, the main difference between the two surfaces is the rebound angle of the ball. The ball kicks up at a steeper angle on clay, partly because of the larger reduction in the horizontal speed of the ball when it bounces and partly because the coefficient of restitution on clay is larger, meaning that the vertical component of the rebound speed is larger. Most players respond by reducing their serve speed substantially (by about 25 km/hr) at the French Open, in order to apply more spin. A ball with topspin and with a reduced horizontal speed strikes the court at a steeper angle and at a higher vertical speed so it kicks up at an even steeper angle.

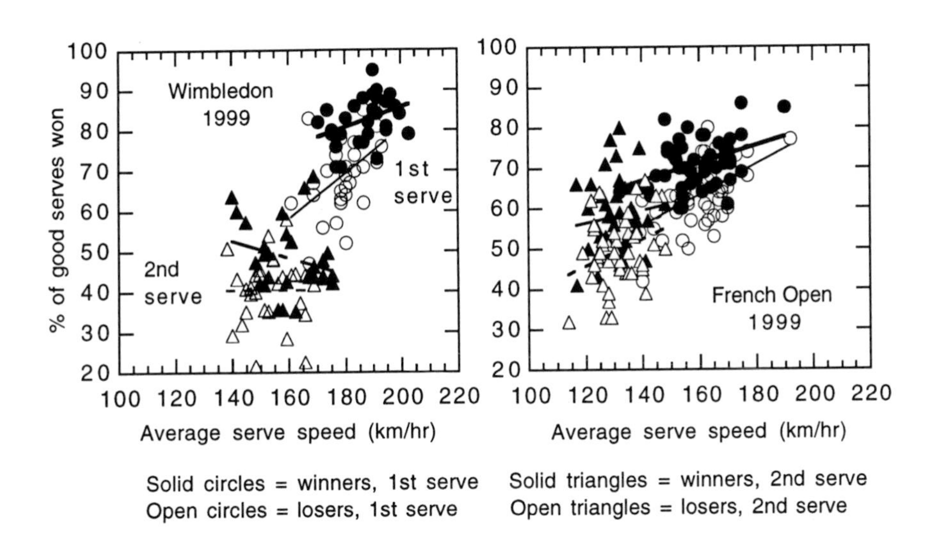

Fig. 2. % of good serves won by the server vs serve speed.

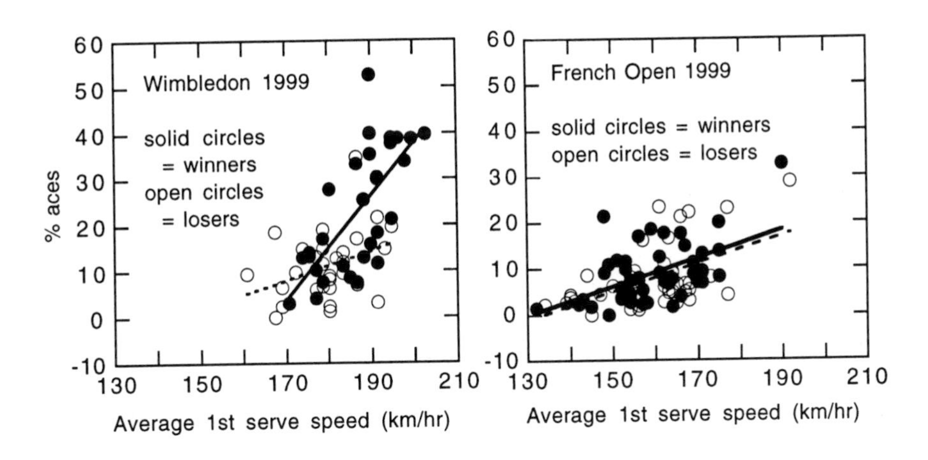

Fig. 3. % of good 1st serves won by aces vs serve speed.

The Wimbledon data in figs. 1-3 shows that the probability of winning a 1st serve increases as the serve speed increases, provided the serve is in. However, the probability that a 1st serve is good decreases as the serve speed increases. The net result, considering all serves, is that players normally win a greater fraction of their 2nd serves than their 1st serves. Taking an average over all players and all serves, regardless of whether the serve was in or a fault, players who won their matches at Wimbledon won 48% of all their 1st serves and 56% of all their second serves. Players who lost their matches won 38% of all their 1st serves and 46% of all their 2nd serves. From this point of view, a reduction in the first serve speed should not be a serious imposition for most players and might even increase the chances of a player winning his 1st serve, depending on the 1st serve %. For example, a larger ball will cause the transit time to increase but it will also increase the 1st serve % due to the larger vertical acceptance angle for good serves. If the net height is raised or if the service court is shortened, the server will be forced to reduce his serve speed in order to maintain the same serve %. Figure 2 shows that the speed of the 2nd serve does not have a significant effect on the chance of winning the second serve, so a reduction in the speed of the 2nd serve should have a relatively small effect.

METHOD 1. CHANGE THE SURFACE

Replacing grass with a slow surface will clearly solve the problem. However, the grass courts at Wimbledon are the heart and soul of "lawn" tennis. This is one place where tradition will clearly take precedence over technology. Some indoor tournaments use a fast carpet for the surface. These should be replaced by materials which slow down the ball on the bounce. Devices which measure the pace (speed) of a court exist and they can be used to evaluate surfaces.

METHOD 2. LIMIT THE RACKET

One of the favorite suggestions, particularly among older players, is to go back to rackets that are made of wood. A related suggestion is that the racket power be limited. Exactly how this is to be done is never specified, and at present, the ITF is investigating whether a practical method of limiting power is possible. Even if a power limit is imposed, it will not produce the results people hope for, as will be explained in the section on tennis balls. Furthermore, a heavy racket may give more power in the hands of a strong player, but it may reduce the effective power of a younger person who may do better with a light racket. Effective racket power is therefore a function of racket speed which depends on both the racket mass and the strength of the player. A change back to wood rackets would mean an increase in weight, and a decrease in racket stiffness. This could mean an increase in racket power for impacts near the centre of the strings, a decrease in power elsewhere due to energy loss in frame vibrations and a decrease in ball control due to the smaller head size.

A major downside to limiting the racket concerns the recreational player. There is no doubt that today's oversize, light, very stiff, graphite reinforced, space age rackets make the game a lot easier to learn and a lot more fun for the recreational player. Unless two sets of rules were adopted (professional rules and amateur rules for example), the recreational players would suffer with a mandated reduction in racket performance.

METHOD 3. ELIMINATE THE SECOND SERVE

It has been proposed that only one serve be allowed per point. This idea might work for an average server, but for the top men on grass, it could prove a disaster. Statistically, more serves are won by the server on the second serve, the first serve often being a "free" point if it happens to go in. Some of the top players can get in 65% of their first serves and may win up to 85% of the points when that serve goes in, giving a 55% overall win rate. If the 2nd serve win rate is also about 55%, it may prove beneficial for that player to hit first serves only as opposed to second serves only. The result would be 35% of the points on serve would end in a fault and the others would be a one or two hit rally. This could really kill tennis as a spectator sport.

METHOD 4. CHANGE THE FOOTFAULT RULE

A number of years ago the serve foot fault rule was changed. At that time, foot contact with the court surface had to be maintained until after the ball was struck. The present rule allows the server to leave the ground and actually be well into the court beyond the baseline as long as contact with the baseline or court is not made until after ball contact. This allows the server to impact the ball about 15 cm higher and up to 60 cm into the court. It also allows the server to get to net quicker.

For a player originally hitting a 190 km/hr serve at 2.4 m above the ground, the additional 15 cm adds 40% to the acceptance window. For a player hitting the same serve originally 2.75 m above the ground the extra 15 cm adds only 20% to the window. Going back to the old foot fault rule may therefore harm the smaller player more than the tall player. The ability to strike the ball on the serve well inside the baseline also increases the acceptance window, but more important, it gives the receiver less time to return the ball. This makes the serve a much more effective weapon. Reverting to the old rule would work well to decrease the effectiveness of the serve and also the serve and volley strategy, but the effect on the shorter or more athletic player might more than compensate for this.

METHOD 5. CHANGE THE BALL

One of the most popular suggestions for slowing down the game is to make the ball less lively, by taking some air out of the ball or reducing the wall stiffness. Changing the weight of the ball also leads to interesting consequences. If the ball is made lighter, it will come off of the racket at a higher speed, but slow down more in getting to the receiver due to air resistance. It the ball is made heavier, it will come off of the racket at a lower speed, but slow down less due to air resistance. For a small change in weight, the two effects tend to cancel each other out and there is no appreciable change in the time the ball takes to reach the receiver. Intuitively, one might expect that a light ball might be slowed more by friction during the bounce, but there is no difference between heavy and light balls in this respect.

The effects of changing the ball mass or its coefficient of restitution (COR), on the serve speed of the ball, can be estimated by modelling the ball as a mass m_1 connected to a spring of spring constant k_1, and modelling the racket as a mass m_2 connected to a spring (ie the strings) of spring constant k_2. An impact of the ball on concrete, or on a hard court, can be modelled with $k_1 = 4 \times 10^4$ Nm^{-1} and a COR $e_1 = 0.75$. If the ball impacts on the strings of a head-clamped racket, and if there is no energy dissipation in the strings, then the COR increases to a value e given by (Cross, 2000)

$$e^2 = \left(k_1 + k_2 e_1^2\right)/\left(k_1 + k_2\right)$$

For example, if $k_2 = 4 \times 10^4$ Nm^{-1} then $e = 0.88$. The COR is unaltered if the head is not clamped, provided the ball impacts at the vibration node in the centre of the strings so that no energy is lost to frame vibrations. In this case, a ball incident at speed v_1 on a stationary racket will rebound at speed $v_2 = e_A v_1$ where

$$e_A = (em_2 - m_1)/(m_1 + m_2)$$

is the apparent coefficient of restitution, typically about 0.4 for an impact near the centre of the strings. This expression for e_A is easily derived from conservation of momentum. Alternatively, if the racket is incident at speed v_R on a stationary ball, then the ball will be served at speed

$$v_{out} = (1 + e_A)v_R$$

The effective mass of the racket at the centre of the strings is typically about three times larger than the mass of the ball. If m_1 = 57 gm and m_2 = 171 gm, then e_A = 0.410 and v_{out} / v_R = 1.410. If m_1 is increased by 10% to 63 gm, then e_A = 0.374 and v_{out} / v_R = 1.374 . For example, if v_R = 40 ms^{-1} then v_{out} decreases by 2.5% from 56.4 to 55.0 ms^{-1} when m_1 is increased from 57 to 63 gm.

A similar small change in serve speed results when the COR of the ball is reduced. For example, if e_1 is reduced by 10% from 0.75 to 0.675 (for an impact on concrete), then *e* decreases from 0.88 to 0.85 (for an impact on the strings). For a 57 gm ball, e_A then decreases from 0.410 to 0.388, corresponding to a 1.6% reduction in serve speed. The ball does not normally bounce very high on grass, and would rebound even less off the grass if the COR is reduced. Lowering the COR would actually speed up the court since the decrease in the horizontal ball speed, $\Delta v_x = \mu(1 + e_1)v_y$ where v_y is the incident ball speed in the vertical direction.

Placing some high friction material (such as rubber fibers) in the ball cover might slow the ball down on the bounce but it would play havoc with the ground strokes. If the ball cover "grabbed" the turf, it would also grab the strings. We would be back to the equivalent of the spaghetti racket, capable of imparting huge spins to the ball, and changing the very nature of the game of tennis.

The proposal that seems to make the most sense is to enlarge the ball somewhat. This will not only give the receiver more time to react and additional time to return a serve, but many of the other consequences of a larger ball may prove beneficial to the game of tennis. When the diameter of the ball is increased, the cross-sectional area of the ball (the parameter that determines the ball's air resistance) increases as the square of the diameter increase. A 6.5% increase in ball diameter will therefore increase the air resistance by 13%. This means that the larger ball will slow down in flight more than a standard size ball. With a larger ball, the extra time the receiver has to return a serve on a grass court is comparable to the extra time the receiver has if the ball bounces on clay (see Table 1). Using larger balls ONLY for the top level men's game on very fast courts (such as grass) would probably be the change that would disturb the tradition of tennis least, yet it may have the desired results. In addition, recreational players might enjoy tennis more with a larger ball and television viewers would find the larger ball easier to see. At high altitudes, where the drag force is reduced by the low density air, a larger ball would be ideal.

It is difficult to predict with certainty the effect of using a larger ball on grass. A 69 mm ball introduces a delay of 0.018 s for a 200 km/hr serve, which is equivalent to serving a standard ball at 194 km/hr. A 72 mm ball introduces a 0.032 s delay, which is equivalent to a standard ball served at 189 km/hr. Figures 1-3 indicate that on grass, this will make a relatively minor difference to the fraction of unplayable serves. As shown in Table 1, the transit time for a 200 km/hr serve on clay, for a standard ball, is 0.026 s longer than on grass. The larger ball will introduce a comparable delay, but one cannot assume that the same delay on grass will have the same effect as on clay.

METHOD 6. SHORTEN THE SERVICE COURT

If the service court is shortened, it would force the big servers to reduce their serve speed in order to get in a reasonable fraction of their offerings. In that respect, the idea works. However, there are several possible drawbacks. First, it might inflict more harm on short players than on tall players. The tall player already has an advantage over the short player, particularly on the serve. This proposal might increase that advantage. In addition, all courts would have to be modified unless it was only done for the fast courts. In that case, players would have to modify their serve just for the fast courts, which is something they might be unhappy about. Nevertheless, they are happy to reduce their serve speed on clay, and to reduce their second serve speed. A related problem is whether it would be necessary to have two sets of service lines, or an interchangeable set, for women's singles or mixed doubles matches.

Calculations of ball trajectories show that if the service court is reduced by 28 cm, then a 1.93 m (6' 4") player would need to reduce his serve speed from 193 km/hr to 175 km/hr in order to maintain the same serve angle window and hence the same percentage of successful first serves. A 1.75 m (5' 9") player needs to serve at 155 km/hr on a normal size court to maintain the same serve percentage as the tall player. If the service court is shortened by 28 cm, the shorter player could maintain the same serve percentage by reducing his speed to 145 km/hr. Given that the necessary reduction in serve speed is less for the shorter player, and that the taller player will hit fewer unplayable serves, short players might even prefer a shorter service court. Figure 1 indicates that the percentage of serves won is almost independent of serve speed when the serve speed is low, especially on the second serve. A short service court should at least be given an experimental trial to gauge the effects.

METHOD 7. RAISE THE NET

Raising the net height will also force players to reduce their serve speed in order to maintain a reasonable serve percentage. Each 1.0 cm increase in net height is equivalent to a reduction in the length of the service court by 7.6 cm, regardless of the initial ball speed or the initial serve height. The disadvantage is that it reduces the window for all shots, not just the serve. However, most groundstrokes normally clear the net by 5 cm or more and will remain unaffected. From a practical point of view, this proposal is the easiest to implement, and should also be subject to experimental trials.

OTHER PROPOSALS

The server could be required to be well behind the baseline when serving. If the server stands 1 m behind the service line, then the transit time to the receiver's baseline is increased by 0.02 s at a serve speed of 180 km/hr (50 ms^{-1}). This would also force players to reduce the serve speed to maintain the same serve percentage, giving the receiver enough time to return most serves because of the extra distance as well as the reduced serve speed. However, the short player may be further disadvantaged and it could mean the end of the serve and volley game.

An interesting proposal has been made to narrow the service box by splitting the center line. This will reduce the number of aces and also allow the receiver to back up a bit and not worry about the serve being swung wide.

CONCLUSIONS

Any suggestion to change the rules of tennis is likely to be met with opposition, particularly by highly ranked players, since their game is so finely tuned. Nevertheless, it is recognized that something needs to be done to improve the professional men's game on fast grass courts, in order to make the sport more interesting for spectators and television audiences. There is no suggestion that anything needs to be done at present for even the top level women's tennis or the recreational game, and the comments in this paper are addressed only to the problem of the serve in the men's game. In that respect, it is recognized that different rules may need to be formulated for different courts and for different levels of competition. Similarly, it is recognized that the grass courts at Wimbledon are sacrosanct and an essential part of the tradition of tennis. What is required is a relatively small change that would be accepted by most players and that would slow the serve down somewhat. The various proposals outlined above need to be tried on an experimental basis and carefully analysed to determine the most satisfactory solution or combination of solutions.

REFERENCES

Cross, R. (2000) The coefficient of restitution for collisions of happy balls, unhappy balls and tennis balls, *Am. J. Phys.* in press.

Haake, S.J., Chadwick, S.G., Dignall, R.J., Goodwill, S. & Rose, P. (2000) Engineering tennis - slowing the game down, *Sports Engineering* in press.

Reaction time testing and grand slam tie-break data

S.J. Haake
Department of Mechanical Engineering, University of Sheffield, UK
P. Rose
The International Tennis Federation, London, UK
J. Kotze
Department of Manufacturing Engineering, Loughborough University, UK

ABSTRACT: Two BOLA projection devices were used to test the reactions of eight good tennis players with simulated serves up to 160 mph. It was found that the number of successful returns decreased with speed and the number of aces increased significantly at speeds over 126 mph. The percentage of tie-breaks at Wimbledon, the US Open and Roland Garros (for 1999) was compared to average first serve speeds for male players. It was found that the number of tie-breaks increased significantly at speeds above 110 mph and that the serve speed and percentage of tie-breaks was related to the surface played upon with faster surfaces having higher serve speeds. It was concluded that the ability to return a ball successfully decreased rapidly above 126 mph, that this was more likely to occur on faster surfaces and that average serves speeds above 110 mph led to an increased percentage of tie-breaks.

INTRODUCTION

The game of tennis has come under increasing pressure from critics in the last decades with claims that it is too fast and is becoming too dominated by the serve (Arthur, 1992). However, there is not much data around to clearly show that serve speeds are increasing or that the serve is indeed dominant. Suggestions have been made that serves above 120 mph tend to result in aces and that as the number of aces increases the potential for tie-breaks also increases (Brody, 1999). If the percentage of tie-breaks for any individual player is related to the speed of his serve (this is predominantly an issue for the male game) then previous scores can be looked at to determine whether serve speeds have increased over time.

This paper aims to determine the serve speed at which players can no longer return the ball, to determine the serve speeds at tournaments and to relate this to the percentage of tie breaks.

REACTION TIME TESTING

Two BOLA ball projection devices, originally created for cricket, have been modified to allow a tennis ball to be served at up to 160 mph (288 kph). The projection machines are capable of imparting spin to the ball in excess of 5000 rpm (Pete Sampras has been observed to serve with spins of around 3000 rpm; Pallis, 1997)

The projection machines were mounted on scaffolding on the service line of an acrylic tennis court. Balls were projected at eight players at the other end of the court with tennis skills ranging from good to semi-professional and the speed of serve slowly increased up to 160 mph and with ten repetitions at each speed. The ball speed was recorded using a Stalker radar gun and the number of good returns and the number of aces were recorded for each player.

Initial results are shown in Fig.1. It can be seen that the number of returns in court decreases with speed. The number of aces starts to increase as the speed of the serve reaches 100 mph

It can clearly be seen that serve speed has an effect on the ability to return the ball and that this starts to take effect at about 100 mph. It is likely that this speed threshold would increase as the player's ability increased.

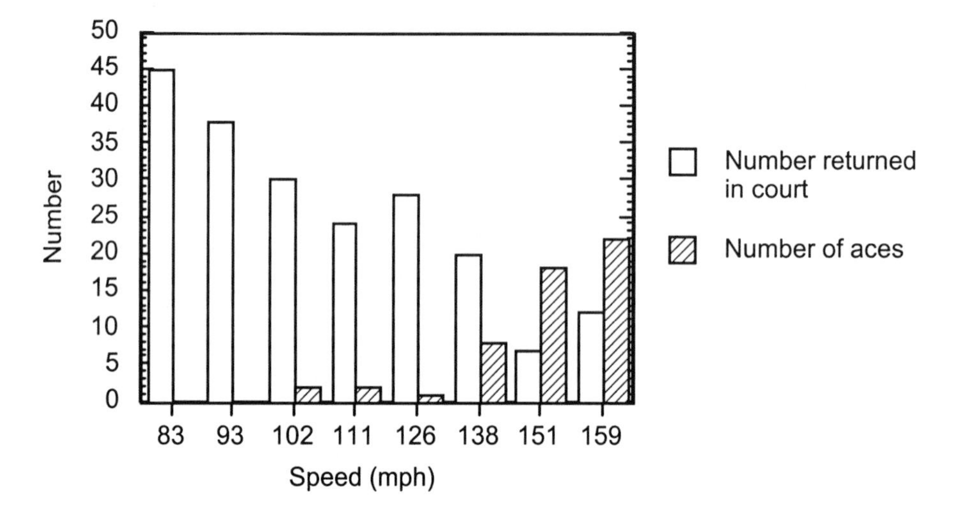

Fig. 1. Number of service returns in court and number of aces vs serve speed using a ball projection machine.

GRAND SLAM TIE-BREAK DATA

SELECTION CRITERIA

Scores were analysed and tie-break data for men for the last 10 years studied for all players at Wimbledon, Roland Garros and the US Open. Sets ending in a 6-6 scoreline were considered to have reached a tie-break. Serve speeds were determined for the same fourteen players from the official tournament internet sites for 1999.

Players were only considered if they had played at all three tournaments and if the total numbers of sets played over the previous 10 years was in excess of 50 (giving approximately a 2% resolution in the number of tie-breaks played).

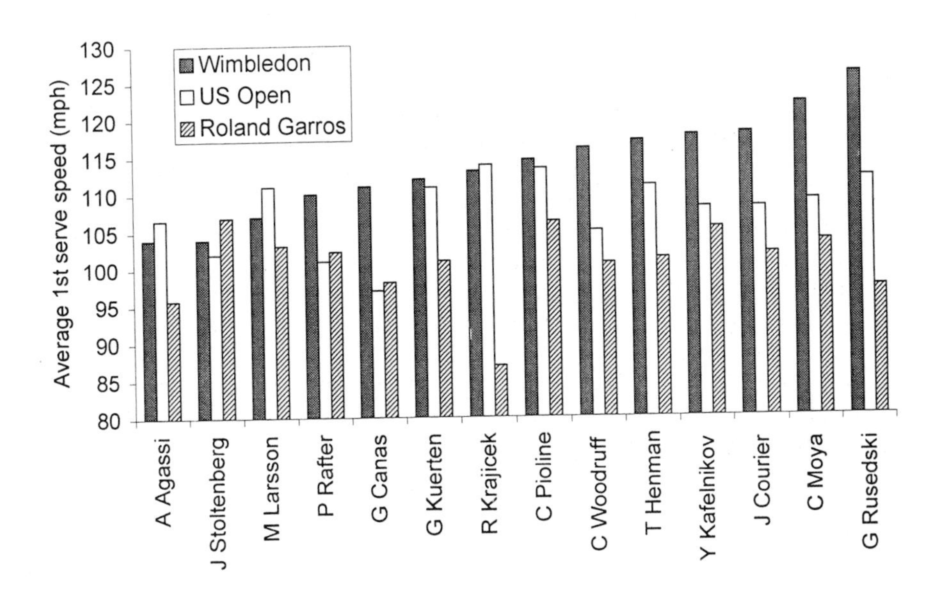

Fig. 2. Average 1st serve speeds for 14 players at Wimbledon, Roland Garros and the US Open for 1999 (in mph).

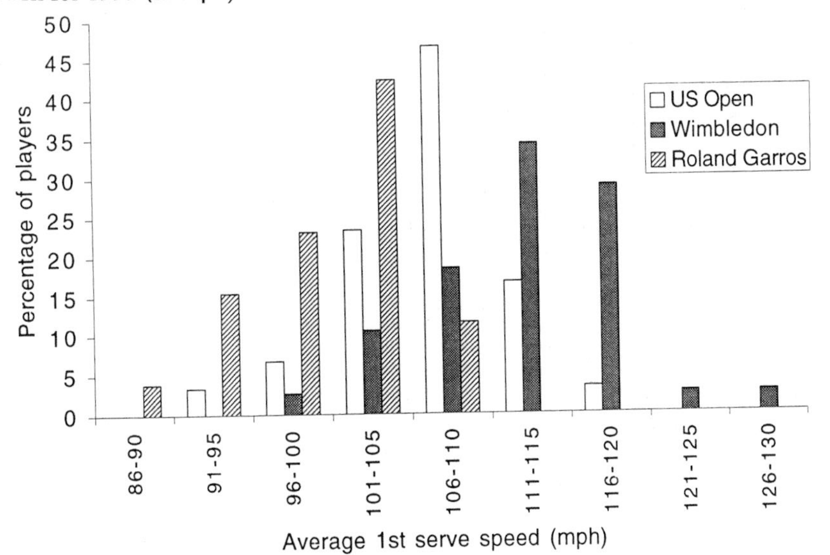

Fig. 3. Histogram of average 1st serve speeds for 14 players at Wimbledon, Roland Garros and the US Open for 1999 (in mph)

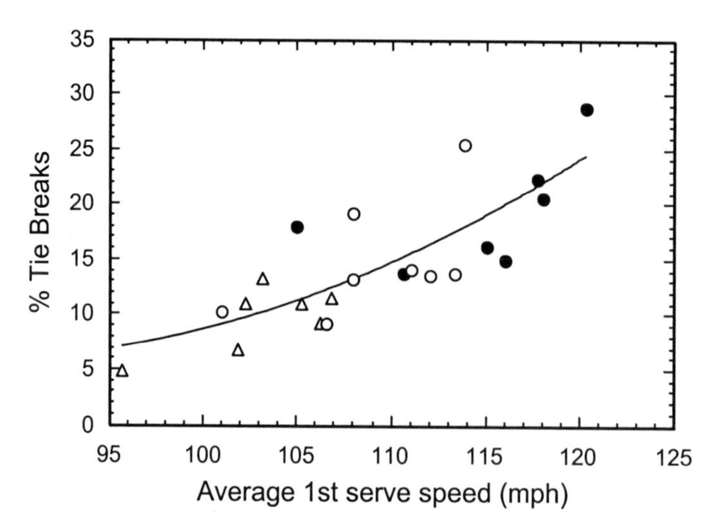

Fig. 4. Percentage of sets ending in a tie-break at Roland Garros (triangles), Wimbledon (filled circles) and the US Open (open circles) vs. average 1st serve speed for a sample of 14 male players.

RESULTS

Prior to analysis, the average 1st serve speeds for each player were analysed (Figs. 2 and 3). The average and standard deviations are shown in Table 1. It can be seen that the average serve speed for all players was fastest at Wimbledon (114 ± 6 mph) followed by the US Open (108 ± 5 mph) and then Roland Garros (101 ± 5 mph). Clearly players have different strategies on different surfaces possibly attempting straight aces on faster surfaces such as Wimbledon while employing spin on slower surfaces such as Roland Garros.

Figure 4 shows the comparison of percentage tie-breaks with the average 1st serve speed for the fourteen players listed in Table 1. It can be seen that the players serve fastest at Wimbledon, then at the US Open and finally at Roland Garros. Corresponding to this, players have the highest percentage of tie-breaks at Wimbledon, then at the US Open then at Roland Garros. It is apparent that the percentage of tie-breaks has a direct relationship with the speed of the serve and the speed of the surface.

Interestingly it can be seen from Fig. 4 that data from all the tournaments follows roughly the same trend with players with faster serves experiencing a higher percentage of tie-breaks. This is probably due to at least two factors; players with a good serve are likely to progress further in the tournament thus meeting players of a similar standard; a match where the serve was dominant would cause games to be won in turn leading to a tie-break. Although there is no historical record of serve speeds, scores exist for all games. Thus percentage tie-breaks can be monitored as an indicator of the dominance of serve speed on the game.

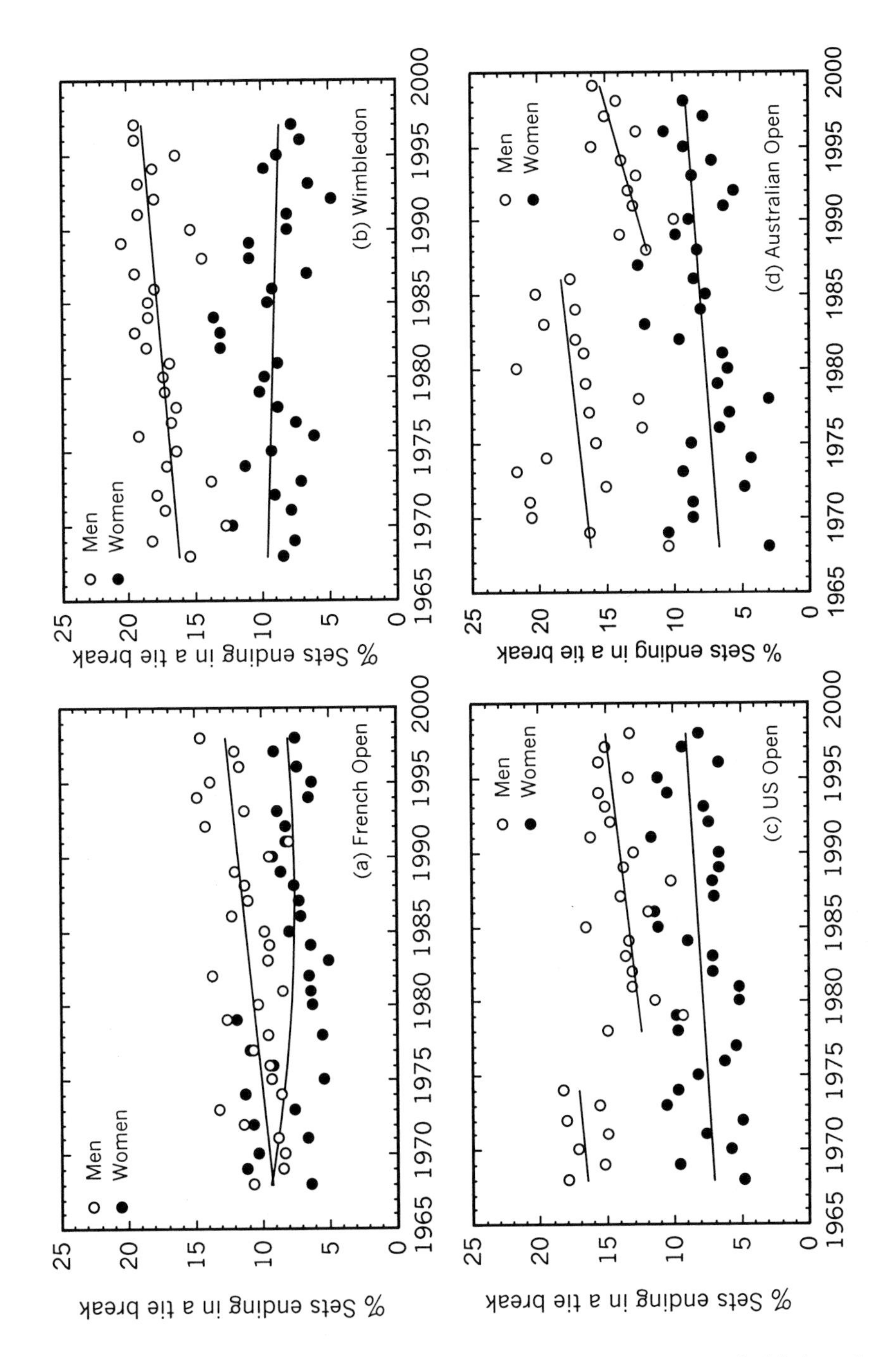

Fig. 5. Percentage of sets ending in a tie-break at Roland Garros, Wimbledon, the US Open and the Australian Open since 1968.

Table 1. Average 1st serve speeds at Wimbledon, Roland Garros and the US Open for 1999 (in mph)

Player	Wimbledon	US Open	Roland Garros
Andre AGASSI	104	107	96
Jason STOLTENBERG	104	102	107
Magnus LARSSON	107	111	103
Patrick RAFTER	110	101	102
Guillermo CANAS	111	97	98
Gustavo KUERTEN	112	111	101
Richard KRAJICEK	113	114	87
Cedric PIOLINE	115	113	106
Chris WOODRUFF	116	105	101
Tim HENMAN	117	111	101
Yevgeny KAFELNIKOV	118	108	105
Jim COURIER	118	108	102
Carlos MOYA	122	109	104
Greg RUSEDSKI	126	112	97
Average	**114**	**108**	**101**
St Dev	**6**	**5**	**5**

DISCUSSION

Figure 5 shows the percentage of sets ending in a tie-break at the French Open, Wimbledon, the US Open and the Australian Open for both men and women since 1965. It can be seen that, at all four grand slam tournaments, the men's game has a higher percentage of tie-breaks than the women's game. For the men's game, it appears that the percentage of tie-breaks is increasing with time at all tournaments which in turn implies that the serve speed is increasing and becoming more dominant. The data for the women's game is more complex and it is possible that serve speeds are not high enough to be a significant cause of tie-breaks.

One issue for the ITF Technical Commission is whether equipment and technology changes have been responsible for changes over time. There is no evidence of a step jump in percentage tie-breaks during the 1980's after the introduction of composite rackets. Thus, lighter rackets are not the sole reason for the increase in serve speed although they may help produce an increase in serve speed through change in technique. Whatever the case, the data indicates that, should the game proceed as it is, then by the middle of this century, the average 1st serve speed at Wimbledon will be 120 mph and a quarter of all sets will end in a tie-break.

An issue that may become apparent as ITF performance standards for surfaces come into place is that of injuries attributed to different surfaces. The causes of tennis elbow are not clear but are generally attributed to peak forces sustained by the arm during play causing injuries to the tendon attaching the extensor muscles to the arm. If this is the case then playing on surfaces such as clay should be much easier on the arm than on surfaces such as hard courts or grass. The serve speed is 13% faster on the grass at Wimbledon than on the clay at Roland Garros while the US

Open appears to be 7% faster than Roland Garros. The implications for arm injuries could be quite significant if peak force is the dominant cause of tennis elbow.

CONCLUSIONS

To conclude, initial testing shows that players have a serve speed threshold at which the number of their successful service returns decreases and at which the number of aces against them increases. It appears that this threshold is somewhere above 100 mph. The serve speed is related to the surface played upon with grass having the fastest serves and clay the slowest. This has implications for arm injuries if it is considered that peak forces are a dominant cause of tennis elbow.

The percentage of tie-breaks experienced by players was proportional both to their average 1st serve speed and the surface on which they played; first serve speeds were 13% faster at Wimbledon than at Roland Garros with the US open somewhere in between.

ACKNOWLEDGEMENTS

The authors would like to thank the International Tennis Federation for all its help and support during this project.

REFERENCES

Arthur, C. (1992) Anyone for slower tennis? *New Scientist*, May, pp. 24-28

Brody, H. (1990) Tennis Pro, March/April, pp. 91.

Pallis, J. M. (1997) Tennis Sport Science, http://wings.ucdavis.edu/Tennis/, NASA's Learning Technologies Project and Cislunar Aerospace, Inc.

How to reduce the service dominance in tennis? Empirical results from four years at Wimbledon

F.J.G.M. Klaassen
Department of Economics, University of Amsterdam, Amsterdam, NL
J.R. Magnus
Center for Economic Research, Tilburg University, Tilburg, NL

ABSTRACT: This paper provides empirical results that will help the discussion on how to reduce the service dominance. Our extensive Wimbledon data set shows that the softer balls at the 1995 Wimbledon championships did not reduce the service dominance. Abolishing the second service, however, will reduce the dominance of service and make the server and receiver more equal. In addition, its implementation is easy and the data show that it will make matches more even. This will make tennis more attractive for spectators. In this sense, allowing for only one service is a good idea.

INTRODUCTION

The service is an important aspect in the game of tennis. In the past its main purpose was only to start a rally. However, nowadays the service is also an important way to win the rally, either directly through an ace, or indirectly through the advantage in the rally following a great serve. Hence, the dominance of the service has increased substantially.

It is often argued that the service dominance is too high, particularly for men's tennis on fast grass courts such as at Wimbledon. This would have a negative effect on the attraction of tennis for spectators. Hence, many proposals have been made to reduce the dominance of service. This paper contributes to this discussion by providing empirical estimates relating to the dominance of service. These are based on an extensive data set from Wimbledon 1992-1995.

Before we can discuss the reduction of service dominance, we need a measure for it. This is the first topic of the paper. We argue that one should use the probability of winning a point on service, not the probability of hitting the first (or second) service in, nor the probability of winning a point on service given that the first (second) service is in. Ironically, television broadcasts inform us about the latter two probabilities, but seldom about the probability of winning a point on service, the better measure.

Next, we discuss how to reduce the service dominance. Such a reduction would make tennis more attractive for spectators, it is hoped. One could make the net of a tennis court higher or the service court smaller. One could also use bigger or softer balls. Softer balls were used in the 1995 Wimbledon championships. However, we find that the use of softer balls had hardly any effect.

Another measure to reduce the service dominance is to abolish the second service, or, in other words, to allow for one service only. This measure is easy to implement and does, of course, lower the service dominance. From the data we can show how large this reduction will be. We find that the probability of winning a point on service will become close to 50%, so that server and receiver have about equal probability to win the point. Moreover, the quality difference between the top players and the others will decrease. These results hold for both men and women. Hence, matches will get more even and this may well make tennis more attractive for spectators.

All results in this paper are based on data from Wimbledon 1992-1995. It is an extensive point-to-point data set on 481 singles matches, resulting in 88,883 observations. In this respect our work differs from existing statistical papers on tennis. Most papers are theoretical and contain no data. If data are available, they are either point-to-point data of one match, as in Croucher (1995), or end-of-match results (6-4/6-3/6-3, say) of several matches (as in Jackson and Mosursky (1997)).

The existing statistical literature on tennis addresses a number of interesting issues relating to the service. Regarding the first/second service strategy, Gillman (1985) concluded that "missing more serves may win more points"; see also Gale (1971), George (1973), Hannan (1976), Norman (1985) and Borghans (1995).

A second category of papers deals with the computation of the probability of winning a game, set, tie-break or match. Such probabilities can be computed if one assumes that each player's probability of winning a point on service is constant during the match (see Hsi and Burych (1971), Kemeny and Snell (1976), and Pollard (1983), among others).

The assumption that the probability of winning a point on service is constant during the match is questionable. Klaassen and Magnus (1999) test this assumption by analysing whether points are independent and identically distributed and reject it. However, more surprisingly given the large data set that is used, the deviations from constancy are not very large. Evidence of dependence at set level is provided by Jackson and Mosurski (1997).

Another category of papers concerns the testing of tennis hypotheses. For example, are new balls an advantage to the server? These and many other hypotheses are considered in Magnus and Klaassen (1998a,b and 1999a,b,c).

In the next section of this paper we briefly describe the Wimbledon data. Then we propose a measure for the dominance of service. The section after that discusses the effectiveness of the use of softer balls and the abolishment of the second service in order to reduce the service dominance. The final section summarizes the results.

WIMBLEDON DATA

We have a data set consisting of 258 matches played in the men's singles and 223 in the women's singles championships at Wimbledon from 1992 to 1995 (the generality of our conclusions may thus be restricted by the fact that the data concern only Wimbledon, a tournament played on fast grass courts). For each match we have detailed information at point level. For instance, we know the winner of the point, whether the first service was in and, if not, whether the second service was in. This leads to a data set of 59,466 points for the men and 29,417 for the women (the reason why we have many more points for the men is that they play for three sets won and the women for two). See Magnus and Klaassen (1999b) for a more extensive description of the data.

The data set accounts for almost one half of all singles matches played during Wimbledon 1992-1995. The matches in our data set are played on one of the five 'show courts': Centre Court and Courts 1, 2, 13 and 14. Usually matches involving top-players are scheduled on the show courts. This leads to an under-representation in the data set of matches with lower-ranked players. To account for this selection problem, we weigh the matches when we compute statistics; see Magnus and Klaassen (1999a) for further details.

MEASURING THE SERVICE DOMINANCE

In this section we discuss how to measure the dominance of service and we compare the measures for men and women. This is necessary for the discussion on the reduction of the service dominance in the next section.

Table 1 provides some characteristics of the service. As in the rest of this paper, we present standard errors in parentheses. To obtain the standard errors, we have treated all points as independent. This is not quite true (see Klaassen and Magnus (1999)), but it is sufficient as a first-order approximation for the purpose of this paper.

Table 1 Service characteristics

Percentage of ...	Men's singles	Women's singles
Points won on service	64.4	56.1
	(0.2)	(0.3)
Points won on 1st service	43.6	37.8
	(0.2)	(0.3)
Points won on 2nd service	51.4	46.6
	(0.3)	(0.5)
1st services in	59.4	60.8
	(0.2)	(0.3)
2nd services in	86.4	86.0
	(0.2)	(0.3)
Points won if 1st service in	73.3	62.2
	(0.2)	(0.4)
Points won if 2nd service in	59.4	54.1
	(0.3)	(0.5)

For the remainder of this paper it is important to get a clear understanding of the relations between the service characteristics. The percentage of points won on service is a combination of the percentage of points won on 1st and on 2nd service. The second part, however, becomes only relevant when the first serve is fault. This leads to:

$$\text{\%points won on service} = \text{\%points won on 1st service} + (\text{\%1st service fault}) \times (\text{\%points won on 2nd service}). \quad (1)$$

For example, table 1 shows that 64.4% = 43.6% + (1-59.4%)×51.4% for the men.

The percentage of points won on 1st service depends on two elements, namely the percentage of 1st service in and the percentage of points won if the 1st service is in:

$$\text{\%points won on 1st service} = (\text{\%1st services in}) \times (\text{\%points won if 1st service in}). \quad (2)$$

For example, table 1 shows that in the men's singles 59.4% of the 1st services is in, and, if the 1st service is in, the men win the point in 73.3% of the time. Hence, the probability of winning a point on the 1st service is 59.4% × 73.3% = 43.6%. A similar formula holds for the 2nd service.

Which statistic in table 1 should we use to measure service dominance? Service dominance is a combination of the dominance of the 1st service and the dominance of the 2nd service. Hence, let us first discuss how to measure the 1st service dominance? Clearly, one should not use the percentage of 1st services in, since this says nothing about the difficulty of the 1st service. Also the percentage of points won if the 1st service in is not appropriate, because this statistic tells us nothing about how often the 1st service is in. The best measure for the dominance of the 1st service is a combination of the two, as given by relation (2). Hence, we measure the 1st service dominance by the percentage of points won on 1st service. Similarly, our measure for the 2nd service dominance is the percentage of points won on 2nd service. Ironically, television broadcasts inform us about the percentage of 1st (or 2nd) services in and sometimes about the percentage of points won if the 1st (or 2nd) service is in, but seldom about the percentage of points won on 1st (or 2nd) service, the most appropriate measure for the 1st (or 2nd) service dominance.

We can now derive the measure for the (total) service dominance by combining the measures for 1st and 2nd service dominance. We use formula (1) for that. Hence, we propose to use the percentage of points won on service to quantify service dominance.

As table 1 shows, we find that the measure for service dominance is 64.4% for the men and 56.1% for the women. Hence, the difference is 8.3%-points. Since the standard error of this difference is 0.4%, the service dominance is significantly larger in the men's singles than the women's singles, as expected (in this paper 'significant' means that the estimate is more than 2 standard errors away from its target). This makes the men's singles a very different game from the women's singles.

HOW TO REDUCE THE SERVICE DOMINANCE?

It is often argued that the service dominance is too high, particularly for men's tennis on fast grass courts such as at Wimbledon. This would lower the attractiveness of tennis for spectators. Hence, many proposals have been made to reduce the dominance of service: making the net higher or the service court smaller, using bigger or softer balls, abolishing the second service. The first subsection discusses the use of softer balls. Then we discuss the abolishment of the second service.

SOFTER BALLS

This measure was implemented at the 1995 Wimbledon championships. The question is whether this has reduced the service dominance. Before we can address this question, we need to know something about the weather, since the weather may also affect the service dominance. The Wimbledon weather has been documented by Little (1995). Because the weather has not been very different in the four years of our observations, the effect of the weather on comparisons between 1995 and the three years before seems negligible.

In the previous section we have explained how to measure the service dominance. In terms of probabilities, this measure is Pr(point won on service). One could estimate this probability for 1995 and compare the estimate with the three estimates based on 1992-1994. However, there may be a downward sloping trend in the probabilities over the years. If one does not correct for that, one may well find that the probability for 1995 is the lowest, even in case of no effect of the softer balls in 1995. Since we are interested in the additional effect of the softer balls, we correct for the time trend in our analysis. More specifically, we use a simple logit model in which the systematic part is a linear function of the year of tournament (year= 92,93,94 or 95) and a dummy for 1995:

$$\Pr(\text{point won on service}) = \Lambda\,(\beta_0 + \beta_1 \times \text{year} + \beta_2 \times \text{dummy95}), \tag{3}$$

where Λ is the logistic distribution function, $\Lambda(x) = \exp(x) / (1+\exp(x))$.

Table 2 Service characteristics depending on the year of the tournament

Probability of ...	Men's singles		Women's singles	
	year	dummy 95	year	Dummy 95
Point won on service	-0.022	0.022	-0.033*	0.030
	(0.012)	(0.031)	(0.016)	(0.042)
Point won on 1st service	-0.033*	0.000	-0.017	-0.043
	(0.011)	(0.030)	(0.017)	(0.043)
Point won on 2nd service	-0.051*	0.045	-0.019	0.049
	(0.018)	(0.046)	(0.027)	(0.068)
1st service in	-0.066*	-0.004	0.021	-0.117*
	(0.011)	(0.030)	(0.017)	(0.043)
2nd service in	-0.125*	0.154*	-0.020	0.013
	(0.026)	(0.067)	(0.038)	(0.097)
Point won if 1st service in	0.029	0.013	-0.049*	0.048
	(0.016)	(0.045)	(0.021)	(0.054)
Point won if 2nd service in	-0.020	0.003	-0.016	0.053
	(0.019)	(0.051)	(0.029)	(0.073)

The top row in table 2 presents the maximum likelihood estimates for β_1 and β_2, along with the standard errors; an asterisk denotes significance at the 5% level. There is no evidence that the softer balls used in 1995 had a negative effect on the service dominance.

This, however, does not mean that the service dominance was equal to the dominance in the years before. From the estimated negative effect of the time trend variable, we observe a gradual decrease in the service dominance over time (significant for the women and almost significant for the men).

Next, we analyse the reasons for this gradual decline. In other words, we examine whether and how the way servers win their points has evolved from 1992 to 1995. Combining relations (1) and (2), we can split up the probability of winning a point on service as

$$\begin{aligned}&\Pr(\text{point won on service}) \\ &\quad = \Pr(\text{point won if 1st service in}) \times \Pr(\text{1st service in}) + \\ &\quad\quad \{1-\Pr(\text{1st service in})\} \times \Pr(\text{point won if 2nd service in}) \times \Pr(\text{2nd service in'}).\end{aligned} \tag{4}$$

For each probability we estimate a similar logit model as in (3). Table 2 contains the results.

For the women the reason for the gradual decline in the service dominance is the decrease in the probability of winning a point on service if the first service is in. This may be caused by an improvement in the return of service by professional players, as is sometimes claimed.

The improvement-of-return hypothesis is also supported by the results for the men in table 2. The significantly negative time trend in both the probability of first service in and second service in shows that the men seem to take more risk on their first and second services. Are they pushed to hit more difficult services, because of the better returns? Nevertheless, the services are still not difficult enough to increase the probability of winning a point on service if the service (1st or 2nd) is in, as table 2 shows. This is again in line with the improvement-of-return hypothesis. Hence, we see this as the main reason behind the gradual decrease in the service dominance.

In summary, we find that the service dominance has decreased over time even without special measures. The use of softer balls had hardly any effect on the 1995 Wimbledon championships. If a faster decrease in the dominance of the service is deemed necessary, then stronger measures are called for.

ABOLISHING THE SECOND SERVICE

An alternative and obvious measure to reduce the service dominance is to abolish the second service, in other words, to allow for one service only. This change in the rules of tennis is easy-to-implement and involves no extra costs, in contrast to measures that change the tennis court, for instance. One can also take this measure for specific tournaments and keep the existing rules for others. For instance, at Wimbledon one could allow for only one service, whereas on the clay courts of Roland Garros or at tournaments for amateurs one could use the existing rule of two services. Finally, abolishing the second service does not directly affect the rally following a service, in contrast to many other measures such as making the net higher or changing the balls.

Although abolishing the second service has appealing effects, it is not clear to what extent it will affect the service dominance. To answer this question it is important to realize that a player having only one service can be seen as equivalent to a player having two services who has missed his/her first service. Hence, with only one service, a player will use his/her current second service. In the language of game theory, the current situation (two services) has an equilibrium which is "subgame perfect" (Selten, 1975, p. 33) and the new situation (one service) is a subgame of the current situation. Hence, the proposed change to one service amounts to actually abolishing the first service, so that for each player the probability of winning a point on service under the new rules equals the probability of winning a point on the second service in the existing situation. (We abstract here from training effects: under the new rule of one-service-only, the players will only have to practice a single service. This will eventually lead to a somewhat better service than the current second service.) The conclusion is that we can use our Wimbledon data set to calculate the effect of abolishing the second service on the service dominance.

To compute the service dominance under the new rule, we could use the estimates for the probability of winning a point on the second service in table 1. These are 51.4% for the men and 46.6% for the women. A note of caution, however, is appropriate. The two percentages of winning a point on the second service in table 1 are only based on points at which the second service occurs (first service is fault). Hence, there is an over-representation of players with a risky first. This is no problem if the percentages are used as estimates for the probability of winning a point on the second service for a random point at Wimbledon for which the second service occurs. After all, such a point indeed often concerns risky servers. However, there is a problem if the percentages of table 1 are used as an estimate for the probability of winning a point on service under the new rule of one service only. After all, under the new rule, *all* players have to hit the single service, not only the risky servers. Hence, using the two percentages of table 1 in the present context of a rule change involves a selection bias.

To avoid the selection bias, we start from the fact that for each player individually the new situation of a single service amounts to the situation of the second service under the existing rule. So, for each player individually we estimate the probability of winning a point under the new rule by the observed percentage of points won on the second service. Then, our estimate of interest is the average of all individual players' estimates. This approach does not suffer from the selection bias discussed above.

The data show that abolishing the second service will reduce the probability of winning a point on service from 64.4% to 51.7% (0.4%) for the men and from 56.1% to 47.3% (0.8%) for the women. (The new percentages do not differ much from the biased ones given above, so the selection effect appears to be small.) The consequence of the change of rule will thus be that the service dominance becomes very much smaller (in the women's singles the service advantage will turn into a service disadvantage!). Furthermore, both percentages get closer to 50%, making the server and receiver more equal.

There is one other appealing consequence of having only one service. Magnus and Klaassen (1999b) demonstrate that top players distinguish themselves from the others particularly by having a better first service. Since it is actually the first service that is abolished by the change of rule, as argued above, the difference between the top players and the others will decrease. Therefore, matches will become more even and thus more attractive for spectators.

CONCLUSION

This paper provides empirical results on the service dominance, more specifically on the question how to reduce the service dominance. We propose to measure the service dominance by the probability of winning a point on service. The data show a gradual decline in the dominance of service over time, probably because the return on service gets better. The softer balls at the 1995 Wimbledon championships did not reduce the service dominance significantly. Abolishing the second service, however, will reduce the service dominance and make the server and receiver more equal. In addition, its implementation is easy and flexible and it will make matches more even. This will make tennis more attractive for spectators. In this sense, allowing for only one service is a good idea.

ACKNOWLEDGEMENTS

We thank IBM UK and The All England Club at Wimbledon for their kindness in providing the data. We also thank Arthur van Soest, Martin Dufwenberg for important comments. This paper is based in part on two of our previous papers, namely Magnus and Klaassen (1999a,b).

REFERENCES

Borghans L. (1995) Keuzeprobleem op Centre Court. *Economisch Statistische Berichten*, **80**, 658-661.

Croucher J. S. (1995) Replaying the 1994 Wimbledon men's singles final. *The New Zealand Statistician*, **30**, 2-8.

Gale D. (1980) Optimal strategy for serving in tennis. *Mathematics Magazine*, **44**, 197-199.

George S. L. (1973) Optimal strategy in tennis: a simple probabilistic model. *Applied Statistics*, **22**, 97-104.

Gillman L. (1985) Missing more serves may win more points. *Mathematics Magazine*, **58**, 222-224.

Hannan E. L. (1976) An analysis of different serving strategies in tennis. In: *Management Science in Sports* (Ed. by R. E. Machol, S. P. Ladany & D. G. Morrison), pp. 125-135. North-Holland, New York.

Hsi B. P. & Burych D. M. (1971) Games of two players. *Applied Statistics*, **20**, 86-92.

Jackson D. & Mosurski K. (1997) Heavy defeats in tennis: psychological momentum of random effect. *Chance*, **10**, 27-34.

Kemeny J. G. & Snell J. L. (1976) *Finite Markov Chains*, in particular pp. 161-167. Springer Verlag, New York.

Klaassen F. J. G. M. & Magnus J. R. (1999) *On the independence and identical distribution of points in tennis.* CentER, Tilburg University, submitted for publication.

Little A. (1995) *Wimbledon Compendium 1995*. The All England Lawn Tennis and Croquet Club, Wimbledon, London.

Magnus J.R. & Klaassen F. J. G. M. (1998a) *On the existence of "big points" in tennis: four years at Wimbledon*. Mimeo, CentER, Tilburg University.

Magnus J. R. & Klaassen F. J. G. M. (1998b) *The importance of breaks in tennis: four years at Wimbledon*. Mimeo, CentER, Tilburg University.

Magnus J. R. & Klaassen F. J. G. M. (1999a) The effect of new balls in tennis: four years at Wimbledon. *The Statistician (Journal of the Royal Statistical Society, Series D)*, **48**, 239-246.

Magnus J. R. & Klaassen F. J. G. M. (1999b) On the advantage of serving first in a tennis set: four years at Wimbledon, *The Statistician (Journal of the Royal Statistical Society, Series D)*. **48**, 247-256.

Magnus J. R. & Klaassen F. J. G. M. (1999c) The final set in a tennis match: four years at Wimbledon. *Journal of Applied Statistics*, **26**, 461-468.

Norman J. M. (1985) Dynamic programming in tennis - when to use a fast serve. *Journal of Operational Research Society*, **36**, 75-77.

Pollard G. H. (1983) An analysis of classical and tie-breaker tennis. *Australian Journal of Statistics*, **25**, 496-505.

Selten R. (1975) Reexamination of the perfectness concept for equilibrium points in extensive games. *International Journal of Game Theory*, **4**, 25-55.

“New balls please”: tennis, technology, and the changing game

A. Miah
School of Physical Education, Sport & Leisure, De Montfort University, Bedford, UK

ABSTRACT: The decision of the International Tennis Federation (July, 1999) to approve trials of different ball types represented a clear admission of the need for tennis to adapt to the enhanced competence of elite athletes. However, such action brings into question to what extent tennis is evolving beyond its modern appearance and how far such change is desirable. Over the last 30 years, advanced technology and athletic capability has resulted in male players having outgrown the structure of the game, which can be seen as having promoted the ITF’s reaction. The need to ensure that tennis remains a challenging game for players at all levels and an exciting game for spectators appears to reflect an interest and concern for the practice-community of the sport. However, it is problematic to conclude that such changes are in the interests of all concerned. This article argues that any such changes to the structure of the game must be preceded with some admission about what future is sought for tennis and thus, where limits might be drawn on the changes made within the game. Furthermore, it is recognised that by invoking the ‘new balls’ proposal, a clear statement is made about what aspects of the game are considered worthy to preserve or not. This article addresses the implications of technological change for tennis, identifying upon what basis such change should take place to ensure a credible future for tennis.

INTRODUCTION

On 12 July 1999, the International Tennis Federation (ITF) announced proposals to introduce two new kinds of tennis ball to tournament level tennis on an experimental basis. The new balls have been designed as a result of a decade’s debate about men’s tennis and how it has become increasingly boring for spectators due to the power of the athletes. This effect, it is argued, underplays the more charismatic aspects of tennis, such as the long rallies (Arthur, 1992; Bierley, 1998). Scientific research has demonstrated that, with the new kinds of ball proposed by the ITF, it is possible to alter the pace of the game on non-hard court surfaces (i.e. grass, clay, indoor grass), primarily from the point of view of the service receive. As is outlined in the ITF announcement in This Week (1999, July 12),

(1) New Ball Type 1 is a faster ball for use on slow surfaces such as clay. These balls will be harder and lower bouncing than standard tennis balls.
(2) Ball Type 2 will be used on medium paced surfaces such as hard courts and will be made to existing specifications.
(3) New Ball Type 3 is a slow pace ball for use on fast surfaces such as grass and some indoor carpets. **Type 3 Balls will be about 8% larger in diameter than standard balls.** (ITF, 1999)

From the introduction of new ball Type 3 it is intended that there will be a slight, though significant, reduction in the speed of a serve, thus allowing the receiving player more time to react to the ball. Presently, service speeds are reaching in excess of 140 mph, which is argued as approaching the limit of human reaction time for the receiving player (Coe, cited in Cislunar Aerospace, 1999). Consequently, if serves start to tend beyond this limit, then the elite game will become merely a serving competition since no player will be able to return a serve. As such, it is argued by the ITF that something must be changed within tennis to try to reduce the dominance of the serve and prevent a future for tennis that could comprise of only serving. Deciding to introduce new balls is but one of numerous proposals that had been made to combat this dominant-serve effect, such as going back to wooden racquets or making the service box narrower (Gray, 1999).

Despite these other proposals, the decision to alter the tennis ball has been chosen to curtail the serve of the male tennis player. This decision is remarkable for numerous reasons. Firstly, the decision reflects the responsibilities of governing bodies to define what shall be considered as the nature of a sport and how it is played on a global scale. Furthermore, the willingness of the ITF to make a change to the rules of tennis sets a pioneering decision for governing bodies of sport around the world. Recognising that tennis is a dynamic enterprise comprising uncontrollable factors that do influence the 'balance' of competition is to the credit of the ITF, where in some sports, tradition and a reluctance for change might well be suffocating the flourishing of a sport. Indeed, this characteristic is typical of the ITF in their interest to address the impact of tennis racquet technology over the last 30 years. Secondly, the ITF's decision is remarkable for what it means historically for tennis and what it means for the future of this changing game. In contrast to the technological development of racquet technology that has resulted from a player's interest to enhance performance, the 'new balls' proposal intends to circumvent performance enhancement and individual choice. Indeed, it might be argued that the proposal hopes to reduce the capabilities of an athletic to perform well. As such, the proposal is explicit in stating that the capabilities of athletes are surpassing the limits of tennis to its detriment.

This paper will address the ITF's 'new balls' proposal by investigating these two kinds of implications and articulating the implications of change more generally. Throughout these analyses, it is important to recognise that, implicitly, they deal with the justifications given by the ITF for making such changes and are thus, directed specifically at the ITF's decision to experiment with new tennis balls. However, the ideas are also more generally interested to identify concerns about technological change in sport and will thus draw upon examples of technology in other sports to inform the appropriateness of the present proposals in tennis. It will be noticeable by its absence that the paper does not articulate what might be given as suitable justifications for introducing change in the structure of tennis. Such a task would extend beyond the limitations of this paper. Rather, the current thesis intends to provide an argument for the need to reconsider what is valuable in tennis.

WHAT IS VALUED IN TENNIS?

In explaining the 'new balls' proposal, the ITF claims to be acting so as to "preserve the nature of the game" (Coe, cited in Cislunar Aerospace, 1999). Yet, it seems uncertain how the ITF can make such claims about the nature of tennis given that what is regarded as valuable within a sport is a function of one's relationship with the particular sport in mind. For example, one might assert the interest of the media to ensure a version of tennis that is exciting or interesting for spectators to watch. Indeed, one might even assert the interests of tennis upon such a premise, given that for tennis to thrive it requires some degree of media exposure and the sponsorship such exposure generates (Gelberg, 1998). This very argument is evident from the ITF's decision to alter tennis and is reflective of the opinion in recent popular press (Arthur, 1992; Bierley, 1998; Blake, 1996). However, it is certainly not the case that such aspirations speak for other parties, such as players, referees, or, indeed, spectators. Nor is it clear that tennis will benefit from such decisions. As such, it is necessary to consider in what sense the term 'benefit' is appropriate. This point is recognised by Kew (1987), where it is discussed how rules and laws are altered within games. Kew identifies how power is distributed within an institution or governing body and the effect of this on how change occurs. If, as might be said of tennis, the authority to alter how a sport is played is determined largely by the legislating governing body, then it cannot be claimed by the governing body that it is able to *know* what is valuable to a sport. This is not to say that its claims are necessarily wrong. Rather, it is to argue that, unless there is the possibility for other parties to be represented in the decision making process, then such change can, at best, only be an approximation of the interests of these other parties.

A similar assertion is made by Morgan (1994) where he argues that decisions about reforming sports on the basis of, for example, a market rationality will inevitably be to the detriment of a sport. Indeed, the very suggestion that such *external* concerns as media exposure are of value can be some guide to identifying how much a sport might be in danger of jeopardising its *internal* goods. Moreover, Morgan advocates that, "all substantive policy matters regarding the conduct and reform of sport be turned over to practice-communities" to ensure that any such reform is for the benefit of the sport (Morgan, 1994, p.237). Morgan argues that it is the practice-community that should be the primary determinants of change in sports, since it is the community who will seek solely to maximise the internal goods of the competition without consideration of external interests such as financial benefits.

That an interest in such external goods is reflective of the present issue is implicated by the comments of the ITF. As Andrew Coe, Head of the ITF Technical Commission states, the new larger ball (Type 3) will "reduce the dominance of the serve, which will make tennis more attractive to spectators. It will also offer greater visibility – both for the players on-court and for television viewers" (ITF, 1999). Clearly, it seems important for governing bodies to ensure media exposure that will generate sufficient sponsorship for the sport. However, this need not imply that simply because such sponsorship might be in jeopardy, that this is a reasonable basis for advocating change. In making such a statement, the ITF must justify why it is that the interests of the media are the primary concern rather than, perhaps, simply ensuring good competition. In so doing, Morgan suggests, governing bodies must ensure that the practice-community is in a position to articulate its interests in a way that will convey influence. As Morgan states, the challenge is to "turn *differends* into litigations" to change disputes where parties are victimised by not being able to make their case, into litigation where "both parties agree on how to phrase the issues that led to the dispute as well as the means for resolving them" (Morgan, 1994, p.238).

Whilst one might accept the need for change, it is necessary to precede changes by first formulating ideas about what kind of game is sought. Why, for example, is it argued by the ITF (Gray, 1999) as a benefit that the new balls would make tennis an easier sport to play? Surely this might also be seen as effectively devaluing the excellence required of an athlete to be proficient in the game and thus, devaluing tennis. In the context of the new specifications in ball design, the ITF does not seem to have embraced discussion between parties of different interests within tennis so as to establish what is important. As such, it does not seem reasonable for the ITF to claim it is in a reasonable position to decide what kind of practice is of interest and thus, what kind of changes (if any) ought to be made to address the dominant serve problem. This is not to say that the decision to introduce new balls to combat the power serve is necessarily a bad one. Rather, it is to recognise that the process by which the decision has been made and the *prima facie* reasons given to justify this choice might be misguided if the practice of tennis is, indeed, the primary concern.

WHAT FUTURE FOR TENNIS?

Asserting a particular kind of ideal about tennis as being valuable to preserve commits one to prioritising that particular kind of game over other, equally valid ideals. Integral of the 'new balls' proposal is an interest to equalise various differences that occur among different kinds of court surface. Thus, new ball Type 1 will hope to speed up such surfaces as clay, whereas ball Type 3 will slow down surfaces such as grass. Yet, it is not at all clear why this is regarded as valuable. Such aspirations will render tennis impoverished of the creativity and variety that it presently has by there being different kinds of surface upon which different kinds of athletes can excel. Consequently, making things too equal might be a problem insofar as it creates uninteresting parity among playing styles in tennis. Whilst the aspirations of the proposal are to equalise the conditions of play and gain a true portrayal of who is the best tennis player, there might be more value in sustaining the differences and simply having a variety of best players on different surfaces. Arguably, variation in surface type is part of the excitement in tennis and is to be cherished rather than removed.

Alternatively, Andrew Coe states that "Tennis is about having rallies" (Chaudhary, 1999, p.9). Yet, it is unclear exactly how prescriptive this statement intends to be. Coe acknowledges that there are limitations to this ideal and that too many rallies can cause problems as well as can too few. However, the very basis of this concern would seem necessary to articulate in order to justify the valuing of rally games over serving games. It would be misleading to assume that this essence is at all static and the ITF does seem to be sensitive to this given that it is recognised how tennis must move with these changing times (Gray, 1999). By identifying what kind of values are purported by the asserting of any specific aspects of tennis as valuable, a more in-depth articulation of the kind of game tennis should be would enable a greater understanding of how the future of tennis should look. Thus, if it is argued that tennis should be about establishing the best serve, then it must be argued why it is that this particular kind of performance is the priority.

Supposing that it is clear what kind of game might be worthwhile to preserve, it remains to be understood what could result from implementing change in tennis. Consequently, the remainder of this paper will articulate some considerations about the effect of changing technology in sport.

PRIORITISING PERFORMANCE OVER PERFORMANCES

Perhaps most significant about change in sport is that any such action can dictate the future of that sport in a way that makes difficult any possibility for reversing such changes. Thus, if tennis is altered on the basis of wanting to preserve some notion of its ideal, such as the rally game, then upon making such a change, it becomes difficult to reappraise such a change. Perhaps most illustrative of this effect is the evolution of the tennis racquet. Fundamentally, tennis racquet technology has been developed to enable a legitimate performance enhancement for athletes. Yet, the rationality of performance enhancement by such means cannot be seen as valuable in itself. Furthermore, its prioritising can have serious implications for what options are available to tennis. Indeed, one might argue that the current predicament facing tennis is partly the consequence of the lack of constraints that had been placed on racquet technology. Had the rationality of performance enhancement been questioned prior to the development of modern racquets, then the dominance of the serve might currently have posed less of a problem.

That the 'new balls' proposal seems merely to lessen the impact of the serve does not seem to challenge this dominant thesis about performance enhancement sufficiently. As such, it would seem that embracing the changes suggested by the ITF merely postpones a further inevitability: that athletes will learn to adapt to these new kinds of balls and, eventually, learn to outplay their constraints. Consequently, presuming a similar rationality as the present circumstances, tennis would then be forced to consider, again, altering other components of tennis to remove the recurrent dominant serve problem. Problematic of this is that if tennis is to continue embracing such changes, then it might find itself on a very slippery slope to a kind of game that is only vaguely memorable as the game of tennis.

A SLIPPERY SLOPE TO TECHNO-TENNIS?

Embracing changes to the constitution of any sport has the potential to bring about an accumulation of changes over a period of time, the end result of which might have been wholly undesirable if it had been known at the beginning of these changes. This is not necessarily to conceive of things as being a slippery slope whereby allowing one kind of alteration (such as new balls), will necessarily render the implementation of further changes to the game of tennis. Rather, it is to recognise the potential for minor alterations to the rules of a game to bring about a major change in the way a game is played. Thus, if seemingly minor changes are made to tennis, such as the proposed alterations to the tennis ball, then an accumulation of such changes could yield a form of tennis that challenges the ideal upon which these changes are based. Tennis has already seen the transformation of the tennis racquet and the enhanced athleticism of elite athletes. The current proposal to alter tennis balls begs the question at what point tennis will become unrecognisable because of such changes.

It seems probable that players will learn to adapt to the new balls and continue to increase their serving speeds. As such, it must be addressed what will be the strategy if such circumstances transpire. Perhaps ball sizes will get bigger and bigger until tennis is played with balls the size of a volleyball in the name of some ideal game that is being preserved. Alternatively (and more likely), the ITF might conclude that, in fact, the size of a tennis ball is also an essential character of tennis – the very characteristic that is being changed in the current proposal. To reiterate, it is problematic of this continuous reappraisal and change that it can lead to a sport that is only vaguely identifiable as current versions of tennis. Furthermore, it is possible that if some future version of tennis were to have been posited as the end result of all such, seemingly small, changes, then it might not have been chosen at all. As such, the ITF must account for the possibility that increasing ball sizes might simply delay the inevitable and must thus, question the ends of such changes. An accumulation of changes, potentially, permits an anything-goes climate for change in tennis. Consequently, it becomes problematic to argue anything as being essential to tennis, which is the critical justification given by the ITF for introducing new balls.

Such determinism might seem improbable of tennis, where decisions about change are neither made too frequently nor all too willingly. Furthermore, the current thesis might appear somewhat traditionalistic in as much as it appears to be rejecting change in favour of inaction that preserves the original form of any sport as if it is timelessly applicable. However, this is most certainly not the case. Rather, this paper questions the reasons given by the ITF in asserting the essence of tennis that it seeks to preserve. Its one question may be summarised thus: If tennis embraces change, then is it also possible to have any sense of essential values that transcends all changes to its structure? It is suggested here that such change is not possible and that, in asserting what is the true nature of tennis, the ITF might find itself simply romanticising what it regards as essential. Moreover, in producing the new balls proposal, the ITF might neither reflect the values of the contingent majority, nor what is independently valuable about the game of tennis.

ANYONE FOR 'NEW' TENNIS?

In addition to difficulties raised by identifying what kind of game is sought, it is important to recognise that change in sports inevitably projects new circumstances and the values they sustain onto future participants. Thus, if new balls become accepted into the formal rules of tennis, then players of the next generation must train with such balls if they hope to reach their optimum level of performance. The significance of this may be understood by considering the tennis racquet and how its specifications have altered present day tennis. Whilst it might be argued that alterations to the tennis racquet (from wooden to the latest composite) accounts for only 4% of the difference in service speed (Gray, 1999), it seem reasonable to suggest that tennis has changed substantially as a result of new technology in racquet design. The wooden racquets of yesteryear weighed nearly double the latest titanium racquets, with Head's T1.S6 weighing in at a mere 8.5 ounces unstrung. Furthermore, contemporary racquets are far stronger than previous designs, which has had an influence on player's ability to improve aspects of his/her game, whether it is ground strokes or service speed. Thus, the technological development of the tennis racquet can be said to have had an irreversible effect upon the way in which tennis is played. Indeed, to some extent one might attribute the ITF's 'new balls' proposal as having being provoked by the way in which tennis racquet technology has influenced the game. It is important to learn from this that any subsequent alteration (such as new balls) will also have an impact on the future playability of tennis. Furthermore, future players inherit the rationality underpinning any alteration in tennis. For example, where the modern racquet has instilled an interest to seek performance enhancement through technological alterations to equipment, the 'new balls' proposal will instil its own values upon the next generation about what is important for tennis. As such, in making a decision that will alter the future of tennis, it is important to consider what will be left for future generations of athletes in providing some possibility of reverting any such change that might not fit with the values of future generations.

UNINTENDED CONSEQUENCES

The significance of unintended consequences is addressed in depth by Tenner (1996), where it is recognised that technology has a tendency to effect change beyond that which was its purpose. In the context of sport, Tenner identifies that endeavouring to provide greater safety the result can be a trade off against a degree of risk that could be argued as constitutive of that sport. For example, the proliferation of new innovations in climbing and mountaineering has provoked a perception of there being much greater safety whilst climbing. Such technologies as Global Positioning Systems has allowing increased certainty of location. However, such 'benefits' can also be argued as detracting from the experience of climbing since climbing is partly about mastering risk in extreme conditions (Miah, 2000). Moreover, that such security is, perhaps, only a perception can have the effect of complacency in dangerous climates.

Perhaps the most probable unintended consequence of the 'new balls' proposal will be a change in the hierarchy of tennis players in the elite, men's game. Thus, in a similar way to how the Németh javelin brought success for the more technically proficient thrower to the prejudice of the power throwers, it could be argued that the more capable ground stroke player will surpass the big hitting server. Indeed, one might even make this same point with reference to how new racquet technology provoked a new emphasis on power, which allowed younger players to adapt and become better than the senior players who were used to older, heavier racquets (Galenson, 1993). Again, this begs the question as to whether such change is fair to athletes and calls for a justification of why the big hitting server, who has spent years developing his serve throughout his career, should be prejudiced by such change.

CONCLUSION

This paper does not wish to suggest that the ITF is mistaken in its decision to experiment with new balls in tennis, far from it. That the ITF recognises the need for an experimental period is evidence of its willingness to reflect upon the impact of change in tennis. Furthermore, that altering the balls within tennis has been a strong proposal for the past decade is testament to the degree of consideration that the proposal has received.

The arguments herein intend to provide some framework for reflecting upon the very consequences of this experimental period that might be useful to guide discussions about the appropriateness of change in tennis. The present thesis does not wish to allow a situation to arise where tennis consists only of serving. However, to justify this perspective on the basis that it would make tennis more interesting to watch, or that it would make tennis easier to learn, does not seem a reasonable position to adopt.

It is paramount that governing bodies are clear about what kinds of experiences are being prioritised when implementing new technology that seeks to alter specific aspects of the game. It cannot be assumed that seemingly inconsequential innovations have little or no effect upon the tennis experience as it has been shown how such innovations do not exist in isolation - it is necessary to recognise each of the altering technologies that combine to change tennis significantly. Historically, tennis has been a game requiring perseverance and determination to do well and by making the game easier it would seem to devalue this technical pursuit. By opting to use various technologies, the ITF is choosing a particular kind of tennis experience over another and might even be infringing upon the experience of another tennis player's experience. Consequently, such decisions would do well to involve the entire practice-community within tennis and must be clear about what kind of game is preserved by such change.

REFERENCES

Arthur, C. (1992, May 2) Anyone for slower tennis? *New Scientist,* **234**, 24.

Bierley, S. (1998, June 29) Tennis: A game in search of a savoir. *The Guardian*, 6.

Blake, A. (1996) *The Body Language: The Meaning of Modern Sport.* Lawrence and Wishart, London.

Chaudhary, V. (1999, June 26) Tennis stars set for revolt over ball change plans. *The Guardian*, 9.

Cislunar Aerospace. (1999) Technology and Tennis - The Balancing Act. *Aerodynamics in Sports Technology,* Hypertext Document: http://wings.ucdavis.edu/Tennis/Features/coe-01.html [Accessed: February 2000].

Galenson, D. W. (1993) The Impact of Economic & Technological Change on the Careers of American Men Tennis Players. *Jon. of Sport History,* **20(2)**, 127-150.

Gelberg, J. N. (1998) Tradition, Talent and Technology: The Ambiguous Relationship between Sports and Innovation. In: *Design for Sport* (Ed. by A. Busch), pp. 88-110. Thames and Hudson, London.

Gray, B. (1999, October) Bigger, but Better? *Tennis*, 12-14.

International Tennis Federation. (1999, July 12) ITF AGM Adopts tennis ball rule change. *This Week: The News Bulletin of the ITF,* Hypertext Document: http://www.itftennis.com/html/new/this_week/archive/1999-o7/12th/newballrule.html [Accessed: 11 July, 1999].

Kew, F. (1987) Contested Rules: An explanation of how games change. *International Review for Sociology of Sport,* **22(2)**, 125-135.

Miah, A. (2000) Climbing Upwards of Climbing Backwards? The Technological Metamorphoses of Climbing and Mountaineering. In: *The Science of Climbing and Mountaineering [CD-ROM].* (Ed. by N. Messenger, W. Patterson, & D. Brook), Human Kinetics, London.

Morgan, W. J. (1994) *Leftist Theories of Sport: A Critique and Reconstruction.* University of Illinois Press, Urbana.

Tenner, E. (1996) Why Things Bite Back: Predicting the Problems of Progress. Fourth Estate, London.

Tennis and the environment

C. Bowers
Ringmer, East Sussex, UK

ABSTRACT: Like any sporting activity, tennis has an impact on the environment. Some of that is sustainable, but much of it may prove not to be in the long term. If the much talked-of threats to the environment materialise or get worse, governments around the world will be forced to take measures that reduce their citizens' consumption of non-renewable energy, maximise the use of raw materials, and reduce ecological damage to a degree at which life becomes truly sustainable. In such a situation, tennis would be forced to rethink some of its current practices. In many ways it is well suited to suriving an era of environmental austerity, as it can be played without any use of energy and using largely natural and biodegradable materials. However, the global tennis circuit would face a restructuring that might see it resemble the circuit of the 1930s with less travel in a more carefully planned schedule. In the light of authoritative reports that say global warming has already begun and that cuts in emissions of polluting and climate-changing gases are becoming more and more inevitable, the tennis world would do well to prepare now for an age in which the energy and raw materials it currently takes for granted are not quite so apparently limitless.

INTRODUCTION

This paper is the expanded version of an article commissioned by ACE magazine on the effect of tennis on the environment. It is by nature speculative, in many instances throwing up more questions than answers, and the journalistic nature of the research accounts for the small number of references. However, it is a starting point in looking at tennis's impact on the environment and how the harmful aspects of that impact could be minimised. It also has a British bias which stems from the original brief for the research.

WHAT IS TENNIS'S IMPACT ON THE ENVIRONMENT?

Just about every activity has some impact on the environment. Even the "greenest" activities, like hiking in the country, can do damage if too many people populate ecologically sensitive areas. Yet some impact is sustainable, in that the damage will easily recover before further damage is inflicted. This main section breaks tennis down into various areas to look at what impact on the environment results from today's practice which is potentially unsustainable.

Racquets

In the days of wooden racquets and strings made from sheep's gut, tennis racquets were a very sustainable commodity, in that the resource base was not a problem and both wood and gut degrade naturally when buried in the ground after their tennis life, with only minimal contamination from the odd bit of glue. Having said that, the last racquet made from a solid piece of wood was used in the 1920s, and the wooden racquets being used as late as the early 1980s at the highest level were a blend of various woods linked in with carbon fibre through the use of toxic and non-biodegradable epoxy resin.

Since then the so-called graphite racquets have taken over, which are essentially a composite construction made mainly from carbon fibre, reinforced with graphite and kevlar, and increasingly titanium. The latest racquets are an offshoot of the aerospace industry, using the know-how gained by defence and military experts.

Though much less labour-intensive, composites require much more energy to produce than wooden racquets and are not biodegradable. Many industries are subject to regulations on what must happen to their product at the end of its useful life, but there are few such rules in the tennis racquet industry. So with eight million racquets sold per year these days, each requiring 200 grammes of composite material, they are piling up, whether in people's garages or in landfill sites. Composite racquets could be incinerated and the fibres recovered, but incineration means a lot of harmful emissions, so that would mean improving one environmental aspect at the expense of another.

Apart from using a racquet for as long as possible, the most environmentally benign approach would seem to be some form of re-use after a racquet is broken or no longer wanted, to make a virtue out of its non-degradability. Yet no obvious use springs to mind. Perhaps some society or philanthropist could sponsor a competition to find the best uses for masses of unwanted tennis racquets which will never degrade, maybe for uses in the Third World?

Nylon strings are also not degradable, but as they are not "cured" they could be recycled as other nylon products (for insulation purposes, or in the manufacture of plastics). A scheme would have to be set up for this purpose, with information on what a club player or stringer wanting to offer used strings for recycling should do.

Grips used to be made of leather, a natural resource which is biodegradable. They are now made mainly from polyurethane, a chemical product which cannot be recycled and does not biodegrade, so old grips could only be re-used (for example for strapping up plants to fences and sticks).

Balls

Around 360 million tennis balls are manufactured each year, most of which end up in the bin, thus in land-fill sites, where most of the constituents will degrade naturally, but over a long time.

A standard ball weighs just under 60g, of which about 11g is made up of cloth and about 47g of rubber and a bit of glue. As long as the rubber trees – mostly in Malaysia – are sustainably farmed, the resource is a natural one and fairly unlimited. In theory, most of the rubber could be recycled, but when A V Syntec, the Australian company which makes Rebound Ace hardcourts, was asked whether it could use old tennis ball rubber instead of the car tyres, the answer was no "as the ball rubber would have to be 'frozen' ground into a suitable size and shape and bound with polyurethane". (But old tennis shoes might be usable – see the section on hardcourts below.) The felt, on the other hand, could be used in preparing the base – an incentive is needed for old tennis balls to be collected for this purpose.

The International Tennis Federation (ITF) ran a scheme called "Bank a Ball" in the mid-1990s, in which club players could hand used balls to their club secretary knowing they would be shipped out to help tennis players in emerging tennis nations. This re-use scheme ran for about three years and contributed well over 100,000 balls to various countries, but it was dependent on two sponsorship contracts for free shipment, and when those expired the scheme fizzled out.

These days balls come in plastic tubes and tins. The plastic tubes can be put in the recycling bins along with PET fizzy drinks bottles and plastic milk bottles, but tins cannot be recycled.

Grass courts

There are obviously much fewer grasscourts now than there were 20 years ago, but in that time the way grasscourts are maintained has undergone major changes. A host of EU regulations affecting weedkillers, pesticides, fungicides and insecticides have changed the rules by which European ground staff work, much of which has put the emphasis back onto court management rather than chemical cures.

Phil Thorn, head groundsman at the tournament in Halle, has to work to stricter standards than most other grasscourt keepers because of heightened environmental awareness in Germany. "Fifteen years ago we would have used a mercury-based poison to treat a worm problem," he says, "which meant we wouldn't have seen a worm for about three years, but the mercury seeped through the soil, causing contamination. These days we use very few chemicals, we have regulations that amount to limiting us to one drop of chemicals in the equivalent of a 50-metre swimming pool of water."

As a living organism, grass is subject to developing a range of diseases, especially if the relationship between the drainage, soil and grass is not quite right. Just as with human diseases, they can be treated with chemicals (like medicines), or through better management (like changes in lifestyle, diet, etc). Among options for better court management are:

- proper court maintenance, for example not letting the layer of thatch get too thick towards the end of summer;
- not using too much fertiliser, even non-chemical, and using the right fertiliser at the right time of year;
- changing the pH of soils (the index to measure acid/alkaline ratio) to make them less attractive to potential bugs
- having "insect-friendly" areas (like weeds and long grass) near but not adjacent to the grasscourts so insects will go there rather than the courts.

There are also nemetodes being developed, living organisms which can be planted into grass to combat certain problems, almost like a form of mini biological warfare, though the science is still relatively new.

"A lot of these alternative techniques are just good groundsmanship," says Thorn, "and the stricter the rules, the more we go back to first principles and good management."

Hardcourts

Given that hardcourts have to stand the rigours of all weathers, they will by definition involve materials which will not naturally degrade, and be made using processes which will involve a certain amount of energy consumption. Yet there are ways of making use of substances which would otherwise have to be disposed of, and tennis bodies could argue it is better to use such substances to make a tennis court than consign them to a land-fill site somewhere near or under a beauty spot.

Despite the numerous different names for hardcourts, they can essentially be broken down into two categories: porous and non-porous or impervious. All hardcourts are built on a base of bituminous macadam (known in the trade as "bitmac"), which can either be built to be porous – ie. letting the water pass through or be soaked up – or impervious to water. On top of this base, a hardcourt surface is added, which is either porous or impervious to match the base.

The bitmac base and many of the surfaces have to be made from quarried materials, or aggregates. This a virgin resource which often involves destruction of the countryside, and there is a limit to how much can be extracted. An alternative to bitmac bases are bases made from compacted rubber, used mainly from old tyres. This is not entirely without environmental problems because to compact or bind the rubber requires polyurethane, an oil-based derivative. However, on the basis that there are constant supplies of tyres which have to be disposed of somehow, there is scope for reducing the pressure on quarrying by finding a constructive use for many old tyres. One of Great Britain's leading experts on artificial sports surfaces Mike Abbott says: "What is holding back the use of compacted rubber is that it's currently more expensive than bitmac, but the industry is bracing itself for a tax on aggregates sooner or later which could change the situation. For example, if the government were to use some of the money from an aggregates tax to fund grant aid for re-using rubber – rather than just seeing the proposed tax as an easy source of revenue – it could encourage a lot more hardcourts to be made with a compacted rubber base."

As for the surfaces, there are too many of them for their respective environmental properties to be examined here. By and large, if a court can be built in such a way that when it comes to the end of its natural life it can be broken up and rebuilt using many of the original materials, that is clearly an advantage.

The Rebound Ace court on which the Australian Open and many of the hardcourt tournaments on the international tennis circuit are played is made up mainly of recycled car tyres. Given that there is a lot of rubber used in tennis (notably in balls and shoes), could this rubber be used, in turn enabling clubs or schools to collect old balls or shoes and thereby reduce the cost of installing a Rebound Ace court? The answer appears to be no, at least for the time being. A V Syntec, the manufacturers of Rebound Ace, say the felt from balls could assist in the court base, but using the ball rubber in the main layer would be too complex. However, they have opened up negotiations with Nike with a view to blending used car tyres with old sports shoes[2] (Nike says research into this is too premature to make any comment).

Claycourts

Claycourts are made from up to six different layers of a complex cross-section, but each of the strata are natural materials, with top dressing we think of as clay sometimes making up just 6 millimetres of a 20-centimetre deep surface. The top dressing on "European" red clay is mostly crushed roof tiles, which begs the environmental question: are these tiles that have been used on roofs and are being recycled as dressing for a claycourt, or are they made specially for tennis courts? If the former, that is good news environmentally. If the latter, then the kiln firing has to count as a minus point. The "American clay" top dressing is made from basalt, which is a natural material which just needs crushing.

The fact that all the constituent parts of a claycourt are natural is good in terms of using natural materials but questionable in terms of their being finite and often virgin resources. In general, a claycourt can be considered more environmentally benign than a hardcourt as its constituent layers are likely to be more easily re-used at the end of its natural life – this could be important if one day environmental reality dictates that we have to use less energy for things like demolition and rebuilding. However, if water is to become a scarce resource, as some scientists believe it might, that is a minus point for clay as it needs regular watering (though using minimal amounts compared with, say, a golf course in a hot climate).

Clay is more susceptible to the weather, making it a viable surface for only eight or nine months of the year in many developed countries. However, the environmentally aware tennis player would aim to fit in with the dictates of the weather rather than try to defy them, which may mean having an off-season during which playing tennis is just not possible.

Floodlights and indoor courts

On the basis that acting environmentally means fitting in with the dictates of nature rather than trying to counter its limitations, strict environmental thinking would preclude the use of floodlights and indoor courts. And in any environmental scoring system, playing outdoors during daylight hours is going to score more highly than using energy for lights and heating.

This is in fact good news for tennis, which was a global sport before the age of apparently limitless electricity, for it means it has the advantage of being able to survive if ever energy becomes such a critical issue that the use of floodlights and indoor lighting become either too expensive or socially irresponsible.

But given that in most European countries there are around 1200 hours of darkness a year during the time the average adult is awake, it is understandable that a tennis club might want to maximise its investment in all-weather courts by providing the chance to play during some of those hours (in some cases it can mean a 35 per cent increase in court usage time). That means floodlights, and there have been several significant improvements in the environmental performance of floodlights which makes using them less intrusive now than it was a few years ago.

There are three broad environmental issues connected with the use and application of floodlights: energy consumption, the visual impact, and transport. The first two are subject to delicate trade-offs, because in certain circumstances lower use of energy can lead to greater glare and light spillage (or "trespass") if it is not matched by the correct choice of light fitting (or "luminaire"). Conversely a set-up which is overspecified in terms of concentrating the light just on the playing area can lead to more energy being consumed, however well controlled it is.

It is important to note that there are various factors at stake in setting up a floodlighting system for one or several tennis courts, with no single optimum formula. Each situation needs to be judged on its merits, and even within one project there may be trade-offs between one environmental impact and another, which can only be answered subjectively. Among the factors to be considered are the height of pylons, the number of pylons, the angle of the light fittings, the type of reflector fitted inside the luminaire, the type and wattage of the light bulbs, and the lie of the land in which the court is situated.

To give examples of a possible trade-off: a court could be lit from "low level" pylons (columns up to 8 metres) fitted with "conventional" type fittings which would use about 75% of the electricity of a court lit from four "high level" pylons (columns up to 10m). However, while less energy would be required, the level of performance would be cut back (in Britain from county standard to club standard), and while the lower height of the column would also minimise the visual impact in daylight there could be more light pollution from this scheme due to increased spillage and glare due to the steeper angles at which the light shines out from the fittings. If higher columns and the accompanying performance level can be tolerated at the site of the court, a planning authority might well insist on the taller column scheme to minimise spillage and glare if it considers those the key issues – even though it would mean more energy being consumed.

A different scheme that fully addresses all the issues of performance, column height, glare control and energy consumption is likely to be one using "low level" (6 to 8 metre) columns fitted with "box" type fittings in place of conventional floodlights. The trade-off here is the additional two columns per court needed by the net. If a club or school is applying for planning permission to install lights in a residential area, the planning authority might well seek to protect residents by insisting on this alternative, as well as a cut-off time for use.

In Great Britain, the Institution of Lighting Engineers has drawn up a four-page factsheet "Guidance Notes for the Reduction of Light Pollution". (Institution of Lighting Engineers, 1997) It breaks down areas in which lighting might be installed into national parks (zone E1), and areas of low (E2), medium (E3) and high (E4) district brightness. It then sets limitations for the specifications of a floodlighting system, which are obviously stricter in an E1 zone than in an E4 zone. Tony Hill of Ayrlect Associates, one of Britain's leading sports venue building services consultancies, says: "While the ILE guidance notes are not mandatory, they are widely accepted among local councils, some of which have incorporated them into their own planning guidelines for floodlighting. If an area can be identified according to the E1-4 classification system, it makes it much easier to develop a design and specification for a lighting project that will minimise damage to the visual environment. Careful application of modern equipment, matched to the correct column height and configuration, increases the chances of low visual impact, and when an applicant uses the ILE guidelines, that in itself sends a signal to planning authorities that the applicant is serious about reducing light pollution. Using such guidelines universally could certainly help reduce the impact of floodlighting on the visual environment."

The impact the energy emissions will have on the environment depends on how the electricity is generated. The large number of fossil fuel power stations in England and Wales means a British floodlit court contributes to global warming, while in France where much of the central electricity comes from nuclear installations, there is virtually no impact on global warming but there would be other environmental problems connected with the nuclear industry. (This syndrome applies to assessing the environmental credentials of electric cars, trams, etc.)

Making tennis courts available after dark also increases traffic. Not only do the extra hours of court use increase the number of cars which will drive into and out of a club, but as the dark hours are often the coldest ones, the incidence of people travelling to tennis by car is likely to be higher by night than by day.

Necessity being the mother of invention, if environmental reality one day begins to bite and floodlights cannot just be turned on and off when two or four players are willing to pay for the time, there could be other ways of lighting courts. For example, clubs might buy their own generator which can be solar or wind powered (though the chances of it impinging on the visual environment are high). Perhaps they might one day even top up the charge by connecting the generator to the club gym, so whenever someone gets on the exercise bike or the rowing machine it sends electricity that either temporarily reduces the need for power from the national grid, or ideally could be stored in the generator for use by floodlights.

Until that time, the best option environmentally, if floodlighting is essential, is to increase the "load factor" by getting four on a court as either a doubles or two singles practice sessions. There are no economies of scale on the use of floodlighting energy that would make lighting two courts less than twice the cost of lighting one (to the same standard). What is a factor is that many floodlight bulbs will have their lifespan shortened considerably if they are turned on when they've just been turned off – this means it is normally better to leave lights on if there is only a 10 or 15 minute gap between court bookings, which is wasted electricity and can only be got round either by additional controls for what is known as "automatic hold-off" or by having strict rules about booking and operating a floodlit court.

The need to provide lighting between the clubhouse or changing rooms and the courts and car parks, plus heating in the changing rooms, when courts are in use under floodlights also adds to the energy consumption of floodlit tennis.

Indoor courts require not only lighting for the tennis hall but heating and ventilation, in addition to the infrastructure of the building or bubble in which the courts are housed. Environmentally this scores relatively badly. While the heating needed for the hall is substantially provided by heat from the lighting, the first unit of energy used for other demands can put indoor courts at a disadvantage compared with playing outdoors under floodlights (if it gets too cold to play outdoors, the environmental approach would be to accept that it is too cold and not play).

On the basis that indoor courts are an important feature of tennis development, measures should be taken to lessen the environmental impact (and are being taken by the more responsible designers). These inlcude using lowest-energy lighting, basic heat recovery equipment, automatic controls to maintain air temperatures at the lowest acceptable level, coupled with environmentally sympathetic materials within the building structure.

Installation of the best technology alone is not enough – owners and operators of outdoor and indoor facilities need to ensure the equipment is monitored and maintained. Only then can the contribution of manufacturers and project design teams to minimising their environmental impact be fully realised.

Tennis clothing

The manufacture of all clothing has environmental implications, and one of the biggest materials used in tenniswear is cotton.

Cotton growing requires more pesticides by volume than any other crop in the world – it uses 11 per cent of the pesticides in use throughout the world, in fact half the cost of growing cotton is the use of pesticides. Pesticides have widespread harmful effects on the environment, and their use in cotton growing has caused the partial drying out of the Aral Sea in Uzbekistan, once the fourth-largest body of freshwater in the world but now a fishless, polluted salt desert (Myers et al, 1999). Cotton produced with fewer pesticides and less harmful chemical treatments enjoyed a mild wave of popularity in the early 1990s, but it did not cause lasting change in the cotton industry. And a product cannot be considered fully organic if it avoids agricultural pollution but is then dyed or chlorine bleached, which is another challenge to the sports and fashion industry.

Approaches to Nike, Reebok and Adidas on their approach to organic cotton elicited contrasting answers. Reebok said: "We do not use organic cotton in our tennis apparel and currently have no plans to do so." Adidas said: "We're trying to figure it out, but you can be sure when one manufacturer has it the others will have it soon afterwards."

Nike appears to be the most forward-looking of the three. It said: "Organic cotton is clearly the better choice [over non-organic]. In 1997 we began using 3 per cent organic cotton in 50 per cent of our 5.4-ounce T-shirts produced in the United States. In 1998 we increased this amount to 75 per cent of our domestic T-shirt volume. Because the organic cotton industry is still slight in comparison to traditional cotton farming, Nike's demand would far outweigh current supply. For example, increasing our usage to 100 per cent organic cotton in just four T-shirt styles would require 7.5 million pounds (3.4m kg) or 42 per cent of the world's total supply. Such a demand would create a shortage in the market. So by increasing our usage as the market and economy allow, we're helping to grow this vital industry, and better our products." Although this response suits Nike's timetable as well as the growth of the organic cotton industry, it is a defensible argument, and assuming the company does follow through its stated intentions, this could be a significant positive development in the environmental impact of all sports, not just tennis.

The only apparent sportswear range for people wanting 100 per cent organic cotton is Patagonia, a company founded in 1957 by an environmentally aware climber, which has sought to be ahead of the clothing industry in its range of sports and camping/outdoor clothes. (Each year it gives 10 per cent of its profits or one per cent of sales to environmental causes.) However, its tennis options are very limited and available only in some seasons.

Regardless whether a tennis shirt is made from organic cotton or not, one of the most environmentally sound things a player can do is to go on wearing it. The tennis fashion industry gives the top players two new collections a year, with the aim of making club and leisure players feel they could not possibly wear a shirt that Agassi was wearing this time last year. But they can, and environmentally they should, otherwise unnecessary and unsustainable amounts of materials – some of them harmful – will get used.

Tennis shoes

Tennis shoe manufacture came in for massive criticism four years ago when the charity Christian Aid exposed widespread abuse of cheap labour in shoe factories in Asia used by the top sportswear names (almost all sports shoes come from Asia these days). Because part of that criticism revolved around exposure to hazardous chemicals and other environmentally harmful substances, the debates about social and environmental conditions in sports shoe factories have gone hand-in-hand.

One of the most problematic ingredients in the manufacture of sports shoes is PVC. The vinyl compound normally associated with home window replacement has been criticised by environmental organisations for many years because of the hazardous chemicals that can leach out during production and disposal of PVC products. Such is the seriousness that in 1998 15 European governments signed the Ospar Convention – this commits all signatories to eliminate emissions of hazardous chemicals into the environment, and in its priority list of the 15 worst chemicals, PVC production & disposal is responsible for eight of them.

Nike announced in August 1998 that it was to stop using PVC in shoes, and also in duffel bags and logos on Nike clothing items. The statement was one of intent, in that research into alternatives is still at an early stage, and Nike says it still does not know when the first pair of PVC-free shoes will be ready. However, Greenpeace toxics campaigner Mark Strutt said: "This is an important move by Nike, and while we have to keep a constant watch on big multinational companies, there seems good reason to believe Nike are serious about this. We were the ones who urged them to publicise their move away from PVC, they weren't going to do anything, and they have also made their new headquarters in the Netherlands totally PVC-free, so the commitment seems to be there."

Reebok and Adidas were less forthcoming. Reebok said it had no plans to eliminate PVC and added: "To the best of our knowledge there are no regulations banning the use of PVC by athletic footwear companies." Adidas said it would like to eliminate PVC but was standing in line with the rest of the sports footwear industry. Its spokesman said: "Most products come from the same place, so if Nike can eliminate PVC, so can we. PVC won't be around for ever, but it's a question of years rather than months, and you can be certain that when the industry finds an alternative we will all have it."

Of the leading sports goods manufacturers, Nike seems to be showing greatest concern and has probably moved furthest in the last couple of years on social and environmental welfare. Its critics might say it had furthest to move after being heavily criticised (and Reebok had a human rights department well before Nike was criticised for its Asian factory conditions), but its answers were much more constructive than the "we're not quite sure" tone of Reebok's and the "we'll wait for someone to do it first" attitude of Adidas. As well as the commitment to eliminating PVC, Nike and Reebok have said they will reduce the amounts of the volatile solvent toluene used in its factories, which is a threat both to the environment and human health. And while there is still much to be done to bring workers' conditions up to those considered minimum standards in western factories, all manufacturers are moving towards agreed minimum standards.

Adidas' spokesman Peter Csanadi added: "I'm afraid environment-friendliness has not been a forerunner in our industry, mainly because so little can be recycled. We are all aware of the environment, and we categorise it together with social and health concerns, but not a lot is happening in our industry at present."

Travel

Just as transport is becoming one of the most visible symbols of the environmental debate, so it is one of the big black marks in tennis. The global tennis circus is made up of 70 tournaments on the ATP Tour, 41 on the Sanex WTA Tour, four Grand Slams, several dozen Davis and Fed Cup ties, plus several hundred Challenger, Futures, Satellite and Junior tournaments. That means thousands of players, coaches, agents, officials, administrators and helpers being flown and driven several thousand kilometres, all of them sending various polluting and climate-changing substances into the air.

Though ships and trains emit harmful gases, the worst forms of transport in terms of emissions and use of land are road and air, which are the two most used in tennis. Tim Henman's schedule in 1999 saw him playing 25 tournaments, one exhibition and two Davis Cup ties. Over the year he flew 68,000 miles, which means he caused emissions of about 21,956 kg of carbon dioxide (the principal gas responsible for climate change through global warming) and 108 kg of nitrogen oxides (one of the main pollutants), plus smaller amounts of carbon monoxide, hydrocarbons and sulphur dioxide.[1]

As for his road mileage, this can only be guessed at, because it won't just be the miles he runs up in his own car (driving to airports, the daily run to London's Queen's Club to practise when he's not at a tournament, plus the three London-based tournaments), but driving to Birmingham twice for Davis Cup ties, and all the miles done ferrying him in official cars at tournaments. If one estimates it at 30,000 in a year, this would make 6,228 kg of carbon dioxide, 72.4 kg of carbon monoxide, and 14.5 kg of nitrogen oxides.

These same figures apply for his coach David Felgate who travels everywhere with him, with slightly lower amounts for his (now former) fitness coach Tim Newenham, his agent Jan Felgate, his then fiancée Lucy Heald, and several journalists who follow the tour with him. Some players and their entourages will have even worse records, notably Pat Rafter who lives in Bermuda but in 1999 had to travel to Harare and Brisbane for Davis Cup ties, and Yevgeny Kafelnikov whose home is on the Black Sea and who has also had to travel to Brisbane for Davis Cup.

With a global tennis circuit, it is hard to see how such pollution can be countered. Improved aircraft technology is making planes a little more fuel-efficient which is helping emissions (and also reducing noise levels), but short of reshaping the calendar to stop the amount of to-ing and fro-ing across the world, and getting players to go by train between tournament cities less than 500 miles apart – which would probably be resisted in the current climate – the options seem limited.

Local travel to tennis – whether it be spectators attending professional tournaments, or club and public parks players having their weekly hit – is where the

---

[1] Calculation based on following assumptions: Swedish airline SAS estimates 214g CO_2 per passenger km and 1.12 g/pkm NOx in a Boeing 747, while Dutch government study puts the figures for Boeing 737 at 146 g/pkm for CO_2 and 0.44 g/pkm for NOx (based on the planes being 71% full). As 88,000 of the 109,400 km travelled were on long-haul (around 80%), 88,000 km calculated on 747 rates and 21.400 at 737 rates. (Figures supplied by Aviation Environment Federation, London)

scope for being more environmentally aware is greater. Many tournaments around the world now have a deal whereby the ticket price includes travel by public transport to the stadium, park & ride schemes are springing up, and tournaments are no longer considering it their duty to provide masses of car parking (and people who have difficulty parking often end up using trains and buses).

As for amateur players, there is something mildly ironic about people taking their weekly or twice-weekly exercise by driving to a tennis or other sports club when walking or cycling could in many cases be part of the exercise (or at least assist with warming-up and stretching). For some, the distance between home and club is so far they would have used up most of their energy by the time they walk on court, and in many cases carrying racquets, balls, shoes, kit, towels, etc is not comfortable for more than a few metres. Yet this latter obstacle has been partly overcome at clubs in Switzerland and Germany, where lockers allow members to leave racquets, shoes and balls in the locker and only bring in and out the sweaty washing. This makes a short walk, cycle ride or bus journey much more plausible.

Recycling at tournaments

Because of the large amount of people who attend tennis tournaments, the potential for recycling is vast. The German Open in Hamburg is believed to have been the first to have recycling banks for bottles and paper on site, and many other tournaments have since copied that (including the Australian and US Opens). Some have introduced paper-only bins in press rooms, though the world's tennis media often seem impervious to such efforts.

CONCLUSIONS AND RECOMMENDATIONS

The fact that this is probably the first research carried out on tennis's impact on the environment suggests the sport has not exactly had the environment at the top of its thinking. There are clearly many things that can be done by everyone from top professionals and administrators, to club players and honorary secretaries. But set against other sports, tennis does not emerge too badly.

When played outdoors *in daylight* it requires virtually no electricity. The move away from grasscourts has meant much fewer pesticides are used (certainly compared to "big grass" sports like golf and cricket), and while most tennis players arrive at the court by car, there is not the intrinsic use of fuel that happens in motorsports or private flying. The fact that four or six new balls last just nine games at the top level and about five sets at club level is very wasteful, but there is ample scope for balls which have had maximum usage to have their felt and most of their rubber recycled – it just needs a financially viable system to be in place.

There is both short- and long-term action that can be taken. An obvious short-term option would be for national tennis associations to issue a simple leaflet on what amateur players can do to minimise their impact of the environment. This could include tips such as:

- Play outdoors whenever possible to conserve energy that would otherwise be used for heating and lighting indoor courts.
- Use balls for as long as possible, and take the old tubes down to the recycling bank.

- Don't get a new racquet unless you really need one – most racquets will last for many years.
- Ask your tennis shop about clothes made from organic cotton and PVC-free shoes – even if they've never heard of either, you're helping spread the word.
- Walk or cycle to the tennis courts if possible, or share a lift if it isn't.

On a longer-term basis, companies involved in tennis will have to make a much greater effort to reduce their impact on the environment. After all, tennis is not of such profound value to society that its harmful aspects could be tolerated in the interests of general well-being, so if a point were reached where limits were put on nations' emissions of polluting and climate changing substances, tennis would only survive if it did negligible damage to the environment.

Of course, tennis bodies could take the view that environmental threats are overstated, that raw materials will not run out, that the adverse effects of pollution and climate change have been exaggerated, etc, and that therefore tennis can carry on as normal. In the light of current scientific thinking, that could well prove a foolhardy approach. For if current environmental predictions prove accurate, there is likely to be a major shift in lifestyles that will make today's tennis – certainly the international tennis circuit – simply unsustainable. There looks to be a limit on how long motoring and flying can continue to grow, and when they are either restricted by international agreements or become much more expensive than they are now, tennis will probably revert to something like it was in the 1930s, when most tournaments were organised on a regional or national level, and only the best players met for a few selected global tournaments (to which they travelled either by very expensive plane, or by train or boat as part of a long trip).

But at least tennis's past proves that it could survive in that environment. It just might be easier to survive if action were taken now than if it were left until the time when the situation becomes critical, action such as building tennis infrastructure which will prove very energy-efficient to maintain and operate.

REFERENCES

Institution of Lighting Engineers, (1997) Rugby – Guidance Notes for the Reducation of Light Pollution.

Myers, D. and Stolton, S., (1999) ed. – Organic Cotton: From Field to Final Product (Pesticides Trust and Intermediate Technology Publications)

Unforced errors and error reduction in tennis

H. Brody
Physics Department, University of Pennsylvania, Philadelphia, PA 19104, USA

ABSTRACT: Unforced lateral (side-to-side) errors can be reduced by returning the ball to the location that it is coming from and by following through your stroke in the direction you want the ball to go. Unforced vertical (depth) errors can be reduced by not hitting the ball as hard, striking the ball when it is higher, going cross-court rather than down-the line, and possibly adding top spin. Rackets with wider heads and tighter strings will reduce errors as will some of the newer rackets with their maximum power region moved up in the head.

INTRODUCTION

If you are a typical tennis player, you will lose many more points by making errors than you will lose because of winners hit by your opponent. If your rate of errors can be reduced, it should lead to an increase in the number of points that you win. There are several ways in which your error rate can be reduced, and some of them will be discussed here.

An improvement in stroke mechanics, lots of practice, and better concentration are obvious ways to reduce your error rate, but these are not the subject of this article. Appropriate shot selection and a different racket or stringing also can lead to a reduction in your error rate, if you are willing to follow advice based on scientific analysis.

LATERAL ERRORS

Errors come about because the ball has hit the net or gone long (an error of depth), or because the ball has gone wide (a lateral error). If you aim your shots so as to have the ball go down the middle, you have an allowable error of almost ten degrees to the right or to the left before the ball goes wide, and this is a sizable margin for error. If your shot bounces well short of the baseline, then your margin for side to side angular error is even greater. The problem is that most of you don't want to play safe and hit every shot down the middle. You want to go for the corners or go down the sideline. You may want to pass an opponent at the net. You want to go for an occasional winner or at least make your opponent run for the ball once in a while. When you do that, you are inviting lateral errors to occur.

CHANGING ANGLES

However, you can reduce the number of errors from shots that go wide, even when you go for corners or side lines by remembering this advice. DON'T CHANGE THE BALL ANGLE! If a shot is coming to you cross court, return it cross court. If a shot is hit down-the-line, return it down-the-line. When you change the ball angle by attempting to return a cross court shot down-the-line or return a down-the-line shot cross court, you are asking for error troubles. The reason for this comes out of the physics of the ball-racket interaction.

When your return does not change the ball angle, the ball's impact direction is perpendicular to the face of the racket at contact. It will then leave your racket in a direction perpendicular to the face of the racket. The ball will go out in the direction of the racket motion whether you swing hard or soft. This is not the case when you attempt to change the ball angle. In this case the direction that the outgoing ball makes relative to your racket face depends on how hard you swing your racket. The higher the relative ball-racket speed, the closer the ball will be to perpendicular to the racket face as it leaves the racket. If your swing is slow, then the ball will leave your racket at a larger angle. Therefore, when your return changes the ball angle, the chances of you making an error and the ball going wide, depend on how hard you swing. When you do not change ball angle, since the ball direction does not depend on how hard you swing, lateral errors should be reduced. It is your choice.

This advice also holds when you volley. If you punch out at your return, the ball will go close to where you aim it. If, on the other hand, you just try to block the ball on your volley, and you are changing angles, the ball will slide off at a large angle, possibly going wide.

However, if you return every shot back to where it came from, your opponent will quickly catch on, and will be there waiting for it. Knowing the facts about the ball-racket interaction, you can reduce the errors that might occur even when you change the ball angle. If you are not going to hit the ball hard, then aim the ball a little closer to the center of the court. If you are really going to swing out, then you can aim your shot a bit closer to the sideline or the corner with confidence.

The famous statement that the angle of reflection equals the angle of incidence holds for light reflecting from a plane mirror, not tennis balls rebounding from a racket.

Often, in a match, when you find yourself well ahead of your opponent, you may ease up a bit, and not hit your shots quite so hard. You do this to reduce the number of errors you are likely to make. This can lead to a problem. If you continue to aim the ball the same way, only not swing as hard, balls that previously went down-the-line will now end in the alley.

A similar problem can be the result of you changing your game plan in the middle of a match. You become concerned about the final outcome, so instead of hitting out and playing your regular game, you decide to play it safer and ease up on your strokes. Again, balls that previously went down-the-line will now end in the alley. You end up making more, not less errors. People will claim that you "choked," when actually all you did was not understand the laws of physics.

TIMING ERRORS

A second type of lateral error is caused by errors in timing your swing. Timing errors cause the contact point of the ball-racket interaction to move forward or backward from the location you are aiming for. If your racket's motion were translational only, this would not be a problem. But the racket is swung (rotated), not translated, so a variation in contact point corresponds to a variation in racket head angle at the time of ball impact. A change in this angle means that the ball has a lateral component of velocity that was not anticipated and the possibility of lateral error.

To reduce this error (assuming the timing error cannot be reduced), the angular change in the racket head due to the timing error must be reduced. This can be done either by swinging the racket slower or by increasing the effective radius of the swing. The former has the disadvantage of reducing the shot speed as well as the lateral error. The latter means swinging from the shoulder or trunk rather than the wrist.

You often hear the advice "Keep the ball on your racket strings longer for better control." What is really being said is to follow through with your racket head in the direction you want the ball to go. If the racket is following the desired ball trajectory, an error in timing will have a minimal effect.

ERRORS OF DEPTH

Where the side-to-side errors in tennis are basically a matter of court geometry, the vertical or depth errors depend on many factors, and your margin for error is often an order of magnitude smaller than the side-to-side error margin. This means, that on occasion, you will be trying to thread a vertical needle with your shot. In order for a shot to be good, it must be hit through a window - either a virtual window in vertical angle as the ball leaves the racket that will allow it to clear the net and bounce in the court, or an effective window just above the net, the upper edge of which corresponds to a ball bouncing right on the baseline. For a given ball speed, if your ball passes over the net above the top of this window, it will bounce beyond the baseline.

ACCEPTANCE WINDOW

The size of this window decreases markedly as the ball speed increases, but increases as topspin is added. When the ball is not hit hard, after it crosses the net, gravity has more time to pull it into the court before it gets to the baseline. Therefore, the vertical window increases in size if you don't hit the ball as hard. The window also gets slightly larger if you increase the height at which the ball is struck. The larger the window, the less likely you are to make an error because of the ball failing to go through it. If the window is large compared to the variation in your strokes from shot-to-shot, you will be known as a steady player who does not beat him or her self. If you select a small window (by hitting the ball too hard, for example), then you will have great difficulty in winning, and most points will end early with you making an error, even if you have reasonable control when you stroke the ball hard. The choice of window size is selected by you and therefore your error rate also is effectively selected by you. If you want to hit the ball hard you will make more errors, both because the window is smaller and because it is more difficult to control the racket on a harder swing. Practice and correct stroke mechanics will reduce the variation of your shots from stroke to stroke, but they will not change the vertical window size. Only shot selection can do that.

The fact that shots that have higher velocity have smaller windows (and lead to more errors) may lead to an unanticipated source of errors. When a ball is coming at you with a very high velocity, if you stroke it back with your normal swing, your resulting shot will be a higher speed ball. This is because the speed of the struck ball is given by the formula:

$$v(\text{struck}) = e * v(\text{incident}) + (1 + e) * V(\text{racket}), \quad (1)$$

where the parameter e is about 0.40. This means that, if the ball coming toward you is hit harder, unless you reduce your swing speed somewhat, your return shot also will have a higher speed. The result will be a reduced window for your shot to be good, and an increase in your error rate. If you don't want this to happen, you should slow down your swing a little when your opponent is blasting the ball at you. As a bonus this also might give you better racket head control.

BALL SPIN

You can increase the small window that comes from hard shots by putting topspin on the ball. Topspin causes the air to provide a downward force, in addition to gravity, that causes the ball to dive into the court. But topspin also has some penalties associated with it. Many players have almost no control of their shots when they attempt to use topspin, so even though the window size increases, their shots spray all over (and often out of) the court. One of the reasons for this may be the following. When a ball bounces, it normally comes off of the court with considerable topspin due to the friction force between the ball and the court surface. When you try to return such a shot with your own topspin, not only must you turn the ball's direction around, you also must turn its spin around. In order to do this you must swing your racket 1 ½ times faster than if you want your return to be flat (with no spin on the ball). If you are already swinging your racket fairly hard (which is why the window is small), the loss of control due to a 50% increase in swing speed may be a major cause of the increase in your errors.

Back spin (under spin or chop) shots have a reduced window because the ball tends to float long after clearing the net. (The interaction of backspin with the air produces an upward force on the ball that counteracts gravity.) Yet, many intermediate level players seem to be able to play a steady game chopping away at the ball and getting it to go through the small window they have set for themselves. This is because they do not have to hit the ball very hard to return it, due to the fact that the incoming ball has acquired topspin because of its bounce. All they have to do is turn the ball's direction around, since the ball is already spinning the way they want it to. Not having to spank the ball really hard means that the player can use a compact stroke, get away with poor racket preparation, and still have good racket head control. This often reduces their shot to shot variation sufficiently to more than compensate the reduced window size.

OTHER METHODS

You also can increase the size of the window on any shot (which reduces the probability of making an error) by hitting the ball when it is high, as opposed to striking it when it is down around your ankles. The incoming ball trajectory does not always give you that option, and many players do not have the control or power to hit shots at shoulder height. But if you can hit the ball when it is higher, vertical errors will be reduced.

Another option you have is the down-the-line versus cross court decision when you are hitting from the corner. This has already been discussed with respect to change of angle, and now it will be analyzed with respect to errors in depth. The vertical window for a successful shot is always greater for a shot hit cross court rather than down-the-line. This is because the court is much deeper (the baseline further away) when you hit at an angle, along the hypotenuse of a triangle. You are effectively hitting into a bigger court when you go cross court, compared to down-the-line. The harder you hit the ball, the greater is the advantage of cross court over down-the-line. If your shot lacks pace, you will not make many more depth errors going down the line. If, however, you really swing out at the ball, then the cross court shot will result in far fewer errors.

NOTE: No mention has been made of the net being lower at the center (where the cross court shot passes over it). That fact is not an important factor in the suggestion to go cross court. This is because for a ground stroke with moderate pace from the corner, the minimum vertical angle needed to clear the net is the same for a cross-court shot and a down-the-line shot.

EQUIPMENT

The racket frame and the stringing in that frame can influence the number of errors that you make. For a long time the literature of tennis has stated that tighter strings lead to better control, and better control means fewer errors. However, nowhere in the literature can you find an experiment, data, or a scientifically valid argument explaining why this is so. An article appearing in the Winter 2000 edition of the International Sports Journal was the first to show a measurement of the correlation between string tension and control. (Brody, 2000) It has long been known that lower string tension does produce greater ball speed (power), but the arguments about the correlation between tension and control are usually anecdotal.

OFF-AXIS IMPACTS

In normal play, you attempt to hit the ball at a specific location on the strings, possibly the center of the head. If you miss that location, you may make an error. This is because the ball comes off of the racket at an unanticipated angle or with a speed higher or lower than desired. The result of this slight miss-hit is a ball trajectory that often ends in the net or over the baseline.

When the ball hits the strings at a location that is not along the long axis of the racket (toward the side of the head), the frame will twist and the rebound will come off with lower speed than the same shot hitting on axis. Assuming the long axis of the racket is parallel to the ground, a ball impact above the axis will result in the racket twisting, a rebound angle greater than desired, and also less ball speed. This combination of effects (larger vertical angle and lower ball speed) will often lead to ball trajectories that still end landing in the court. If, on the other hand, the ball strikes below the axis, the rebound ball angle will be lower than desired and the ball speed again will be lower. This combination often leads to shots that end in the net. (An error reducing strategy would be to attempt to always hit the ball slightly above the axis. However, the unanticipated consequence of such a strategy might result in tennis elbow from the constant twisting of the racket due to off-axis ball impacts.)

The fractional reduction of rebound ball speed depends on how far off axis the impact is and a physical property of the racket called the moment of inertia about the long axis. The larger this moment of the racket, the less the rebound ball speed will be effected by off axis hits and the more stable the racket will be against twisting. Therefore, by going to a racket with a larger polar or roll moment of inertia (the moment about the long axis) you will make fewer errors due to the miss-hits, some of which are almost inevitable. As a general rule, rackets with a wider head (as opposed to wide-body, which is thicker) will have a larger moment of inertia and be more stable. A frame that is 25% wider (10 inches instead of 8 inches) can result in a 50% greater moment. This means that using a racket with a wider head usually will result in fewer errors due to this type of miss-hit. Howard Head knew what he was doing when he designed the original Prince oversize racket.

BALL IMPACTS ALONG THE AXIS

If the ball hits the racket along the long axis of the racket, but at different locations along that axis (closer to the tip or to the throat), errors of depth can again occur. For many rackets and classic ground stroke swings, the ball rebound speed decreases as the ball impact point moves from the throat area toward the racket tip, all other parameters held constant. If you are attempting to strike the ball at the center of the head and trying to hit a deep shot, ball impacts that occur closer to the throat are likely to go long because the ball will come off of the racket with too great a speed. This can be compensated for by consciously hitting your normal (central impact) shots so they bounce short. Now the balls that inadvertently impact closer to the throat will land in the court. This is a very conservative strategy, and is comparable to hitting every ball softly to insure a large window and few errors.

Some of the modern, very stiff, handle-light frames now available, have succeeded in moving the power region further up into the head of the racket. When one of these frames is combined with a flexible wrist swing (not the classic long, flowing stroke), the effective maximum power point of the racket can move up so that it is at or near the center of the head. This is because the center of the head is moving faster than the throat on these type of swings. Calculations show that when the power maximum is at the center of the head, the variation of rebound ball speed as the ball impact point moves away from the center, is greatly reduced, compared to the classic racket used with a classic stroke. In other words, the "power spot" has been enlarged and made more uniform. Clearly, this will reduce errors of depth.

In addition, there is a built-in error correcting mechanism that occurs when the maximum power point is at the center of the racket head. Any shot that misses the center will come off the racket with slightly less rebound speed than a shot that hits at the center. If you then attempt to hit your shots so that balls that strike the racket at the center of the head land deep in the court, depth errors (balls going over the baseline) due to variation in impact location will automatically be reduced. Any shot that hits the strings slightly closer to the throat or slightly closer to the tip than the center of the head, will come off of the racket with slightly less speed, and therefore will land a bit shorter. These types of slight miss-hits will not result in balls that land long, and therefore depth errors are reduced. Because this effect is working in your favor, you can hit your shots deep, go for the baseline, and play a more aggressive game.

CONCLUSIONS

A reduction in the error rate should clearly mean an increase in the number of points won. There are methods other than the traditional "lots of practice" that can lead to a lowering of the error rate. A knowledge of the science behind ball trajectories and the ball racket interaction can lead to a number of strategies and tactics which will reduce errors. While these recommendations lead to a more conservative style of play, they also lead to more points won, which should translate into more matches won. Equipment can be selected and tuned to reduce the rate of unforced errors.

REFERENCES

Brody, H. and Knudson, D. (2000) A Model of Tennis Stroke Accuracy Relative to String Tension. International Sports Journal, 4, 38-45

Brody, H. (1987) Tennis Science for Tennis Players. University of Pennsylvania Press, Philadelphia, PA

5 Sports science

Muscle activity in tennis

E. Paul Roetert
American Sport Education Program, Champaign, IL, USA
Todd S. Ellenbecker
Physiotherapy Associates, Scottsdale, AZ, USA

ABSTRACT: This paper outlines the physical demands of tennis focusing specifically on the components of flexibility strength and agility. Specific muscle involvement in each of the tennis strokes is outlined. Available research has given us a profile of the modern tennis player from a physiological standpoint, however more research is needed for different age groups, levels of play and both genders. In addition, there is a real lack of longitudinal data available in the current literature related to tennis. Future research is recommended in these areas.

INTRODUCTION

Due to the popularity of the sport, tennis has received a great deal of attention from scientists. In the United States alone, over 2000 junior tournaments are held each year. In addition, many tournaments are available for adult and senior players. Researchers have been interested to learn more about how the game should be played, the psychological aspects of competition, how players learn specific motor skills, how human physiology plays a role in tennis and a number of other related topics (Groppel and Roetert, 1992). This purpose of this paper is to focus on the tennis player from a physiological perspective and specifically on the muscular actions involved in each of the tennis strokes. To be able to compete effectively against progressively more elite opponents, requires higher levels of physical fitness related to such factors as strength, power, muscular endurance, flexibility, coordination and agility. Therefore, standardized testing might provide a useful addition to subjective coaching appraisals (Quinn, 1989). By testing junior tennis players early in their careers, physical training programs can be implemented to strengthen weaknesses and lessen the risk of injury (Roetert, et al, 1995). The exercise prescription for individual players depends on a number of factors, including the initial fitness level of the player, the goals of the player, the injury history of the player, and the level of play of the participant (Chandler, 1995).

MUSCULAR DEMANDS OF TENNIS

Tennis has often been characterized as a sport in which players must respond to a continuous series of emergencies. (Groppel, 1986). Among others, USTA (1998) identified the following physical characteristics related to successful tennis play: sprinting to the ball, changing directions, reaching, lunging, stretching, stopping and starting and balance. All this while maintaining balance and control to hit the ball effectively. These characteristics make it clear that speed and agility are important components to successful tennis performance. In fact agility and speed were found to have a higher correlation to tennis performance than any other physical performance factor (Roetert, et al. 1992). Throughout a tennis match players may hit hundreds of balls while running from side to side. Since tennis requires players to make shots that place the body in extreme positions, proper flexibility is paramount. Flexibility is the degree to which the soft tissue structures surrounding a joint – muscles, tendons and connective tissues – stretch (Sobel et al. 1995). Proper flexibility for both upper and lower body are critical for tennis performance and injury free play, therefore testing and training programs should be designed with muscular balance for the whole body in mind. Muscular endurance (applying force and sustaining it over time) can help players hit the ball just as hard at the end of a match as the beginning. Even though in tennis, upper body strength deserves attention for achieving optimum stroke potential and injury prevention, coaches should include a proportionately significant amount of lower body strength and endurance conditioning in their tennis training design (Bergeron, 1988). Ellenbecker and Roetert (1995) showed that, in contrast to the upper body, symmetrical lower extremity strength can be expected in tennis players. In addition to muscular endurance tennis requires explosive movements. Players with explosive first steps get into position quickly, set up well, and hit effective shots. In addition, an explosive first step, will give players the speed necessary to get to balls hit farther away (Roetert, et al 1995). Therefore both upper and lower body power are necessary in tennis.

MUSCLE INVOLVEMENT IN THE STROKES

If researchers can investigate which muscles are involved in each of the strokes, when they are active, what the predominant types of muscle action are, what the motor patterns are, and what the speeds of motion are, then they can provide this useful information to coaches and players. In addition, trainers and strength and conditioning experts can then help in the design of exercise programs that are tennis-specific. These training programs, if they appropriately incorporate flexibility and strength components, can also minimize the degree of structural and functional asymmetry. All this information can and should be used for the purposes of performance enhancement as well as injury prevention. Based on the available research to date, the table below highlights the major muscle groups involved in each of the tennis strokes.

MUSCLE GROUPS USED

Table 1 Muscle groups used in tennis strokes

Forehand Drive and Volley	
Action	**Muscle Used**
Push-Off	Soleus, gastrocnemius, quadriceps, gluteals
Trunk Rotation	Obliques, spinal erectors, abdominals
Forehand Swing	Anterior deltoid, pectorals, shoulder internal rotators, elbow flexors (biceps), serratus anterior

One-Handed Backhand Drive and Volley	
Action	**Muscle Used**
Push-Off	Soleus, gastrocnemius, quadriceps, gluteals
Trunk Rotation	Obliques, spinal erectors, abdominals
Foreward Swing	Rhomboids and middle trapezius, posterior deltoid, middle deltoid, shoulder external rotators, triceps, serratus anterior

Two-Handed Backhand Drive	
Action	**Muscle Used**
Push-Off	Soleus, gastrocnemius, quadriceps, gluteals
Trunk Rotation	Obliques, spinal erectors, abdominals
Backhand Swing (Non-dominant side)	Pectorals, anterior deltoid, shoulder internal rotators
(Dominant side)	Rhomboids and middle trapezius, posterior deltoid, middle deltoid, shoulder external rotators, triceps, serratus anterior

Table 1 contd. Muscle groups used in tennis strokes

Serve and Overhead	
Action	**Muscle Used**
Trunk Rotation	Obliques, spinal erectors, abdominals
Knee and Hip Extension (before impact)	Quadriceps, gluteals
Arm Swing	Pectorals, shoulder interior rotators, latissimus, triceps
Arm Extension	Triceps
Wrist Flexion	Wrist flexors

Reprinted with permission from "Complete Conditioning for Tennis", United States Tennis Association, Human Kinetics, Champaign, IL.

These muscle groups have to accommodate explosive movement patterns and highly intensive maximal concentric and eccentric muscular effort. To get an accurate picture of the high velocity demands of tennis, testing procedures should emulate the type of concentric and eccentric muscular contractions during the game (Ellenbecker and Roetert, 2000; Dillman, 1991; Shapiro and Stine, 1992; Roetert, et al 1992). Similarly, because of the repetitive upper and lower extremity muscular activity, muscular endurance testing will also assist in the evaluation and design of optimal conditioning programs (Ellenbecker and Roetert, 1999). Proper flexibility testing protocols should focus on the imbalances and other inherent stresses of the game. Much research is available in this area for different levels and age groups of tennis players (Chandler, et al, 1990; Chinn, et al, 1974; Ellenbecker, 1992; Ellenbecker, et al, 1996; Kibler, et al, 1996; Kibler, et al, 1988), but more longitudinal data is needed (Roetert, et al, 2000).

CONCLUSIONS

Testing procedures should provide tennis coaches and players with better information regarding performance and training methods. A current body of physiological research related to strength, flexibility, agility and speed is available based on results of tennis-specific field tests, goniometric measurements, isokinetic measurements, and EMG data. This information should be the basis of future research and needs to be expanded to different age groups, levels of play and both genders. In addition, players should be studied over time so that more longitudinal data becomes available. All this information should lead to the design of appropriate and sport-specific training methods. Since tennis requires good footwork and speed of movement in positioning oneself for effective strokes, training techniques should mimic the demands of the sport (Roetert, et al, 1997). The translation of scientific data to practical training methods and applications is the most important step in assisting players with both the enhancement of performance and the prevention of injury.

REFERENCES

Bergeron, M.F. (1988). Conditioning the Legs for Tennis. *National Strength and Conditioning Association Journal*, 10: 40-41.

Chandler, T.J. (1995). Exercise Training for Tennis. *Clinics in Sports Medicine*, 14 (1), 33-46.

Chandler, T.J., Kibler, W.B., Uhl T.L. et al. (1990). Flexibility Comparisons of Junior Elite Tennis Players to Other Athletes. *American Journal of Sports Medicine*, 18: 134–136.

Chinn, C.J., Priest, J.D., Kent, B.E., (1974). Upper Extremity Range of Motion, Grip, Strength, and Girth in Highly Skilled Tennis Players. *Physical Therapy*, 54: 474–82.

Dillman, C.J., (1991). The Upper Extremity in Tennis and Throwing Athletes. Paper presented at the United States Tennis Association Annual Meeting (Tucson, AZ).

Ellenbecker, T.S., (1992). Shoulder Internal and External Rotation Strength and Range of Motion of Highly Skilled Junior Tennis Players. *Isokinetics and Exercise Science* 2: 1–8.

Ellenbecker, T.S. & Roetert, E.P. (1995). Concentric Isokinetic Quadricep and Hamstring Strength In Elite Junior Tennis Players. *Isokinetics and Exercise Science*, 5, pp. 3-6.

Ellenbecker, T.S., Roetert, E.P., Piorkowski, P.A., Schulz, D.A., (1996). Glenohumeral Joint Internal and External Rotation Range of Motion in Elite Junior Tennis Players. *Journal of Orthopedics and Sports Physical Therapy*. 24: 336 – 341.

Ellenbecker, T.S., Roetert, E.P., (1999). Isokinetic Muscular Fatigue Testing of Shoulder Internal and External Rotation in Elite Junior Tennis Players. *Journal of Orthopedics and Sports Physical Therapy*, 29: 275–281.

Ellenbecker, T.S. & Roetert, E.P. (2000). Isokinetic Testing and Training in Tennis. *Isokinetics in Human Performance*, L.E. Brown, Ed. pp. 358-377. Human Kinetics, Champaign, IL.

Groppel, J.L. (1986). The Biomechanics of Tennis: An Overview. *International Journal of Sport Biomechanics*, 2: 141-155.

Groppel, J.L. & Roetert, E.P. (1992). Performance Profiles of Nationally Ranked Junior Tennis Players. *Journal of Applied Sport Science Research*, 6 (4): 225-231.

Kibler, W.B., Chandler, T.J., Livingston, B.P., Roetert, E.P., (1996). Shoulder Range of Motion in Elite Tennis Players. *American Journal of Sports Medicine*, 24(3): 279 – 285.

Kibler, W.B., McQueen, C., Uhl, T., (1998). Fitness Evaluation and Fitness Findings in Competitive Junior Tennis Players. *Clinics in Sports Medicine*, 7, 403 – 416.

Quinn, A.M., (1989). Evaluate Your Players' Tennis Fitness. In: *Science of Coaching Tennis*, J.L. Groppel, J.E. Loehr, D.S., Melville and A.M., Quinn (Eds.) Champaign, IL: Leisure Press. pp. 131 – 248.

Roetert, E. P., Garrett, G. E., Brown, S.W., & Camaione, D. N., (1992). Performance Profiles of Nationally Ranked Junior Tennis Players. *Journal of Applied Sport Science Research*, 6, (4), pp. 225 – 231.

Roetert, E. P., Piorkowski, P.A., Woods, R.W. & Brown, S.W. (1995). Establishing Percentiles For Junior Tennis Players Based On Physical Fitness Testing Results. *Clinics in Sports Medicine*, 14, (1), pp. 1-21.

Roetert, E.P., Ellenbecker, T.S., Chu, D.A. & Bugg, B.S. (1997). Tennis-Specific Shoulder and Trunk Strength Training. *Strength and Conditioning*, June: 31-39.

Roetert, E.P., Ellenbecker, T.S., Brown, S.W. (In Press). Shoulder Internal and External Rotation Range of Motion in Elite Junior Tennis Players: A Longitudinal Analysis. *Journal of Strength and Conditioning Research.*

Shapiro, R., Stine, R.L. (1992). Shoulder Rotation Velocities. Technical Report Submitted to the Lexington Clinic, Lexington, KY.

Sobel, J., Ellenbecker, T.S. & Roetert, E.P. (1995). Flexibility Training for Tennis. *Strength and Conditioning.* December: 43-50.

United States Tennis Association, Roetert, E.P. & Ellenbecker, T.S. (1998). *Complete Conditioning for Tennis*, Human Kinetics, Champaign, IL.

Rackets, strings and balls in relation to tennis elbow

B.M. Pluim
Royal Netherlands Lawn Tennis Association, Amersfoort, The Netherlands

ABSTRACT: What characteristics of the rackets, strings, and ball help to minimise vibrations and possibly, the load on the tennis elbow? This presentation covers the latest information on the effects of equipment modification on vibration, and the expected effects on tennis elbow. All aspects of the racket will be discussed, including material and composition, size, length, weight, balance, grip size, and grip material. The material, size, tension, and pattern of the strings will be critically analysed, as well as the effect of so-called vibration stoppers. Pressurised and non-pressurised tennis balls are compared, as well as the new 'slow-type' and 'fast-type" tennis balls. Practical recommendations regarding the ideal equipment choice to reduce impact-induced stresses on the arm will be given.

INTRODUCTION

Tennis elbow, or lateral epicondylitis, can be a very serious and debilitating condition that can keep a player sidelined for many months. It is a common condition among tennis players, although only about 5% of those who experience problems with tennis elbow are actually tennis players. The peak incidence of tennis elbow is found in the 35 to 50 age group. A study of 2,633 average tennis players showed that 31% of all players either had current pain or a history of elbow pain (Priest et al, 1980). Slightly higher frequencies of 35% and 41%, respectively, were found in local league players (Carroll, 1981; Gruchow, 1979). Kamien, who studied 260 long-time tennis players, reported an even higher frequency of 57% (Kamien, 1988). Professional tennis players have a lower incidence of tennis elbow than recreational athletes and more often suffer from medial epicondylitis.

TENNIS ELBOW

What are the symptoms of tennis elbow? Players typically complain of pain over the outer aspect of the elbow, which is aggravated by radial extension of the wrist and resisted supination or passive pronation of the forearm (Kamien, 1990). The onset of pain may be sudden or gradual, and may radiate into the forearm. The most painful tennis stroke for a player with tennis elbow is usually the backhand. Sometimes the pain is so intense that the arm cannot be used for the normal tasks of daily life. Physical examination reveals tenderness at the insertion of the m. extensor carpi radialis brevis tendon at the lateral epicondyle, pain during resisted extension of the wrist and pain with passive wrist flexion with extended elbow. Pathologic specimens show that this is a degenerative condition, more like a tendinosis than a tendinitis, resulting from chronic repetitive microtrauma. The extensor carpi radialis brevis tendon is involved in all cases of tennis elbow, and the extensor digitorum tendon in approximately one third.

TREATMENT

Treatment generally consists of rest, heat, massage, non-steroidal anti-inflammatory drugs, the application of ice after play, and the use of an elbow brace around the forearm. Rehabilitation exercises are started once the patient's symptoms begin to subside. The goal is to increase the strength, endurance, and flexibility of the extensor muscle group. Corticosteroid injections have been shown to be effective in certain cases (Verhaar, 1992). However, because of the negative influence of corticosteroids on the strength of the tendon bone junction, injections remain a controversial subject. Treatment should preferably be limited to two injections. Surgical intervention should be considered only in patients whose injuries fail to respond to an appropriate rehabilitation programme.

Various factors are thought to increase the likelihood of developing tennis elbow: age, frequency of play, strength and flexibility deficiencies in the forearm extensor muscles, faulty technique, grip tightness, and equipment. Hatze (1976) first proposed the idea that high frequency vibrations generated by ball-impact may have a causal relationship to tennis elbow.

IMPACT-INDUCED STRESSES ON THE ARM

Let's take a closer look at the impact-induced vibration of the racket-and-arm system at ball contact. If vibration is an important factor, it may be worth the time and effort to carefully investigate modification of equipment. If a tennis player's chances of suffering tennis elbow are ultimately more than one in two, surely it must be worth spending time to select the best tools!

What characteristics of the racket, strings and ball help to minimise vibrations and possibly, the load on the elbow? This presentation covers the latest information on the effects of equipment modification on vibration, and the expected effects on tennis elbow. There is surprisingly little consensus in the literature about equipment variables such as the composition, weight, grip size and string tension of the racket. However, armed with the necessary background information, it is possible to reach agreement on most aspects on the basis of common sense and logical deductions.

1) MATERIAL AND COMPOSITION

Wood

From the end of the 15th century until the late 1960s, all tennis rackets were made of wood (Kuebler, 1995). The advantages of wooden rackets are good playing characteristics and good damping qualities. However, production of laminated wooden rackets is very labour-intensive. The wood has to be bent into the right form after being made supple in water or steam, and then several layers (between 7 and 11) of wood must be glued together. Some contemporary wooden rackets are strengthened by adding carbon fibres to the frame. However, because of the success of the cheaper, stronger, lighter and more durable composite frames, wooden rackets are rarely made today.

Metal

Around 1970, the first metal rackets became available, Jimmy Connor's Wilson T2000 is one of the most famous examples (Kuebler, 1995). The most commonly-used metals were steel and aluminium. Steel rackets never became very popular, because the metal is quite heavy. Aluminium is relatively cheap, light, strong, and easy to process. The basic profile only needs to be bent into the correct form. The aluminium is rigid and light, which makes it more manoeuvrable and suitable for players with an attacking game. The biggest disadvantage is the transmission of vibrations to the hand and arm, which makes these rackets very uncomfortable to play with. Gruchow and Pelletier (1979) found that play with metal tennis rackets is an independent risk factor for the development of tennis elbow. Metal rackets are not, therefore, recommended for players with elbow problems.

Composite materials

In the early 1980s, the first composite rackets became available. These generally consist of a combination of graphite and fibreglass, to which other materials such as kevlar, boron or ceramic are added. The purpose of mixing these materials is to create a racket with certain flexibility, strength, weight and other properties that no single material possesses.

Kevlar, for example, is durable and light, and has good vibration damping properties (it is also used in bullet-proof vests!). Boron makes the racket rigid and is very strong and light. Graphite is light and fragile, but has good vibration damping qualities, and is therefore very often used in rackets. Short carbon fibres are more effective in damping vibrations than the long fibres, which offer a continuous path for the transmission of vibrations (Kamien, 1988). Silicon carbide is strong and has good damping qualities. Fibreglass adds flexibility to the racket.

It is possible to manufacture composite rackets that produce less vibration when striking the ball by increasing the ratio of the matrix to fibres. A nylon matrix also dampens vibrations more effectively than the commonly-used epoxy resin.

To reduce the load on the arm, choose a composite racket with good damping qualities.

(2) THE SIZE OF THE RACKET HEAD

Every racket face has an area where the shocks transmitted to the arm are minimal and where the ball rebounds with highest velocity - the so-called `sweet spot' (Brody, 1988). Actually there is not one, but three sweet spots, but that is of no further relevance here. The size and location of the sweet spot depends on factors such as the distribution of the weight over the racket, the flexibility of the racket, the size and shape of the head etc. In general, the larger the racket head, the larger the sweet spot and the lower the chances of hitting the ball off-centre, which results in loss of control, loss of ball velocity and an increased load on the arm and hand. This has been shown in an elegant study by Elliott et al. (1980), who demonstrated lower vibration levels and higher rebound velocities in oversized rackets, in comparison with their conventional counterparts.

Another reason for using a racket with a large head is to prevent the racket from twisting and turning. When the ball is hit off-centre, the racket tends to twist in the hand, which may result in a miss-hit. The resistance of a tennis racket to twisting is known as the polar or roll moment of inertia. This polar moment of inertia is proportional to the mass and to the square of the distance to the axis of rotation.

If this moment of inertia is increased, the racket will be more stable. This can be achieved either by increasing the mass at the edges, or by making the frame wider. It is much more effective to make the frame wider than to increase the mass, because the added mass only increases the moment linearly, while the width is squared. Increasing the width by 6 cm increases the resistance to turning by 40%! A wider head provides more stability when the ball is hit off-centre than a standard head, when the axis is missed by the same amount. However, with a wider racket one can miss-hit by a greater distance. A miss-hit with a wider racket will create a greater twisting force, while the same shot with a standard racket could hit the frame or miss the racket completely.

We can illustrate this with the example of a figure skater performing a pirouette. If she wants to turn faster she tucks in her arms. To slow down, she spreads her arms wide.

To reduce loads on the arm, choose a racket with a large `sweet spot', such as a mid-size or oversize racket.

(3) FLEXIBLE AND STIFF RACKETS

A flexible racket bends a great deal (has a high level of head displacement) on impact. A racket that bends little is described as `stiff'. It has been suggested that a flexible racket provides more power, as reflected in the slogan `Flex means power'. However, this is not true. If the ball hits a flexible racket, the racket head deforms quite considerably. The time taken for the racket to deform and return to its original shape is approximately 15 ms. This is longer than the dwell time for the ball on the strings (about 4-6 ms) (Groppel, 1987). Therefore, the ball has already left the strings before the frame has straightened again and the energy fed into racket deformation is lost to the ball. Consequently, there is a loss of kinetic energy. For more power, choose a stiff racket.

However, a flexible racket is kinder to the arm, because the flexion will absorb some of the shock and spread it over a longer period. This hypothesis is supported by a study conducted by Stone et al. (1999), who compared the incidence of tennis elbow among players using first generation and graphite composite rackets. Eighty-one entrants into the Men's National 40's Tennis Tournament (Monroeville, PA, USA) were surveyed for history of tennis elbow and racket usage. Thirty-one of the players had never developed tennis elbow, 17 had developed it more than 10 years ago, when wide body rackets were introduced, and 33 developed it during the last 10 years. The data suggest that the newer, stiffer, composite rackets may increase the rate of developing tennis elbow.

A possible compromise is a racket with a very stiff head and some degree of flexion in the shaft, since flexion in this location does not degrade the racket's power as much as in the head. This is realised in a racket with a so-called tapered profile: thin (and therefore flexible) in the shaft and wider (and therefore stiffer) in the racket head, such as a wide-body.

To reduce loads on the arm, choose a racket with a stiff racket head and a flexible shaft.

(4) LONG-BODY AND STANDARD SIZE RACKETS

A conventional racket is 68.6 cm long, but several years ago, longer rackets were introduced, with bodies of up to 73.7 cm. Long-bodied rackets increase the player's reach slightly. A long-body also enables a player to serve from a greater height, which decreases the risk of serving into the net. This will increase the power of the serve, because the player can take more risks. However, the longer fulcrum means that the momentum of the racket increases, which will increase the load on the arm. This is not a good idea if one is suffering from elbow problems.

Furthermore, players who switch from a standard-size racket to a long-body have to get used to the increased length of the racket. This means that quite a few balls will be hit off-centre at first. Since off-centre ball impacts increase the load on the arm approximately three-fold in comparison with centre impacts, a switch to a long-bodied racket is inadvisable for players suffering from tennis elbow.

To reduce loads on the arm, do not switch to a long-body if you have elbow problems

(5) THE WEIGHT OF THE RACKET

The weight of the racket varies between 275 and 360 grams. In recent years, the introduction of new materials, better design, and improved construction has led to a trend away from heavy rackets. Nowadays it is possible to make a lightweight frame that is both stiff and durable - something that would have been impossible in the past. Both racket weight and the speed of the racket head are factors that determine the velocity of the ball (Brody, 1988). The extra racket-head speed that can be achieved with a lighter racket head may more than compensate for its lighter weight and lead to higher ball velocity.

For a player with arm problems, however, a heavier racket is preferable. Even though more effort is required to generate racket speed, the shock transmitted to the hand and arm is less. The greater the mass of the racket, the greater its ability to absorb shock. A heavier racket also promotes better control, since it minimises twisting and turning in the player's hand. Strokes therefore tend to be longer and more fluid.

To reduce loads on the arm, choose the heaviest racket you are able to handle without sacrificing technique.

(6) BALANCE

The balance of a racket depends on the distribution of the weight over the racket. The balance point, or centre of gravity, is the point where the racket remains in balance when resting on a support. A racket that has its centre of gravity at its geometrical centre is balanced. A head-heavy racket has its balance point on the head side of the racket's centre. A head-light racket has its balance point closer to the grip than to the head (Brody, 1988).

Which type is best for a player with elbow problems? A head-light racket *feels* lighter than a head-heavy racket, although they actually weigh the same. This is because the *first moment* of a head-light racket - the weight of the racket multiplied by the distance from the balance point to the location of the hand - is less than that of a head-heavy racket. A head-light, low-moment racket is generally easier to manoeuvre than a head-heavy, high-moment racket, and may therefore be more suitable to someone with elbow problems.

To reduce loads on the arm, choose a head-light racket

(7) SMALL OR LARGE GRIPS

The grips of most rackets range between 4 1/8 inches and 4 7/8 inches in circumference. The sizes are usually labelled from 1 to 7. In the past, players used thicker grips. Grip sizes 6 and 7, which used to be very common, are very hard to find today.

A grip that is too small or too large may cause problems. In both cases, the player may have to grip the racket too tightly to prevent it from twisting, and high grip force may increase the risk of elbow injury (Brody, 1989). Increased grip tension at the moment of impact dampens the vibration of the racket. When the racket is held tightly at the moment of ball-racket contact, the magnitude of racket vibration is reduced. However, due to the increased mechanical coupling between the racket and the hand, the player experiences increased arm vibration (Hennig, 1992; Knudson, 1991). Researchers conclude that most of the oscillation energy has to be taken by the hand to dampen the vibrations of a racket in a short time.

Using electromyography, Adelsberg showed that racket torque is best controlled by using the largest comfortable grip size (Adelsberg, 1986).

Also, the larger the grip size, the greater the dorsal flexion of the wrist and the lower the tension on the origin of the forearm extensor muscles (Murley, 1987). Therefore, to reduce the chances of the racket twisting in the hand and developing tennis elbow, use a good, high-friction, non-slip grip covering, squeeze the handle firmly and use the largest grip that is comfortable. An easy way to determine the correct grip is by measuring the distance from the long crease in the palm (the second one down from the fingers) to the tip of the ring finger.

To reduce loads on the arm, choose the largest grip that is comfortable for you.

8) GRIP MATERIAL

Most rackets come with a leather grip. There are a number of materials that one can wrap over the grip to prevent the racket handle from becoming slippery. Many types of synthetic materials are quite porous and can absorb a good deal of moisture. But the material of the grip is also important when it comes to the prevention of tennis elbow. A cushioned grip tape can increase racket damping by up to 100% (!) and may reduce the grip reaction force by 20% (Wilson, 1995).

To reduce loads on the arm, choose a cushioned grip tape.

9) STRINGING MATERIAL

Is it better to string the racket with gut or with synthetic strings?

In the past, all rackets were strung with gut made from the intestines of cattle or sheep. The intestines of two cows are needed to produce one set of gut strings. The fabrication process is very labour-intensive, and gut strings are therefore more expensive. They are also more fragile, less durable and less resistant to damage from moisture and humidity than synthetics. So, why play with gut? A comparison of gut and first-grade synthetic-gut strung at the same tension suggests that gut strings give slightly higher post-impact ball velocities, improved control and better racket `feel' (Gropppel, 1987; Ellis, 1978). These are the reasons why most top players favour gut strings. Gut strings also have greater resilience and better damping qualities than synthetic strings.

What about the qualities of synthetic strings? The first synthetic strings, made of nylon and polyester, appeared in the 1930s. They became really popular after 1976, when the first oversized rackets appeared, because the lower elasticity of these strings was less of a problem in these rackets (Groppel, 1987). Two types of strings can be distinguished: mono-filamented and multi-filamented. Mono-filamented strings (made of nylon) are the cheapest, but also have the poorest playing qualities. They are hard to the touch, not very flexible and uncomfortable to play with. The playing characteristics of multi-filamented strings resemble those of gut more closely. They are also cheaper and more durable than gut.

Gut should be your choice if money is not a problem and you do not mind bringing several extra rackets to the courts. A high quality multi-filamented string is a good second choice.

To reduce loads on the arm, choose gut or high-quality synthetic string.

10) STRING SIZE: THICK OR THIN?

The size of a tennis racket string is measured in `gauges'. A lower gauge (for example, 16) indicates a thicker string. Most strings are 16 to 18-gauge. The advantage of a thicker string is its durability. A thicker string will last longer and not break as quickly as a thinner string. A thinner string however, is more elastic and has better shock-absorbing capacities. Therefore, thinner strings are easier to play with and reduce stresses transmitted to the player.

To reduce loads on the arm, choose a thinner string.

11) TIGHT OR LOOSE STRINGS?

Tennis racket strings are designed to return 90 to 95% of the energy that they receive from the incoming ball. Balls themselves are much less efficient: they return only about half the energy when they bounce on a hard surface. Actually, we should all know this, because it is in the rules! The rules of the International Tennis Federation require that a ball have a bounce of more than 53 and less than 58 inches when dropped 100 inches. This means that the ball loses about 45% of its energy after the bounce. If we drop the ball on our racket instead of the hard floor, it will bounce back to a considerably higher level, showing that less energy has been dissipated. Thus, the strings return almost all of the energy that they store when they deform, while the ball only returns about half the energy when it hits and deforms.

At the point of ball-string contact, therefore, the ideal situation is that the ball is deformed as little as possible and the strings as much as possible (Baker, 1978). Lower strings tensions encourage such a relationship, because the strings will deflect more, and the ball will deform less. Tighter strings therefore mean less power and looser strings more power. Reducing the tension of the racket strings also reduces the load on the arm, because less swing is needed.

There is another reason why looser strings are better for the arm, and that is because they increase the dwell time of the ball on the strings. The longer contact time means that the shock of the ball impact is spread over a longer period of time. The magnitude of the force at any given time is therefore reduced. In other words, when the strings are softer (at lower string tension), the arm does not feel the shock of hitting the ball as much as with harder strings (at higher string tension).

However, the string tension should not be too low. If string tension is lower than 40 lbs., energy is lost because of excessive string movement (Elliott, 1982).

To reduce loads on the arm, string the racket at a lower tension.

12) THE STRING PATTERN

The density of a string pattern is measured in strings per square meter. The closer together the strings, the stiffer the racket plays. The effect of a high-density pattern is similar to that of a racket that is tightly strung. As a result, rackets with high string densities play better with thinner, more elastic strings.

To reduce loads on the arm, a high-density string pattern should be avoided.

13) VIBRATION STOPPERS (STRING IMPLANT DEVICES)

Vibration stoppers are small inserts made of rubber or plastic. They come in different shapes and sizes, and are placed between the two main longitudinal strings close to the balance point of the racket. Many players use this vibration absorbing material in their strings. It is sold under brand names such as Vibratrap, Vibrazorp, Nonvib etc. However, since these damping devices only weigh 1 to 2 grams, and the racket may weigh up to 360 grams, there is not a lot of mass available for shock absorption. Such a small device cannot absorb and damp the vibrations from the frame in a finite time. So, what is the actual effect of these devices on frame vibrations?

In a study by Tomosue et al (1995). it was shown that the damping material was effective in reducing impact shock. The researchers concluded that the double-layered sponge appreciably reduces the amplitude of string vibrations, which has an apparent effect on frame vibrations. In another study, however, the influence of the damping device on vibration damping of the frame was found to be negligible (Wilson, 1995). Perhaps these differences can be explained by the fact that when the ball is hit exactly at the vibrational sweet spot of the racket, the vibration stopper works quite well at damping down the high frequency vibrations. However, when the ball is hit off-centre, the vibration stopper makes no appreciable difference to the vibrations transmitted to the hand.

More research is needed before we can give a definite answer to the question of whether vibration stoppers really do play a significant role in the prevention of tennis elbow.

To reduce loads on the arm, choose a vibration damper. It may help, and it will not hurt.

14) A PRESSURISED OR NON-PRESSURISED TENNIS BALL

In the game of tennis, two types of balls can be distinguished: a pressurised and a non-pressurised ball. A non-pressurised ball bounces because of the thickness and suppleness of the rubber that is used. A pressurised ball bounces not only because of the rubber, but also because the internal pressure is greater than the external pressure. The durability of a pressurised ball is mainly determined by the bounce. Once the ball looses internal pressure, it does not bounce as well any more and feels dead. The height of the bounce remains constant for longer with a non-pressurised ball than with a pressurised ball. Therefore, the durability of a non-pressurised ball is generally longer than of a pressurised ball and is determined mainly by the wear of the fabric cover.

There is a general feeling that players with elbow problems should use pressurised tennis balls in preference to non-pressurised ones. Is there any scientific evidence to support this notion? An elegant study was performed Caffi et al. (1995) who investigated rebound characteristics of pressurised and non-pressurised balls, and of new and used balls. The researchers determined the coefficient of restitution for all balls tested. The coefficient of restitution parameter is the ratio of the post-impact ball velocity to the relative velocity of the ball before impact with respect to the target. The study demonstrated that, with new balls tested in the laboratory, the coefficient of restitution of non-pressurised balls is lower than the corresponding value of pressurised balls, for the whole range of velocities considered. That means one has to hit a non-pressurised ball harder to acquire the same speed as with a pressurised, which will put more load on the arm. The researchers also investigated the influence of ball wear. As expected, used balls have a slower rebound than new ones. Therefore, players suffering from elbow problems should preferably use new tennis balls.

Use new, pressurised tennis balls.

15) THE SIZE AND DEFORMATION OF THE TENNIS BALL

From 1 January 2000 to 31 December 2001, two further types of tennis balls may be used on an experimental basis. The `fast-type' ball will have more less deformation and less return deformation than the regular `medium-type' ball, i.e. it is stiffer and harder, and should only be used for play on slow pace surfaces. The ball rebounds less than the medium ball and slips across the surface more. This causes the ball to rebound shallower and faster, which means that less energy is lost as vertical momentum. Therefore, the forward velocity of the ball is higher than of a regular type tennis ball. Also, since there is less ball deformation, less energy is lost when the ball hits the strings. This allows the player to put in less effort, thereby reducing stress on the elbow.
The `slow-type' ball is slightly larger than the regular ball, with a diameter of 2.825 to 2.85", and should only be used for fast-pace court surfaces. This ball is supposed to have the same stiffness as the standard size ball in compressions tests. It is likely that the peak force will be lower and the contact time longer when the ball hits the racket, resulting in less peak load on the arm.

On a fast pace surface, try the 'slow-type' ball. On a slow pace surface, try the 'fast-type' ball.

CONCLUSIONS

The following equipment modifications may be tried in order to reduce loads on the arm and possibly prevent tennis elbow. Choose a composite racket with good damping qualities and a large `sweet spot', such as a mid-size or oversize racket. The racket may have a stiff head, but the shaft should be flexible. Do not switch to a long-body if you have elbow problems. Go for the heaviest racket you are able to handle without sacrificing technique, but make sure it is head-light. The grip should be the largest one that is comfortable for you. Try a cushioned grip tape. String the racket with a thinner gut or high-quality synthetic string at a lower tension. A high-density string pattern should be avoided. Try a vibration damper, since it may help, and will not hurt. Make sure you use new, pressurised tennis balls. On a fast pace surface, try the 'slow-type' ball. On a slow pace surface, try the 'fast-type' ball.

REFERENCES

Adelsberg, S. (1986) The tennis stroke: an EMG analysis of selected muscles with rackets and increasing grip size. *American Journal of Sports Medicine*, 14, 139-142.

Baker, J. and Wilson, B. (1978) The effect of tennis racket stiffness and string tension on ball velocity after impact. *Research Quarterly of Exercise and Sport*, 3, 255-259.

Brody, H. (1988) Tennis science for tennis players. *University of Pennsylvania Press*, Philadelphia.

Brody, H. (1989) Vibration damping of tennis rackets. *The International Journal of Sport Biomechanics*, 5, 451-456.

Caffi, M. and Casolo, F. (1995) Ball dynamic characteristics: a fundamental factor in racket dynamic optimization. In: *Science and racket sports*. (eds Reilly, T., Hughes, M., Lees, A.), pp 146-152. E&FN Spon, London.

Carroll, R. (1981) Tennis elbow: incidence in local league players. *British Journal of Sports Medicine*, 15, 250-256.

Elliott, B.C., Blanksby, B.A.., Ellis, R. (1980) Vibration and rebound velocity characteristics of conventional and oversized tennis rackets. *Research Quarterly of Exercise and Sport*, 51, 608-615.

Elliott, B. (1982) The influence of tennis racket flexibility and string tension on rebound velocity following a dynamic impact. *Research Quarterly of Exercise and Sport*, 53, 277-281.

Ellis, R., Elliott, B., Blanksby, B. (1978*)* The effect of string type and tension in jumbo and regular size rackets. *Sports Coach*, 4, 32-34.

Groppel, J.L., Shin, I., Thomas, J.A., Welk, G.J. (1987) The effects of string type and tension on impact in midsized and oversized tennis racquets. *The International Journal of Sport Biomechanics*, 3, 40-46.

Gruchow, H.W., Pelletier, D. (1979*)* An epidemiologic study of tennis elbow. *American Journal of Sports Medicine*, 7, 234-238.

Hatze, H. (1976*)* Forces and duration of impact and grip tightness during the tennis stroke. *Medicine and Science in Sports*, 5, 88-95.

Hennig, E.M., Rosenbaum, D., Milani, T.L. (1992*)* Transfer of tennis racket vibrations onto the human forearm. *Medicine and Science in Sports and Exercise*, 24, 1134-1140.

Kamien, M. (1988) Tennis elbow in long-time tennis players. *Australian Journal of Science and Medicine in Sport*, 20, 19-27.

Kamien, M. (1990) A rational management of tennis elbow. *Sports Medicine*, 9, 173-191.

Kuebler, S. (1995) Buch der Tennisrackets von den Anfängen im 16. Jahrhundert bis etwa 1990. Kuebler Gmbh, Singen.

Knudson, D.V. (1991*)* Factors affecting force loading on the hand in the tennis forehand. *The Journal of Sports Medicine and Physical Fitness*, 31, 527-531.

Murley, R. (1987) Tennis elbow: conservative, surgical, and manipulative treatment (letter). *British Medical Journal*, 294, 839-840.

Priest, J.D., Braden, V., Goodwin Gerberich, S. (1980) The elbow and tennis, Part 1: An analysis of players with and without pain. *Physician and Sports Medicine*, 8, 81-91.

Stone, D.A., Voght, M., Safran, M.R. (1999) Comparison of incidence of tennis elbow in first generation and graphite composite racquets. *Sportsmedicine and Science in Tennis*, 4, 3 (abstract).

Tomosue, R., Sugiyama, K., Yamamoto, K. (1995) The effectiveness of damping material in reducing impact shock in the tennis forehand drive. In: *Science and racket sports*. (eds Reilly, T., Hughes, M., Lees, A.), pp 140-145. E&FN Spon, London.

Verhaar, J.A.N. (1992) Tennis elbow. Thesis. *Universitaire Pers Maastricht, Maastricht.*

Wilson, J.F., Davis, J.S. (1995) Tennis racket shock mitigation experiments. *Journal of Biomechanic Engineering*, 117, 479-484.

Fatigue, carbohydrate supplementation and skilled tennis performance

P.R. McCarthy-Davey
School of Applied Science, South Bank University, London, UK

ABSTRACT: Tennis is a typical multiple sprint sport whereby players perform a multitude of multi-directional movements, including sprinting, running, jumping, lateral movements, cross over steps, split steps and lunges interspersed with periods of recovery. Tennis is a sport which uses the whole body for stroke production unlike some other sports which predominantly use the lower body musculature e.g. distance running. Furthermore, the demands of the tournament schedule require players to be engaged in competition and or training for the majority of the year. Thus maintenance of skill, power, speed and endurance throughout a competitor's year makes tennis a physiologically and psychologically tough sport to participate in at the 'elite' level.

The aim of the current paper is to outline current research in three main areas: fatigue, carbohydrate supplementation and their influence upon skilled tennis performance. Limited research exists to date on the aforementioned parameters however it is of utmost importance that players are educated in and understand the necessity of applying nutritional support and research to their game in order to be able to optimise their performance.

INTRODUCTION

Tennis is classified as a multiple sprint sport, demanding from its participants the ability to perform both short-term high intensity exercise and a high endurance capacity. Matches can be as brief as an hour or extend to five hours or more. Throughout this time players need to be able to react to the opponent's shots and move rapidly around the court. The state of play however is constantly changing and high degrees of strength, power, speed and flexibility are required for the repetitive execution of the groundstrokes and service.

DEMANDS OF THE GAME OF TENNIS

The physical demands of the game are dependent upon many influential factors including the playing surface, styles of play, environment and techniques utilised. Schönborn (1984) made observations regarding the length of a point, relative to the playing surface. As expected, it was observed that points on faster surfaces such as grass were consistently shorter than those played on slower surfaces such as clay. Further analyses were made by Brabanec (1994) illustrating that approximately 67%

of points in a match are of less than 5s duration. Muscle metabolism during such short rallies is predominantly fuelled by Phosphocreatine (PCr) which is in limited supply in the body; and the anaerobic breakdown of glycogen (anaerobic glycolysis). The periods of low intensity and the short recovery between points (20s) and games (30s) and changeovers (1 min 30s) provides an ideal opportunity for the replenishment of PCr and the removal and oxidation of the by-products of glycolytic metabolism. Furthermore, myoglobin (iron containing pigment present in muscle) stores of oxygen may also be replenished during the recovery periods.

Statistical analyses of tennis matchplay are presented in Table 1.0. Previous research, which has quantified the 'actual' time in and out of play, can be criticised for the short duration in which the match is analysed. In some instances, such studies have been known to have drawn conclusions from match analyses with a sampling period of only 10 minutes (Seliger *et al.*, 1973). Longer periods of observation and or video analyses provide far more useful information on the 'actual' work to rest ratios (work:rest) of the players. Christmass *et al.,* (1993) reported work to rest ratios of 1:1.8. Once again such parameters are highly dependent upon the styles of play utilised, surface type and individual player behaviour.

Coupled with the physical demands that are imposed, players are also required to perform at different extremes of psychological pressure. Furthermore, tennis players not only require athleticism, but athleticism combined with the application of skilled performance. Thus tennis, alongside other racket sports requires good cognitive ability, peripheral and spatial awareness and physical and skill attributes.

SKILLED TENNIS PERFORMANCE

Most tennis coaches and players would agree that in order to be successful, players must be able to perform both consistent and accurate shots (Sailes, 1989). For the purpose of this paper, consistency shall be referred to as 'hitting the ball back repetitively without error, without any specific regard to the length or placement of the ball.' Accuracy however shall be defined as 'the ability to direct the ball deep in the court and vary its direction, whilst keeping the ball in court.'

It must be emphasised however that players producing high levels of accuracy may not necessarily be highly consistent. Similarly it must not be assumed that high levels of consistency enable a player to possess 'pin point' accuracy. Thus players will place themselves at a disadvantage if they lack accuracy, consistency and depth in any of their strokes (Knapp, 1963).

FACTORS INFLUENCING SKILLED TENNIS PERFORMANCE

Many factors influence tennis skill and performance, including the type of warm up (Anshel and Wrisberg, 1993) and the playing environment (Dawson et al, 1985). Stroke velocity has also been reported to influence an 'open skill' such as the groundstrokes (Caraugh *et al.*, 1990, Siegal 1994) more than that of a 'closed skill' such as service. Maximum accuracy in an 'open skill' was observed to be associated with maximum velocity of movement. This finding was not revealed however with the service ('closed skill').

Table 1. Statistical analyses of tennis matchplay

Study	Christmass *et al.*,(1993)	Bergeron *et al.*,(1991)	McCarthy (1997)	Seliger *et al.*,(1973)
n	8	10	14	16
Match duration (min)	90	85	67.7 ± 21.3	10
Mean time of point (s)	10.2 ± 0.3		2.9 ± 0.1	
Time between points (s)	16.8 ± 0.2		19.4 ± 0.3	28.9 ± 3.1
Mean no. of shots per point	4.6 ± 0.1			62 ± 10 (total no. of points)
Mean no. of rallies				29.9 ± 3.1
% Heart rate max.	86 ± 1	74%		60
% VO_2 max	86.2 ± 1	60-70%		
Work:rest ratio	1:1.8			
Blood glucose conc. (mM)		4.5-5.2		
Blood lactate conc. (mM)		1.5 ± 0.5 (rest) 13.2 ± 3.2 (play)		
Plasma volume change (%)		0.7 (pre-play) 2.3 (post-play)		

Certain physiological situations have been reported to have a detrimental effect on skill and performance. Such factors include lactate and the resultant abnormally high acidity levels in the muscle and blood due to its accumulation - 'metabolic acidosis'; (Liesen, 1983; Green, 1979); dehydration (Craig and Cummings, 1966; Saltin, 1964) and fatigue (Ivoilov *et al.*, 1981, Hoffman *et al.*, 1992). Motor performance and reaction time have also been negatively affected by fatiguing exercise (Alderman, 1965; Wrisberg and Herbert, 1976). Liesen (1983) reported peak blood lactate concentrations of 7 to 8 $mmol.l^{-1}$ to have been associated with a decline in both technical and tactical performance, even at the 'elite' level in tennis and handball. Similarly, Green (1979) associated such high lactate levels with a loss of control and a decline in skill in ice hockey players. In all of the aforementioned sports, upper body movements are superimposed with lower body movements. In the study of Green (1979) it was suggested that these dual demands may have imposed a decline in blood flow to the legs resulting in an increased lactate production from anaerobic glycolysis. Comparisons between studies are difficult, especially due to the lack of specified sampling sites.

THE INFLUENCE OF FATIGUE ON SKILLED TENNIS PERFORMANCE

Fatigue can be caused by many factors and is still not fully understood. The mechanisms of fatigue are thought to be both peripheral (inadequate force development resulting from the breakdown in the excitation / contraction process in the muscle) and central (inadequate activation of the muscle by the central nervous system). In depth discussions on the possible causes and suggested mechanisms of fatigue are beyond the scope of this paper.

McCarthy *et al.,* (1995) studied the influence of fatigue on maximal skilled tennis hitting performance. Eighteen 'elite' tennis players performed a groundstroke and service skill test before (ST1) and after (ST2) an intermittent hitting performance test of 4 minutes with maximal hitting bouts against a ball machine with 40s seated recovery repeated to exhaustion. Mean time to exhaustion was 35.4 ± 4.6 min. Mean hitting accuracy declined by approximately 70% from the start of the performance test to the point of exhaustion and from the start of the performance test to 75% way through ($P < 0.01$). Certain groundstroke and service skills declined from ST1 to ST2. Thus tennis players performing a skill test, which simulated the metabolic demands of matchplay, experienced fatigue, which had a detrimental effect on skilled tennis performance. The decline in skill observed may have been associated with many underlying causes of fatigue such as metabolic acidosis, poor hydration status, central fatigue or local muscular fatigue. Certainly some subjects complained of sore forearms at the point of volitional fatigue in this study.

In a further study conducted by McCarthy (1997) grip strength was observed to decline in both the dominant ($P < 0.01$) and non dominant hand ($P < 0.05$) following a simulated tennis match and performance test consisting of tennis hitting to volitional fatigue. Associated with this decline in grip strength was the expected increase in grip fatigue index. This was an interesting observation considering the findings of Nwuga (1975), that tennis hitting exerts a force of only 45% and 60% of maximum grip force in the forehand and backhand at impact respectively. The observed decrease in grip strength and increase in fatigue index may display the local muscular fatigue experienced by tennis players, when not only performing in 'actual' matchplay, but also at the point of fatigue. Although players are not using maximal gripping forces when playing, the accumulative effect over the duration of a match would also seem to be important. Such declines in grip strength observed might have a detrimental effect on tennis hitting performance and certainly players who are able to sustain their performance into the later hours of a match are at an advantage. This also has implications for both training and testing the tennis player as currently the majority of protocols include only grip strength as an objective index of the functional integrity of the upper extremity. Kibler and Chandler (1989) and Elliot (1982) stated that in order to fully understand the action of the muscles used to grip the racket in tennis it is essential that both grip strength and endurance are measured.

CARBOHYDRATE SUPPLEMENTATION AND ITS EFFECT ON SKILLED TENNIS PERFORMANCE

Considering that a decline in skill has been observed when players become fatigued (McCarthy *et al.*, 1995) and that a well documented link between low carbohydrate stores and the onset of fatigue exists (Bergstrom *et al*, 1967), nutritional intervention may play a decisive role in the maintenance of skilled performance parameters.

The ingestion of a carbohydrate beverage may help to prevent low blood glucose levels (hypoglycaemia) and provide an alternative fuel source for immediate use by the working muscle; possibly maintaining performance and extending endurance capacity. The repetitive and explosive nature of the game of tennis over a prolonged period constitutes considerable proportions of energy derivation from muscle glycogen. Vollestad *et al.*, (1984) have reported significant depletion of muscle glycogen in all muscle fibres when subjects performed exercise at varying intensities, especially in the type II muscle fibres. Furthermore, the research of Nicholas *et al.*, (1996) reported that the ingestion of a 6.9% carbohydrate-electrolyte (CHO-E) solution throughout intermittent running resulted in a sparing of muscle glycogen, occurring predominantly in the type II muscle fibres. The predominant utilisation of type II muscle fibres is expected in tennis when tennis skills are performed explosively and repetitively.

Tennis players are dependent upon CHO as a fuel source for the powerful actions of the game such as the overhead, service, sprints, jumps and other rapid and explosive movements performed to elicit an attacking game. Even as early as 1963, Lundqvist; a Swedish player in the Davis Cup, was so aware of the high physical demands of the game of tennis that he made no effort on return of service, in order to save energy for the execution of his own service! Thus even 37 years ago elite players realised the need to recover and maintain a skilful and explosive game.

Few good nutritional studies have been reported in peer reviewed journals in the specific field of tennis physiology (Maclaren, 1998). The limited research on the positive effects of carbohydrate supplementation upon skilled tennis performance shall now be discussed.

Keul *et al.*, (1995) observed the positive effects of supplementing CHO throughout 6.5 hours of Davis Cup matchplay. Players reported improved alertness and the maintenance of concentration and co-ordination throughout matchplay. Burke and Ekblom (1982) also showed the benefits of CHO supplementation upon various tennis performance parameters. Two, 2 hour trials of simulated competitive matchplay were performed under 5 conditions: i) CHO polymer ($75g.l^{-1}$), ii) water, iii) thermal dehydration ($-1.5kg^{-1}$.BW), iv) no fluids and v) control (neutral environment). Two performance measures of 'total points' of stroke accuracy (service and groundstrokes) and power output (vertical jump) were recorded respectively. Plasma glucose concentrations were observed to increase in the CHO trial by 8.9% in contrast to a decline in blood glucose concentration in all other trials. Furthermore, players' error rates increased in all trials except the CHO and control trials. Thus Burke and Ekblom (1982) recognised that through maintaining a players' hydration levels and plasma glucose concentration with a CHO beverage, the quantity of error was minimised and skill was maintained.

More recently, Vergauwen *et al.*, (1998) have shown skilled performance benefits when CHO was supplemented. Players underwent a 'Leuven tennis test' consisting of 50 rallies (1st and 2nd serves) followed by returning 5 balls from a ball machine to a random selection of neutral, offensive and defensive tactical situations. The 'Leuven tennis test' was repeated following 2 hours of strenuous training. A velocity precision index (VP) and velocity precision error (VPE) index was used to account for velocity, precision and error respectively. Players on the placebo trial experienced a greater decline in VPE, first service and an increase in the number of errors when compared to the CHO trial respectively. Furthermore, players performed slower shuttle run times at the end of the placebo trial and reached fewer feeds from the ball machine in comparison to the CHO trial.

McCarthy *et al.*, (1997) performed a similar study to that of Vergauwen *et al.*, (1998), whereby blood glucose and blood lactate concentrations were obtained and skilled tennis performance was observed following simulated 'actual' matchplay as opposed to a strenuous training session. In a study of McCarthy (1997) players ingested $3 ml.kg^{-1}$BM of either a 6.9% CHO-E solution or a flavoured water placebo. Following the simulated 'actual' tennis matchplay (92 min 46s) players performed a tennis hitting performance test to the point of volitional fatigue. Players aimed at targets placed in the rear singles court area and were asked to aim at the targets throughout the performance test. Mean times to volitional fatigue of 22.4 ± 8.2 min and 18.5 ± 4.8 min in the carbohydrate and placebo trials did not differ respectively. Nevertheless, hitting accuracy declined in the placebo condition from the start of the performance test to 50% way through and at volitional fatigue ($P < 0.05$), whereas hitting accuracy was maintained throughout the performance test in the carbohydrate condition. Blood glucose concentration was also maintained above that in the placebo trial throughout simulated matchplay. Thus the ingestion of a carbohydrate-electrolyte beverage throughout simulated matchplay provided players with a maintained hitting accuracy and elevated blood glucose concentration to the point of volitional fatigue, whereas the same performance and metabolic advantages were not exhibited with the ingestion of a placebo.

Some studies have reported limited benefits of nutritional intervention upon skilled tennis performance. The studies of Mitchell *et al., (*1992) and McCarthy *et al.,* (1995) showed no skilled tennis performance benefits when a carbohydrate beverage was ingested during simulated matchplay. Nevertheless, in both studies blood glucose concentration was maintained in the carbohydrate trials at a higher level than those in the placebo trials respectively. Research conducted by Owens and Benton (1994) also showed that individuals, who ingested a carbohydrate beverage, experienced elevated levels of blood glucose. Owens and Benton (1994) also associated the elevated blood glucose concentration with faster reaction times and improved mental speed and faster and more efficient information processing levels, as opposed to when the placebo was administered. Thus, maintenance of a positive metabolic profile even without measured performance benefits may have a positive effect on the cognitive aspects of a player's game, especially in games of long duration. It would seem that the type of protocol utilised and the performance parameters measured play a substantial role in the sensitivity of performance alterations as a result of the nutritional intervention.

CONCLUSION

At the elite level, tennis players are constantly attempting to improve their performance and gain an advantage over their opponents. Players will not be able to achieve such an aim nor perform adequate training levels if attention is not given to nutritional support. The current paper has reported the benefits of nutritional ergogenic aids, such as carbohydrate-electrolyte beverages to skilled tennis performance. Such nutritional interventions may prove decisive in underpinning success in a highly competitive skilled racket sport such as tennis.

REFERENCES

Aldermann R. B. (1965) Influence of Local Fatigue on Speed and Accuracy in Motor Learning. *Res. Q. Exerc.* **(2)**, 131-140.

Anshel M. H. & Wrisberg C. A. (1993) Reducing Warm Up Decrement in the Performance of the Tennis Serve. *J. Sport Exerc. Psych.,* **15**, 290-303.

Bergeron M. F., Maresh C. M., Kraemer W. J., Abraham A., Conroy B., and Gabaree C. (1991): Tennis: A Physiological Profile During Matchplay. *Int. J. Sports Med.,* **12** (1), 474-479.

Bergstrom J., Hermanssen L., Hultman E. & Saltin B. (1967) Diet, Muscle Glycogen and Physical Performance. *Acta Physiol. Scand.,* **71**, 140-150.

Brabanec J. (1994) *Creating Efficient Training Sessions.* Coaches Review (December), pp. 1-3.

Burke E. R. & Ekblom B. (1982) Influence of Fluid Ingestion and Dehydration on Precision and Endurance Performance in Tennis. *Ath.Train,* 275-277.

Caraugh J. H. (1990) Speed-Accuracy Trade Off During Response Preparation. *Res. Q. Exerc. Sport,* **61** (4), 331-337.

Christmass M. A., Richmond S. E., Cable N. T. & Hartmann P. E. (1993) A Metabolic Characterisation of Singles Tennis. *J. Sports Sci,* **11**, 543.

Craig F. N. & Cummings E. G. (1966) Dehydration and Muscular Work. *J. Appl. Physiol.,* **21** (2), 670-674.

Dawson B., Elliot B. C., Pyke F. & Rogers R. (1985) Physiological and Performance Responses to Playing Tennis in a Cool Environment and Similar Intervalised Treadmill Running in a Hot Climate. *J. Hum. Mov. Stud.,* **11** (1), 21-34.

Elliot B. C. (1982) Tennis: The Influence of Grip Tightness on Reaction Impulse and Rebound Velocity. *Med. Sci. Sports Exerc.,* **14** (5), 348-352.

Green H. J. (1979) Metabolic Aspects of Intermittent Work With Specific Regard to Ice Hockey. *Can. J. Sports Sci.,* **4** (1), 29-34.

Hoffman M. D., Gilson P. M., Westenburg T. M. & Spencer W. A. (1992) Biathlon Shooting Performance After Exercise at Different Intensities. *Int. J. Sports Med.,* **13**, 270-273.

Ivoilov A. V., Smirnov Y. G., Chikalov V. V. & Garkavenko A. G. (1981) Effects of Progressive Fatigue on Shooting Accuracy. *Teoriya-i-Practika-Fizicheskoi-Kultury*: **7**,12-14.

Keul J., Berg A., Konig D., Huonker M. & Halle M. (1995) Nutrition in Tennis. In: *TENNIS: Sports Medicine and Science,* (Ed. by W. Hollmann, K. Struder, A. Ferrauti, & K. Weber). pp. 219-226. Rau publishers, Dusseldorf.

Kibler W. B. & Chandler T. J. (1989) Grip Strength and Endurance in Elite Tennis Players. *Med. Sci. Sport Exerc. Sci.,* **65** (suppl. 1.12), S65.

Knapp B. (1963) *The Attainment of Proficiency and Skill in Sport.* London: Routledge.

Liesen H. (1983) Schnelligkeitsauadauer Im Fussball aus Sportsmedizinscher Sicht Fussballtraining. **1**(1): 27-31.

Maclaren D.P.M. (1998) Nutrition for Racket Sports. In: *Science and Racket Sports II,* (Ed. by A. Lees, I. Maynard, M.Hughes & T. Reilly). pp. 43-51. E. & F.N. Spon, London.

McCarthy P.R. (1997) *Influence of Fatigue and Dietary Manipulation Strategies on Skilled Tennis Hitting Performance.* PhD. thesis. Department of Physical Education, Sports Science and Recreation Management, Loughborough University.

McCarthy P. R., Thorpe R. D. & Williams C. (1995) The Influence of a Carbohydrate-Electrolyte Beverage on Tennis Performance. In: *TENNIS: Sports Medicine and Science,* (Ed. by W. Hollmann, K. Struder, A. Ferrauti, & K. Weber). pp. 210-218. Rau publishers, Dusseldorf.

Mitchell J. B., Cole K. J., Grandjean P. W. & Sobczak R. J. (1992) The Effect of a Carbohydrate Beverage on Tennis Performance and Fluid Balance During Prolonged Tennis Play. *J. Appl. Sport Sci. Res.*, **6** (2), 96-102.

Nicholas C.W. (1996*) Influence of Nutrition on Muscle Metabolism and Performance During Prolonged Intermittent High Intensity Shuttle Running in Man.* PhD. thesis. Department of Physical Education, Sports Science and Recreation Management, Loughborough University.

Nwuga V. C. (1975) Grip Strength and Grip Endurance in Physical Therapy Students. *Arch. Phys. Med.,* **56,** 296-299.

Owens D.S. & Benton D. (1994) The Impact of Raising Blood Glucose on Reaction Times . *Neuropsychobiology*. **30**, 106-113.

Sailes G. A. (1989) A Comparison of Three Methods of Target-Oriented Hitting on Baseline Groundstroke Accuracy In Tennis. *Appl. Res. Coach. Athl.*, **4** (1), 25-34.

Saltin B. (1964) Aerobic and Anaerobic Work Capacity After Dehydration. *J. Appl. Physiol.*, **19** (6), 1114-1118.

Schönborn R. (1984) *Principles of Kinetic Theory.* In, L.T.A. European. Tennis Association Coaches Symposium Proceedings. Marbella, pp. 1-35.

Seliger V., Ejem M., Pauer M., & Safarik V. (1973) Energy Metabolism in Tennis. *Int. Z. Angew. Physiol.*, **31**, 333-340.

Siegal D. (1994) Response Velocity Range of Movement and Timing Accuracy. *Percept. and Mot. Skills,* **79**, 216-218.

Vergauwen L., Brouns F. & Hespel P. (1998) Carbohydrate Supplementation Improves Stroke Performance in Tennis. *Med. Sci. Sports Exerc.,***30**, (8),1289-1295.

Vollestad N.K., Vaage O. & Hermansen L. (1984) Muscle Glycogen Depletion Patterns in Type I and Subgroups of Type II Fibres During Prolonged Severe Exercise in Man. *Acta Physiol. Scand.,* **122**, 433-441.

Wrisberg C. A., & Herbert W. G. (1976) Fatigue Effects on the Timing Performance of Well Practised Subjects. *Res. Q. Exerc.*, **47** (4), 839-844.

Nutritional status of performance-level junior players

S.R. Parsonage
United States Professional Tennis Registry (UK), London,
University of Greenwich, London, UK

ABSTRACT: The diets of junior players identified as potential performance players need to be optimised to meet the dual demands of tournament tennis and a high growth rate. Dietary analysis, and Perceived Exertion Ratings (PER) of matches, of juniors in the Kent Performance scheme showed carbohydrate and fluid intakes were often inadequate for physiological needs, and that intakes of several micronutrients were below recommended levels. Players, parents and coaches received nutritional information packs aimed at increasing the proportion of energy from carbohydrate, and their diets were re-assessed. There was no significant change in either carbohydrate or nutrient intakes as a result of the nutrition intervention, but PER values recorded by players were higher. The best mode of delivery of sports nutrition information, in timing, approach and content, plus the need for diet assessment as an indicator of success, is discussed.

INTRODUCTION

The importance of an appropriate diet for high level performance is well-understood by professional players, some of whom are reaching top 100 world rankings in their teens. This has created a degree of pressure on junior players who aspire to the professional game, and has resulted in players as young as 10 or 11 years old participating in training programmes and competition schedules which impose considerable physical and mental demands on them. Technical, tactical, fitness and mental skills are all practiced in these programmes, yet relatively little attention is given to the nutritional status and dietary intakes of players who are also going through a period of rapid physical growth and development.

It is common practice to coach juniors in stroke production and tactics to prepare them for early tournaments, and as they become successful to introduce other areas such as fitness. Sports nutrition relies largely on the coach involved - if he/she has an interest or knowledge-base in this field then some will be passed onto the player, but this is the exception rather than the rule. So many juniors can reach county or even a national rankings without receiving any information or education on basic or sports-specific nutrition. This raises the questions as to what information should be given, at what age or stage of the players development, and what is the most effective mode of delivery? These can only be answered by diet assessment before and after nutritional interventions to ascertain whether players simply ignore information, understand it but do not put it into action, or actually change their diets short- or long-term in response to it

METHODS

A pilot project was set up involving 46 junior players aged 9-16 (mean 11.0 years) who were part of the Kent County Performance training programme. Depending on age and individual development, this programme encompasses 2-3 individual lessons, 3-4 training sessions and a fitness regime during midweek, plus a full tournament schedule competing throughout holiday periods and most weekends (Kent County LTA Council Report 1999)

All players and their coaches met the same sports nutritionist at one of the designated squad training sessions, when the project was outlined and players given introductory packs, which included letters to parents who were unable to be present. Detailed instructions were given for recording three day dietary intakes and Perceived Exertion Ratings (PER) for future matches. All the players, parents and coaches were given open access to the sports nutritionist for assistance with any problems during the project. The players' diets were analysed using Dietplan 5.2 (Forestfield Software) based on data from McCance & Widdowson (1990). PER values for matches were noted for warm-up, 1st, 2nd and 3rd (when played) sets.

Following the initial assessment, players and parents were sent information packs aimed at improving the energy profile of their diets, and specifically to increase the proportion of energy derived from carbohydrates. Coaches also received the same nutritional information packs so they were aware of what nutritional changes were being advised. This took place early in the summer tournament season to allow time for any changes to be made and for the players to adjust to new eating habits. During the main tournament season players were again asked to record dietary intake over a three day period, and to note PER values for matches as above.

All players completing the final stage of the project were offered an individual consultation and long term support from the sports nutritionist. The County Performance officer and coaches involved were also given feedback regarding the results of the project.

RESULTS

The profiles of the junior players starting the project, and of those completing the initial and final stages of the project, are presented in Table 1.

Table 1 Age and sex profile of junior players at each stage of project.

		Boys:	Girls:	All:
Starting	Number	32	14	46
	Mean age	11.2	10.8	11.1
	+/- SD	2.1	1.9	1.9
Initial stage:	Number	14	7	21
	Mean age	10.6	11.0	10.7
	+/- SD	1.7	2.3	1.5
Final stage	Number	5	1	6
	Mean age	10.8	13.5	11.2
	+/- SD	1.5	0	1.6

The number of players failing to complete each stage of the project reflects the varying levels of interest, motivation, coach support and parental involvement with sports nutrition as an adjunct to enhanced performance (Table 1). Although there are no statistical differences in the age/sex profile of players completing each stage, there was an observable trend for players whose parents still had greater influence over their eating habits to remain active participants in the project. Overall 13% of the players who received starter packs completed all the stages in the project, 16% for boys and 7% for girls.

Energy intakes (Table 2) calculated from the initial diet assessments were, on average, 17% higher than Estimated Average Requirements (EAR) for boys and girls in the same age range as the sample (Report on Health & Social Subjects 1991). The energy profile of the diets i.e. the contributions to total energy intake made by protein, carbohydrate and fat, were similar to the Dietary Reference Values (DRV) for those dietary components. As would be expected of a population sample such as this, energy intakes and profiles showed a wide variation between individuals. Sources of carbohydrate intake varied - nearly half the sample recorded soft drinks, squashes and fruit juices as the major source of carbohydrates in their diet, while breakfast cereals (25%) or bread (25%) were the next most common food sources. Potatoes (including chips), rice and pasta all made less significant contributions as did chocolate and confectionery.

A significant proportion of the sample (69%) had dietary intakes of 2 or more micro-nutrients below Reference Nutrient Intake (RNI) levels. Only 27% of this group took multi-vitamin and mineral supplements on a regular basis. 25% of the whole sample showed sub-RNI intakes of 6 or more micro-nutrients, and were associated with energy intakes in the lower portion of the distribution range. Nutrients most commonly showing low intakes were iodine, calcium, iron, magnesium and zinc, but in a smaller number of individuals potassium, phosphorus, selenium, copper, thiamine, riboflavine and folic acid were also involved.

Most players derived around 50% of their food energy from carbohydrates, which is close to the recommended energy profile for the normal population. This figure did not alter significantly even after receiving specific nutritional advice on increasing carbohydrate intake to meet sports-specific demands. In fact the average % energy from carbohydrate was slightly lower in the final than in the initial assessment. Only one individual recorded a significant increase in carbohydrate intake between the two assessments

Table 2 Energy intakes and profiles of diets at initial and final dietary assessments (means +/- SD)

	Initial	Final
Total energy intake Kcals/day	2435 +/-597	2634 +/-422
Protein energy % total	13.5 +/-2.6	13.2 +/-2.6
Carbohydrate energy % total	50.0 +/-4.4	48.5 +/-2.9
Fat energy % total	36.4 +/-3.9	38.3 +/-2.0

As an indication of exertion levels and fatigue encountered during matches, the players recorded PER values (Borg 1982) for matches played during the initial and final stages of the project. Values were recorded courtside by the players for a minimum of 3 matches (Table 3), and the opponents playing rating and the prevailing weather conditions were also noted.

Table 3 Perceived Exertion Rating values recorded by players during initial and final assessment periods

	1st set	2nd set	3rd set
Initial	11.9	13.1	16.2
Final	14.6	15.3	16.8

During the initial phase most of the players were involved in area ratings tournaments which involve playing opponents of a similar standard, at least for a few rounds. During the final phase later in the summer, many of these performance players were involved in county, regional or national championships, the results from which would determine their ranking for the next season. Additionally, the weather was considerably warmer at this time of the year. There is a clear trend towards players playing at a higher level of perceived exertion during the final phase of assessment.

DISCUSSION

The energy intakes of the junior players was consistently higher than the Estimated Average Requirement for boys and girls of the same age. This is entirely expected and is a reflection of the significantly higher level of physical activity regularly undertaken by this group. The proportion of energy intake provided by carbohydrates (50%) while adequate for a normal population, would generally be regarded as low compared with the recommended 60% or more for athletes training and competing in an intermittent sprint sport such as tennis (Costill et al 1971). However none of the participants had received any prior sports nutrition information other than what was available through the general media. The information, in the form of verbal discussion and printed leaflets, given to the players and their parents gave simple dietary guidelines for increasing carbohydrate intake both on a regular basis and specifically immediately before matches. Although being extremely well received and apparently understood, this sports nutrition information did not seem to make any positive impact on the energy profile of their diets. The main food sources of carbohydrate for many of the players were fruit juices, soft drinks, and squashes, with sugared breakfast cereals coming a close second, i.e. a relatively high proportion of simple rather than complex carbohydrates. Most consumed some bread, but little rice, pasta, potatoes or other vegetables. This has short term implications for maintenance of blood glucose levels during play, for replenishment and maintenance of muscle glycogen levels, and long term for dental health.

The high consumption of foods rich in simple carbohydrates, combined with the low intake of fruit, vegetables and cereals, is a major factor for the high incidence of sub-optimal intakes of a number of minerals and a few vitamins. While the shortfalls were not of the magnitude associated with frank deficiency states, they do raise doubts about nutritional status for supporting growth, high level physical performance, and long term good health. Sub-optimal intakes of micro-nutrients tended to be associated with lower energy intakes, below EAR for some individuals. It would not be unreasonable that an increase in total energy intake would correct these shortfalls in most cases. Individuals recording low energy intakes were not underweight for age or height, so it is possible that an element of under-recording had occurred during the 3 day dietary assessment (Livingstone et al 1990)

A further area of concern was the fluid intake of the players, which appeared to be surprisingly low considering the level of physical activity undertaken regularly, often in warm or hot conditions. More than half the players recorded total fluid intakes of less than 1000mls per day, although there was an increase in fluid intake (16%) between the initial and final assessments. This was probably due to the warmer playing conditions experienced during July and August. An earlier pilot survey (unpublished) had shown that 25% of junior players failed to take any fluid at all during change-of-ends breaks, while a further 40% either drank too little or an inappropriate type of drink for fluid replacement. So two-thirds of junior competitors at this level may be playing with some degree of inadequate hydration. This failure to remain properly hydrated during competitive matches could account for the 15% average increase in Perceived Exertion Rating values for matches recorded by the players during the final assessment (see Table 3), since earlier onset of fatigue is an inevitable result of dehydration (Maughan 1994). Other factors to be considered here are reduced muscle glycogen stores resulting from only an inadequate proportion of energy from carbohydrate, plus the fact that the later matches were of far greater importance to the individuals in terms of deciding their ranking for the next season so that they played with greater effort and intensity.

The assumption, often made by sports nutritionists, that giving the information to the athlete will ensure that it is put into practice, is erroneous. Athletes diets are rarely recorded before and after nutrition education, and while verbal enquiry might show that they have a good understanding of what is required in terms of changing eating habits, a dietary assessment may show a completely different picture. Junior athletes eating habits are largely decided by their parents, who may be unaware of the performance-enhancing effects of good sports nutrition. Parental and coach involvement was a strong factor in compliance with the project protocol and in tangible effort to improve the players diet. It is clear that teaching basic sports nutrition as early as possible in playing careers needs to involve more than just giving the information, however attractively. Working on a one-to-one basis with the players and their parents may prove to be more efficient that the group approach. At present this only happens when a specific problem arises with a player.

Sports nutrition requires continual reinforcement and assessment from coaches and nutritionists, and the use of top professional players as role models to illustrate the performance benefits of good sports nutrition.

REFERENCES

Borg G.A.V. (1982) Psychophysical bases of perceived exertion. *Med Sci Sports Exerc* **14,** 377-381

Costill D.L., Bowers R.W., Branam G., and Sparks K. (1971) Muscle glycogen utilisation during prolonged exercise on successive days. *J. Appl. Physiol.* **31,** 834-838

Kent County Lawn Tennis Association, (1999), *Report of Council.*

Livingstone M.B.E., Prentice A.M., Strain J.J., Coward W.A., Black A.E., Barker M.E., McKenna P.G. & Whitehead R.G. (1990) Accuracy of weighed dietary records in studies of diet and health. *British Medical Journal* **300,** 708-712.

Maughan R.J. (1994) Fluid and electrolyte loss and replacement in exercise. In *Oxford Textbook of Sports Medicine*, [Ed. by M.Harries, C.Williams, W.D. Stanish and L.L. Micheli], pp. 82-93, Oxford University Press.

Paul A. A.& Southgate D.A.T. (1978) *McCance & Widdowson's The Composition of Foods,* 4th ed. HMSO, London.

Report on Health & Social Subjects 41 (1991) - Dietary Reference Values for Food Energy and Nutrients for the UK, HMSO, London.

ACKNOWLEDGMENTS

This project was funded by a grant from the PTR Foundation (USA). With thanks to Kent County LTA Performance Officer, players, parents and coaches for their cooperation.

Sport-specific conditioning for tennis: science and myth

T.J. Chandler
Marshall University, Huntington, USA

ABSTRACT: Training for elite tennis performance must be based on the sport specific metabolic and muscular demands of the sport. Two aspects of training for tennis will be discussed, cardiorespiratory training and muscular training. A common error in conditioning the cardiorespiratory system of elite tennis players is to perform extensive amounts of long slow distance running. Sprint/interval running is likely the most sport specific method of training the cardiorespiratory system for tennis. A common error in conditioning the muscular system of elite tennis players is to perform resistance training movements intentionally slow. Resistance training for tennis is likely most effectively performed with a moderate to moderately light intensity at a near maximal rate of speed.

The sport of tennis has evolved over the past 20 years to a sport requiring speed, quickness, and power. Changes in the conditioning programs of elite tennis players have evolved more slowly than changes in the game. There are many possible reasons for this. One may be a lack of recognition of the changing physiological demands of the sport. Scientifically, there is also relatively little research that deals specifically with tennis conditioning. There are some traditional training methods that may be counterproductive to enhancing performance in tennis. This paper will look at two specific aspects of training tennis players; cardiorespiratory training and muscular training for elite performance in the sport of tennis. Two traditional training ideas will be evaluated specific to tennis. They are 1) the value of traditional long slow distance training, and 2) the speed of movement during resistance training.

CARDIORESPIRATORY TRAINING FOR TENNIS

CardiorespiratoryCardioreaspiratory ndurance is the capacity of the heart and lungs to provide oxygen and nutrients to working muscles, and to remove metabolic waste products that result during cellular metabolism. Because tennis matches can last for a long period of time, cardiorespiratory endurance is a performance factor in the sport. The work bouts during a tennis match, however, are typically of a higher intensity and a shorter duration than traditional aerobic training.

Training the cardiorespiratory system for performance in tennis has been an area of some disagreement in recent years. Traditional long slow distance training utilizing the aerobic energy systems is at times recommended for elite tennis players. Because the activity demands in tennis are intermittent, the specificity of training principle would suggest that training for the sport primarily involve an intermittent cardiorespiratory demand. In intermittent activities, anaerobic metabolism is primarily utilized during high intensity bouts of activity with aerobic metabolism predominating during the recovery periods. The primary goal of the sport scientist in terms of training for tennis should be to determine the type of cardiorespiratory training and muscular training that will best promote performance in tennis.

There is some overlap in terms of cardiorespiratory training and muscular training, and this must be considered. The muscular system is an integral part of any cardiorespiratory activity. It is important to understand that metabolic changes in muscle are likely specific to the musculature being trained (Kraemer, 1994). Typically, the most often used method of improving aerobic fitness utilizes the large muscles of the lower extremity. The metabolic changes related to running/jogging primarily effect the specific muscles utilized during training. These changes are not conducive to improving strength and power of the lower extremity

AEROBIC VS ANAEROBIC TRAINING FOR TENNIS

Both traditional aerobic training and interval training will improve VO_2 max, and will enhance the ability of the tennis player to recover between points, games, and during changeovers. Long slow distance training over time may cause greater increases in VO_2 max over an extended period of time, but interval training is most effective at producing relatively rapid changes in aerobic power (Gorostiaga, 1991). Sprint training with relatively short rest intervals may produce the most rapid gains in aerobic power. With long slow distance training, other adaptations may occur that may be detrimental to performance in tennis (Kraemer, 1994).

Traditional aerobic training produces adaptations to the cardiovascular system that improve the efficiency of the body to work aerobically. Positive changes occur related to heart rate, stroke volume, blood pressure, and oxygen utilization. In skeletal muscle, long slow distance training produces metabolic changes primarily in the slow twitch muscle fibers, including an increase in enzymes necessary for aerobic metabolism, a small degree of hypertrophy in the slow twitch muscle fibers, and an increase in the long-term endurance in the trained muscles.

It is possible that these changes to skeletal muscle will negatively affect power output in the trained muscles. The changes that might occur with distance running, for instance, would primarily affect the lower extremity musculature. Kraemer demonstrated interference in strength and power specifically in the lower body musculature when combining aerobic training with strength training (Kraemer, 1994). This means that extensive distance running combined with lower body speed, strength, and power training will likely cause a decrease in speed, strength, and power.

SIMULATING WORK/REST INTERVALS IN THE SPORT

Research evaluating the work/rest intervals in the sport of tennis indicate relatively brief work intervals (5-10 seconds) followed by brief rest intervals (20-25 seconds between points, 90 seconds on a changeover)(Chandler, 1997). The specificity principle would suggest that training should be performed using similar work/rest intervals. In fact, interval training has been demonstrated to produce considerable gains in VO_2 max (Gorostiaga, 1991). Simulating the work/rest intervals in tennis does not mean that every work bout should last for 8 seconds and every rest bout should be 25 seconds. On the contrary, a variety of work/rest intervals should be utilized, but the majority of training, particularly in the inseason phases of training, should be in the range of the average work/rest intervals.

CARDIORESPIRATORY TRAINING AND FOOT SPEED/QUICKNESS

Foot speed and quickness is essential to the sport of tennis. As stated, tennis has evolved into a power game which, not only requires the players to hit the ball harder, but also requires them to react quicker to their opponents shot. Foot speed, quickness, and agility is critical to performance at the elite level.

As mentioned, long slow distance training recruits primarily slow twitch muscle fibers and will cause metabolic adaptations to occur in those fibers that facilitate aerobic metabolism. Running faster, i.e., sprint and interval training, recruits fast twitch muscle fibers in addition to the slow twitch muscle fibers. Additionally, the neural stimulus to the slow twitch muscle fibers is altered such that they are now being recruited to perform rapidly. The metabolic and enzymatic changes in the muscle occur with sprint/interval training to facilitate the development of speed and quickness.

SUMMARY OF CARDIORESPIRATORY TRAINING

Since sprint-interval training can provide essentially equal improvements in VO_2 max and is most likely to facilitate increases in speed/quickness, it should be the preferred method of training the cardiorespiratory system. Extensive aerobic training may cause changes to occur in the muscle that may negatively affect speed and power. This point is discussed to further under skeletal muscle training.

MUSCULAR TRAINING FOR TENNIS

To draw a distinct line between cardiorespiratory training and skeletal muscular training is difficult because cardiorespiratory training does affect changes in skeletal muscle. This section of the paper primarily deals with resistance training and other forms of anaerobic training associated with changes in strength and power.

POWER OUTPUT IN SKELETAL MUSCLE

Power output in skeletal muscle involves the force of contraction and the time interval over which that force is applied. There is an inverted U relationship between power and velocity (which is determined by the resistance when speed of movement is explosive) such that moderate resistances produce the highest power outputs with both very heavy and very light resistances demonstrating a decreased power output (Enoka, 1994). Figure 1 demonstrates this relationship. Considering the specificity principle and the fact that the tennis racquet provides a light resistance, it would be reasonable to perform a majority of the resistance training for tennis with a moderate to moderately light resistance.

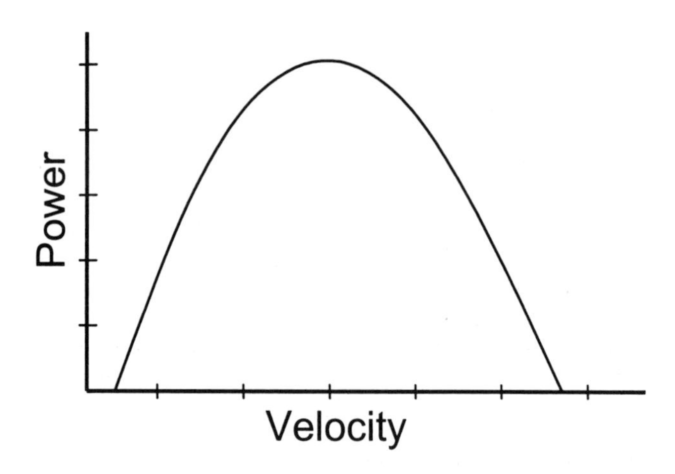

Fig. 1 The inverted "U" relationship between power and velocity illustrates that power output is greatest in the mid-ranges of velocity (adapted from Enoka, 1994).

SPEED OF MOVEMENT IN RESISTANCE/ANAEROBIC TRAINING

One myth with resistance training in general is the premise that training should be performed "intentionally slow" in athletes training for a high power output. This discussion will be limited to speed of movement in the concentric phase of muscle action, and does not apply to an eccentric muscle action.

Intentionally slow movements decrease the power output, and thus may fail to provide a stimulus to improve performance specific to increasing the power output (Force velocity curve ref). When performing a "power" stroke in tennis, the athlete must hit the ball as hard as possible. It should be obvious that moving intentionally slow when hitting a tennis ball will drastically decrease racquet speed and thus the speed of the ball. Again considering the specificity principle, concentric action with resistance training should be performed explosively. Athletes should train at a variety of speeds, but never intentionally slow. The resistance, as demonstrated in Figure 2, controls speed of movement. As the resistance increases, movement speed decreases in an inverse relationship. Speed of movement also affects the recruitment of fast twitch and slow twitch muscle fibers.

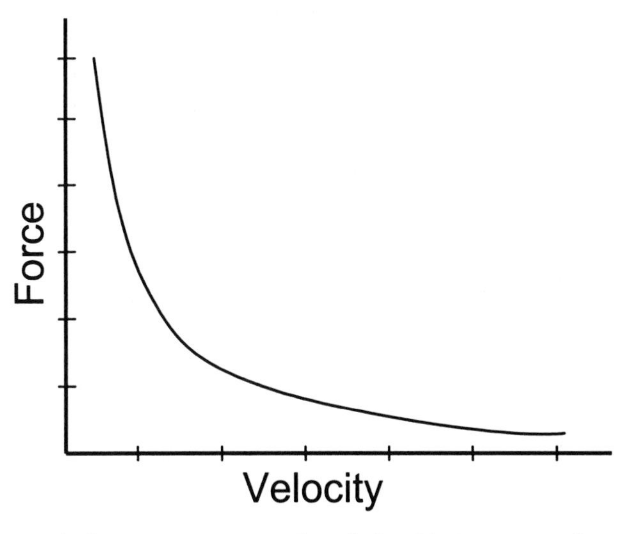

Fig. 2 The force-velocity curve represents the relationship between resistance (force) and the velocity of movement. Training at near maximal movement speeds with a variety of resistances will theoretically facilitate a shift in the curve to the right (i.e., increase movement speed) (adapted from Enoka, 1994).

FAST TWITCH AND SLOW TWITCH MUSCLE FIBERS

As discussed, tennis is a power/endurance sport. An elite tennis match is characterized by intermittent bursts of activity from both the lower and upper body. Both fast and slow twitch muscle fibers are recruited during a power activity. Slow twitch muscle fibers are always recruited first because they have a lower threshold of stimulation (Costill, 1986). This order of recruitment is depicted in Figure 3, and is referred to as the "size principle" of muscle recruitment. But because fast twitch muscle fibers develop tension more rapidly than slow twitch muscle fibers, they are likely responsible for the majority of the force development a higher movement speeds.

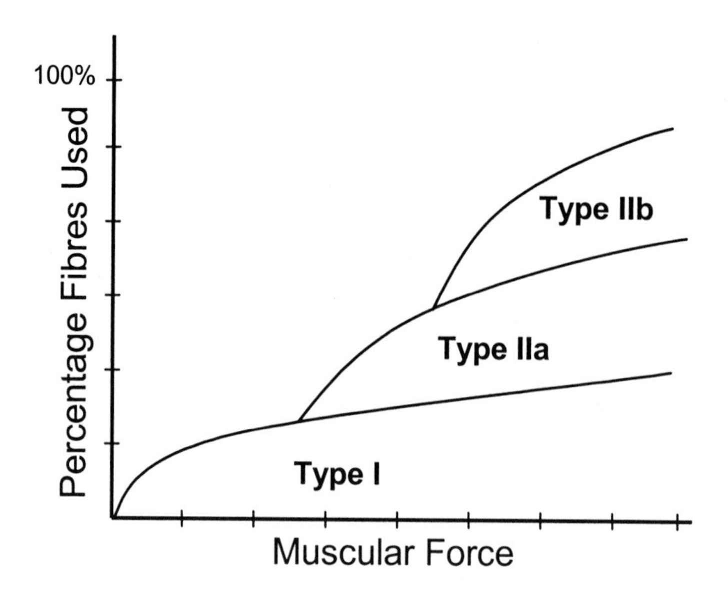

Fig. 3 The size principle of muscle recruitment demonstrates that with low force movements, Type I muscle fibers are recruited. As the force requirements increase, Type IIa fibers are recruited, and lastly Type IIb fibers are recruited (adapted from Enoka, 1994).

There are different ways to classify muscle fiber types, with the most common classification system using Type I, Type IIa and Type IIb nomenclature for slow, intermediate, and fast twitch muscle fibers respectively. Although research fails to support the conversion slow twitch to fast twitch, or fast twitch to slow twitch, it has been suggested that conversion from IIb to IIa may be possible (Kraemer, 1994). This would be most likely to occur with slow training, and may compromise the overall ability of the fast twitch muscles to exert maximal power. This point may hold true for both slow resistance training and slow running. If the neural recruitment pattern requires a rapid response, both fast and slow twitch muscle fibers are recruited. Slow twitch muscle fibers are recruited first, but the intensity of the neural input also recruits fast twitch muscle fibers. Because fast twitch muscle fibers develop tension more rapidly than slow twitch muscle fibers, they are responsible for the rapid tension development in the muscle. To train fast twitch muscle fibers in a sport specific manner, movement speed should be near maximum depending on the resistance.

TRAINING FOR MUSCULAR ENDURANCE

The concept of muscular endurance must be considered, as tennis matches can last several hours. The difference between training for long term and short term endurance must be considered. Long term endurance requires continuous energy production for continuous muscular activity, as in running a long distance. Long-term endurance requires aerobic metabolism to supply this energy over a long period of time. Short-term endurance, on the other hand, requires a burst of energy, followed by a rest period, and then followed by another burst of activity. It should be clear that endurance for tennis is primarily of the short-term variety, but may be repeated for up to several hours.

Short-term endurance can be enhanced with resistance training and sprint/interval training with work/rest intervals specific to the demands of the sport. Thus both cardiorespiratory training and muscular training can be performed in such a manner as to promote short-term muscular endurance.

SUMMARY OF MUSCULAR TRAINING

Muscular training at intentionally slow speeds recruits slow twitch muscle fibers and may interfere with the ability of the muscle to develop peak tension rapidly. Training at near maximal speeds relative to the amount of resistance used recruits fast twitch muscle fibers, and is conducive to producing muscular speed and power. Resistance training for tennis involves choosing the appropriate movement speeds, intensities, and work/rest intervals to promote muscular adaptations that are specific to the sport.

CONCLUSIONS

Conditioning for tennis provides a challenge to the sport scientist developing a conditioning plan because of the varied demands of the sport. Even within the sport, the demands depend on the characteristics of the game of the player and the opponent, as well as the surface. A baseline player, for example, requires more short-term endurance, as the points are likely to be longer. A serve and volleyer would require more speed and quickness. These designations are slightly artificial, however, as the style of play is to some degree dictated by such factors as the style of game of the opponent and the court surface.

In terms of the cardiorespiratory system, the demands of the sport of tennis indicate a need for moderately high levels of oxygen consumption. Traditional aerobic training, however, may interfere with the power output in the trained muscles, which is also important to performance in the sport. To train the cardiorespiratory system for tennis, high intensity interval training will produce the necessary levels of oxygen consumption allowing the athlete to also produce superior gains in muscular power. Additionally, there may be a shift from type IIb (fast twitch or fast glycolytic) to type IIa (intermediate or fast oxidative glycolytic) muscle fibers, which may have a negative effect on speed and quickness.

In terms of the muscular system, tennis requires both power and endurance. Muscular power is a function of force output per unit of time. The sport of tennis requires repetitive use of the muscular system to produce a high power output in repeated bouts over a long period of time. Therefore, training the muscular system to improve power output is beneficial to performance in tennis. The inverted U relationship between power output and resistance means that a moderate training resistance will likely produce the greatest increases in muscular power. Because tennis utilizes a relatively light resistance, training for tennis should likely be performed on the moderate to moderately light end of the power continuum. Muscular power is dramatically decreased in concentric muscle contractions with intentionally slow movement. For this reason, in a majority of resistance training for tennis, concentric muscle contractions should not be performed intentionally slow.

Appropriate training of both the muscular system and the cardiorespiratory system will likely have the most positive effect on tennis performance. The most common errors in training tennis players is performing a large quantity of traditional aerobic training and performing anaerobic/resistance training exercises intentionally slow. Both may lead to a decreased power output in a sport that requires a high power output for success at the elite level. These recommendations appear sound based on the training data that is currently available. However, very little data is available on elite tennis players. Additional research needs to be done to evaluate these concepts in elite tennis players.

REFERENCES

Chandler, T.J. (1998). Sport Specific Conditioning for Tennis. *Science and Racket Sports II*, A. Lees, I Maynard, M Hughes, T Reilly, Eds, E and FN Spon, New York, NY.

Costill, D.L. (1986) Inside Running: Basics of Sport Physiology, Benchmark Press.

Enoka, RM (1994). *Neuromechanical Basis of Kinesiology.* Human Kinetics, Champaign, IL.

Gorostiaga, E.M., Walter, C.B., Foster, C. and Hickson, R.C. (1991). Uniqueness of interval training at the same maintained exercise intensity. *European Journal of Applied Physiology*. 63(2):101-107.

Kraemer, W.J. (1994). Neuroendocrine Responses to Resistance Exercise, In *Essentials of Strength Training and Conditioning*, Baechle, TR, Ed., Human Kinetics, Champaign, IL.

The biomechanical fundamentals, the commonalities, and the preferences of tennis strokes

D. Van der Meer
Van Der Meer Tennis Center, Hilton Head, SC, USA
W. Ben Kibler, M.D.
Lexington Sports Medical Center, Lexington, KY, USA

ABSTRACT: Coaching techniques for optimum performance should be based on knowledge of the biomechanics that are required for stroke production, but should be tailored to the preferences and abilities of the individual. This can be best achieved by understanding the biomechanical points common to all strokes, and constructing coaching techniques to make sure the strokes conform to these common points even though the strokes may vary in other points.

BIOMECHANICAL FUNDAMENTALS – "WHAT MAKES THE BALL GO, AND WHAT HAPPENS WHEN IT DOESN'T"

In its basic essence, coaching high-performance or effective tennis is concerned with how to "make the ball go"—faster and more efficiently. This requires that the tennis coach formulate teaching methods that allow maximum efficiency of stroke production and allow maximum effectiveness on the court. To produce maximum efficiency, the coach must have a knowledge of the biomechanics that underlie stroke production, and how to apply this knowledge in the individual athlete. Also, the coach must be able to analyze and correct problems "when the ball doesn't go"—when there is inefficient stroke production.

THE KINETIC CHAIN

Efficient force to create maximum ball velocity is produced through a coordinated activation of the segments of the body, starting with the ground reaction force of the feet on the ground, and ending with acceleration of the racquet through the ball. This activation sequence is termed a kinetic chain.

Although the kinetic chains in tennis may be slightly different in the serve and the groundstrokes, they have several points in common. First, the largest portions of kinetic energy or force generated in the stroke are developed in the legs and trunk. 51% of the kinetic energy, and 54-60% of total force is produced in this way (Kibler, 1995; Schonborn, 1999). Second, the kinetic chain is oriented to converting linear momentum to angular, or rotational momentum, around a stable post leg (Schonborn, 1999). Third, each segment has a cocking, or stabilization phase, and an acceleration phase (Kibler, 1995). Fourth, large and rapid motions are required in the joints, especially the shoulder (Kibler 1993). Finally, segment dropout or kinetic chain breakage decreases the ultimate force or energy available to make the ball go, and puts abnormally large strains on the surrounding segments (Kibler, 1995). A 10% reduction in kinetic energy from the hip/trunk to the shoulder in the serve requires a 14% increase in shoulder rotation velocity or 22% increase in shoulder mass to create the same kinetic energy at the hand and racquet. There are several reasons for kinetic chain breakage, but the most common include muscle weakness, muscle inflexibility, joint injury, or poor mechanics of the strokes.

COACHING FUNDAMENTALS

To facilitate optimal biomechanical application, the coach should be able to differentiate between limiting factors and personal preferences. Limiting factors are those that are either detrimental to the production of the variables of the ball control, primarily spin and speed or that may increase the risk of injury. Personal preferences are those idiosyncrasies that account for individuality in style, but that do not necessarily have an impact on physical performance.

THE SERVE

BIOMECHANICAL FUNDAMENTALS AND COMMONALITIES

Efficient kinetic chain force production for the serve appears to require four common points in the sequence. First, the leg/hip segment requires some degree of knee flexion in cocking. Knee extension from flexion then provides the upward linear momentum, transferring the ground reaction force to the trunk (Elliott et al., 1995). The back leg provides most of the upward and forward push, while the front leg provides the stable post to allow rotational momentum. Second, the trunk and scapula must rotate and retract, to allow shoulder/arm positioning in cocking (Kibler, 1998). This is the stable position to allow rapid acceleration into the ball. Third, the shoulder must externally rotate and horizontally abduct to achieve cocking, and internally rotate in acceleration. Shoulder internal rotation from cocking is the most important single kinematic variable in the serving kinetic chain, having the highest velocity, occurring closest to the ball impact, and allowing maximum acceleration through the ball (Kibler, 1995). Finally, the forearm must pronate to accelerate the racquet through the hitting zone. This motion is coupled with shoulder internal rotation (Elliott et al., 1995; Schonborn, 1999).

Kinetic chain alterations can occur from weak leg muscles, especially on the front leg. This will create a "pull-through" rather than "push-through" movement of the hips and trunk with either lateral trunk flexion or forward trunk flexion. Decreased trunk rotation will decrease cocking. Tightness of the shoulder rotation will affect both shoulder motion and forearm pronation.

COACHING FUNDAMENTALS AND COMMONALITIES

In the serve, there are a number of commonalities that are prerequisite for the most efficient and effective execution of the stroke. These are the fundamentals that should be evident in every serve.

(1) Stable base – Feet aligned, spread and positioned sideways allowing for knee flexion, and the development of rotational momentum of both the lower and upper body.
(2) Simultaneous movement of both arms – Tossing arm follows the line of the body rotation as the hitting arm goes into the cocking position (closed racquet face). This closed racquet position is best achieved with a proper service grip (continental). With the arms working together in this manner, the athlete will be better able to achieve the desired trunk and shoulder rotation.
(3) The hitting phase – The beginning of the hitting phase is highlighted by the rotation of the hips, trunk, and hitting shoulder. This is followed by a deceleration of the right hip (blocking) which allows for the transfer of energy to the shoulder and arm. The shoulder internal rotation (leading elbow) influences the racquet drop continuity and the acceleration of the elbow forward.
(4) Extension phase – The racquet begins its rapid ascent on edge. This permits optimal internal rotation of the shoulder and pronation of the forearm.
(5) Contact – At this point, a complete extension of the hitting side is demonstrated. The pronation of the forearm and outward rolling of the wrist continues through impact.
(6) Deceleration – After the powerful action of hitting the ball and the ensuing extended followthrough, the player recovers balance.

Note: Any jumping that occurs is incidental and is a result of the upward thrust of the legs pushing against the ground.

THE FOREHAND

BIOMECHANICAL FUNDAMENTALS AND COMMONALITIES

The open stance forehand allows more rapid rotational momentum and quicker reaction to a hit ball (Schonborn, 1999). In contrast to the serve, most of the push-off and rotation force generation occur through the same leg, the back leg. This requires strong hip muscles. Maximum racquet acceleration through ball impact is accomplished mainly by shoulder internal rotation, with hand motion, a combination of forearm pronation, and wrist flexion, as a late contributing factor (Kibler, 1995; Schonborn, 1999).

Kinetic chain alterations include inadequate push-off with back leg, creating a "pull-through" movement from the non-dominant side muscles, and lack of shoulder rotation, requiring a reliance on forearm pronation with wrist flexion to create a "slapping" motion to accelerate the racquet.

COACHING FUNDAMENTALS AND COMMONALITIES

In the forehand, there are a variety of actions that fall into the realm of sound biomechanical execution. These adaptive technical actions result from the reception differences of each ball being hit—height, spin, speed, direction, and depth combined with many possible sending intentions. Just the same, there are certain commonalities that form the basis of the modern forehand.

(1) Movement to the ball – Though getting into position to hit the ball may require variations in footwork, ideally the last step should produce a relatively wide base.
(2) Stable base – Having the feet well spread in the hitting stance allows for the center of gravity to move while staying within the base of support. Add to this an appropriate knee bend and the player is able to maintain balance for a longer period during the hitting phase. This wide, low stance provides the base of support that is needed to keep the upper body and head in balance.
(3) Grip/Stance relationship – The most efficient and effective forehands in today's game are hit with a Semi-Western grip and from an open stance. This permits the contact point to be well forward of the center of gravity. For the right-handed player, the right leg acts as the drive leg and provides most of the push-off for the stroke. This stance facilitates rotation of the upper body allowing for optimal pre-stretch.
(4) Preparation – The backswing is initiated with a sequential backward rotation of the hips and shoulders (turn). This results in a coiling effect in which the shoulders rotate more than the hips and which increases the energy transferred from the legs. During the preparation, the non-hitting arm follows the line of rotation and is brought back parallel to the baseline.
(5) The hitting phase – Preparation without timing is inefficient. This means that the backswing should flow into the stroke. As the racquet is still moving backwards, two things occur: the forward rotation of the right side begins, and there is a relaxed lowering of the arm. These two cause a pre-stretching of the wrist. The racquet acceleration through the ball is a result of the force produced by the summation of rotations—hips, trunk, shoulders, arm, and wrist.
(6) The contact point – The contact point should always be ahead of the center of gravity, but will vary according to the grip that is used.
(7) Followthrough – The followthrough is dependent on the path of the racquet prior to impact and the amount of rotation of the forearm and wrist. Most modern forehands finish with the hitting shoulder well in front and rotated over 200 degrees from the preparation phase. As well, the elbow is flexed and the racquet wraps around the body.

THE BACKHAND

BIOMECHANICAL FUNDAMENTALS AND COMMONALITIES

The backhand stroke can be divided into one-handed and two-handed styles, and the two-handed style can be further broken down into non-dominant arm prime mover and dominant arm prime mover patterns.

The one-handed backhand requires trunk rotation around the stable post of the lead leg and shoulder external rotation to accelerate the racquet through ball impact.

The dominant arm two-handed backhand is similar to the one-handed pattern, with reliance on the dominant leg for stability, and trunk rotation and dominant shoulder external rotation for racquet acceleration.

The non-dominant arm two-handed backhand is more similar to the forehand, with most of the acceleration contributed by non-dominant leg push-through, non-dominant shoulder internal rotation, and non-dominant wrist flexion.

Kinetic chain alterations for the one-handed backhand include lack of trunk rotation and shoulder muscle weakness. For the two-handed backhand, the major alterations include lack of trunk rotation or a "pull-through" movement if back leg "push-through" is diminished.

COACHING FUNDAMENTALS AND COMMONALITIES

For the backhands, some of the commonalities of the forehand also exist such as the stable base. In the case of the one-handed backhand, we see the following:

(1) Stance – A square stance is advocated although there is allowance for particular situations in which a closed or open stance may be preferable. The square stance facilitates torso and shoulder rotation (pre-stretch).
(2) The grip – The eastern backhand grip permits a more stable wrist and an extended arm at impact and during the followthrough.
(3) The hitting phase – The forward action of the racquet is initiated by the trunk and external shoulder rotation. Because the hitting shoulder is positioned ahead of the body, the degree of hip rotation is limited. The angle between the shaft of the racquet and the forearm is maintained throughout the hitting phase.
(4) The contact point – As a result of the grip and the position of the hitting shoulder, the contact point is well ahead of the body.
(5) Followthrough – The followthrough is again a function of the path of the racquet prior to impact and the degree to which the stroke is a rotating shoulder socket vs. a rising shoulder socket.

In the case of the two-handed backhand, we see differences depending on the grip used. An eastern backhand grip with the right hand produces a right arm dominant stroke. The left arm is more of a guide arm, but has the effect of bringing the back shoulder and hips into a square position at contact. An eastern forehand grip with the right hand produces a left arm dominant stroke. In this stroke, we generally see a more aggressive rotation of the hips and shoulders.

In both the forehand and backhand groundstrokes, in which the players are hitting lifted strokes, the push against the ground (drive leg extension) combined with centrifugal force often leads the player to leave the ground. This is not the case with underspin or slice shots.

REFERENCES

Elliott BC, Marshall R, Noffal G (1995): Contributions of Upper Limb Segment Rotations During the Power Serve in Tennis. *J Appl Biomech*, 11:433-442.

Kibler WB (1993): Analysis of Sport as a Diagnostic Aid. In: The Shoulder: A Balance of Mobility and Stability (ed by RF Matsen, F Fu, and R Hawkins).

Kibler WB (1995): Biomechanical Analysis of the Shoulder During Tennis Activities. *Clin Sports Med,* 14:79-86.

Kibler WB (1998): The Role of the Scapula in Athletic Shoulder Function. *Am J Sports Med*, 26:325-337.

Schonborn R (1999): Advanced Techniques for Competitive Tennis. Meyer and Meyer, Aachen Germany, pp 11-73.

Modern day tennis coaching: the impact of the Sport Sciences

M. Crespo, D. Miley and K. Cooke
Development Department, International Tennis Federation, UK

ABSTRACT: The purpose of this review is to examine some of the developments that have occurred in sport science specific to tennis coaching and to determine what implications there have been for the modern tennis coach. Advances in Biomechanics, Physiology, Psychology and Teaching Methods are examined in relation to the practical applications for tennis coaching.

INTRODUCTION

Tilden (1925) stated "There are two general rules of body position so elemental in tennis... 1. Await a stroke facing the net, with body parallel to it. 2. Play every stroke with right angles (sideways) to the net. This is true for service, drive, chop, volley, smash, half-volley and lob". The previous statement represents some of the earliest tennis coaching literature available, however the game of tennis has evolved to such an extent that the present day game is hardly recognisable to that of the 1920's and 1930's. To a similar extent the coaching literature has also evolved. The progressive accumulation of information in the field of sport science means that the modern author of coaching materials has an expanding wealth of resources from which to develop the guidelines for player and coach practice. The sport sciences have helped to develop a better understanding of almost every aspect of the game. Therefore the purpose of this review is to highlight some of the major scientific contributions to the development of modern coaching and playing guidelines. The purpose is not to produce a comprehensive or complete review of the literature but rather to select some of the more influential scientific advances that have had an impact on tennis coaching.

THE PHYSIOLOGICAL CHARACTERISTICS OF TENNIS

The physiological demands of tennis have been addressed by an increasing number of scientific studies examining many of the metabolic and time-motion characteristics of actual match play. Jacobsen (1992) found that the mean time taken to complete a point during actual tennis match play was 6.5 seconds for men and 6.6 seconds for women whereas McCarthy (1997) found that the mean time taken to complete a point was 2.9 +/- 0.1 seconds. The variation in these findings may be explainable by the fact that the studies may have used different surfaces or standards of player, never the less the duration of the majority of points reported in actual match play appears to be in the range of 5 to 10 seconds. Brabenec (1994) reported that 67.7 +/- 3.8% of the points lasted less than 5 seconds and only 14.5 +/- 2.8% lasted greater than 10 seconds.

Tennis play is comprised of a variety of movements with repeated accelerations, decelerations, turns and jumps with activities involving both the upper and lower body. The intensity of tennis match play has been reported in a number of studies the results range from 60% to 86% of maximum heart rate. The calculated work to rest ratio also reported in a number of studies is 1:2 (Elliot and co-workers, 1985).

Peak blood lactate levels have been shown to reach 5.86 +/- 1.33 mmol / litre. The blood lactate concentrations during match play have been shown to correlate to relative exercise intensity (r = 0.71, Christmass and co-workers 1993). Due to the effect of large upper body movements (as in tennis) it has been suggested that this may result in reduced blood flow to the lower limbs and consequently lead to an increase in anaerobic glycolysis and lactate production (McCarthy, 1997). The level of metabolic acidosis as reflected by blood lactate concentrations has been reported to lead a decrement in technical and tactical function in tennis. The maintenance of these functions is paramount to successful performance (Schonborn, 1999).

The impact of such activity also has thermoregulatory connotations for the tennis player. The metabolism of the tennis player during match play generates considerable heat that must be dissipated in order to maintain homeostasis. The primary mode by which this heat is dissipated is the evaporation of sweat; the impact of this is significant fluid losses if fluid consumption does not match fluid losses. It has been reported that if total body water is not maintained, premature fatigue and significant performance decrements will most likely occur (Bergeron and co-workers, 1995). There is clear evidence that the consumption of a fluids such as a carbohydrate electrolyte solution can improve tennis performance parameters (McCarthy, 1997).

THE IMPLICATIONS FOR TENNIS COACHING

The metabolic and time-motion studies done in tennis allow the tennis coach to have a scientific basis for the development of physical conditioning and training regimes that are specific to tennis and will help to maximise performance. Tennis is a game of bouts of moderate to high intensity exercise predominantly utilising the high-energy phosphate systems to regenerate adenosine tri-phosphate for repeated muscular action (Groppel and Roetert, 1992). Training should match the moderate to high intensity nature of the game in order to induce adaptations that will maximise match play performances (Chandler, 1998). The intensity of the training should consider the importance of a tennis specific work-rest ratio (known to be 1:2). The tennis coach should ensure that his or her players consume adequate fluids to match fluid losses especially when competing in a hot environment. Players should consume fluids such as water or an isotonic carbohydrate electrolyte solution in order to maximise performance and recovery in tournament play (Schonborn, 1999).

Crespo and Miley (1998) have indicated that the coach needs to plan the drills that will be performed during a session in order to respect the physiological demands of tennis. In order to be systematic in the training, easier drills should be performed before more difficult drills, and technical drills (with no decision making) should be done before tactical ones. Besides, learning or correction drills should be done before drills aimed at stabilising the skills of the player, drills which involve practising co-ordination or fine motor skills (e.g. serve, drop-shot) should be performed before those which involve practising other skills (e.g. groundstroke endurance, etc.). Furthermore, tough drills should be mixed up with fun or competition drills to ensure motivation.. Besides, the coach has to schedule recovery drills towards the end of the session to maximise the quality of the training. On the other hand, adequate warm up routines to avoid injury and to prepare the body to exercise and cool down procedures to help eliminate any build up of lactic acid should be emphasised.

THE PSYCHOLOGICAL CHARACTERISTICS OF TENNIS

The development of science has not been limited to the physical field, recent advances in the psychological domain have given a enlightened insight to mental factors which influence the general performance and learning of tennis.

MOTIVATION, BURN-OUT AND PERCEPTION OF SUCCESS

The ability to undertake the demands required to compete at the highest level in tennis requires high levels of motivation and the maintenance of this motivation. Balaguer and Atienza (1994) studied the motives of youngsters for playing tennis and if these motives differ depending on players' age and gender. Results confirm previous investigations and reveal that the main motives of youngsters for playing tennis include increase the playing level, keeping physically fit, improve skills; make new friends, and keep in shape.

Gould and co-workers (1996a; 1996b) examined burnout in competitive junior tennis players. Analyses revealed that the burned out in contrast to comparison players had significantly higher burnout scores and less input into training. They were more likely to have played high school tennis and more likely played up in age divisions. Besides, they practised fewer days, were lower in external motivation, were higher in amotivation and reported being more withdrawn. They were less likely to use planning coping strategies, and were lower on positive interpretation and growth coping.

In their study on perceived causes of success among elite adolescent tennis players, Newton and Duda (1993) concluded that there were two conceptually coherent personal goal-belief dimensions. They were: Ego orientation and the beliefs that ability and maintaining a positive impression were the primary causes of success, and Task orientation coupled with the effort and a de-emphasis on external factors and deceptive tactics would lead to tennis accomplishment.

Balaguer, Duda and Crespo (1999) studied the relationship of goal orientations and the perceived motivational climate created by the coach in relation to competitive tennis players' perception of improvement, satisfaction and coach rating. Results supported the motivational advantages of a task-involving atmosphere. When the environment created by the coach was more task involving and less ego involving, the players were more satisfied with their coach, with their results, and they felt that they were progressing more.

CONTROL OF EMOTIONS AND CRITICAL SITUATIONS

The competitive tennis environment produces a great deal of stress factors that need to be coped with and one of the keys to maintaining performance in such an environment is the control of emotions. Wughalter and Gondola (1991) developed a psychological profile of professional female tennis players and to compared it to that of other college-age athletes. Results showed that professional tennis players were less tense, depressed, fatigued and confused, and had more vigour than college-age individuals. Younger players tended to display a psychological profile more similar to the norming sample of college-age women than older players, older players exhibit the 'iceberg profile', i.e. they scored higher on the vigour mood state and lower on all other mood states than younger players and college aged-women.

Weinberg, Richardson and Jackson (1981), and Weinberg, and Jackson (1989) studied the effects of situation critically on tennis performance of males and females. Results showed that after losing the first set, males came back to win the match significantly more often than females. In contrast the results of Ransom and Weinberg (1985) showed that there was no difference between males and females in coming from behind. They concluded that self-confidence appears to be one variable that consistently differentiated successful from less successful tennis players.

The evaluation of stress levels in junior tennis by Pandelidis and co-workers (1997) showed that the high levels of stress described in adult tennis players are not found in young tennis players. The article stated that children tend to lose concentration when under constant pressure and the loss of concentration interferes with performance, leading to injury. Some experts suggest that highly stressed players may use injury as an acceptable alternative to risking failure in high-pressure competition or to quitting. The effective management of stress in adolescent players can reduce injury, enhance performance, and prevent premature burn out.

CONTROL OF THOUGHTS, SELF-EFFICACY AND SELF-TALK

The highly intermittent nature of tennis match play has been suggested as an additional complication in maintaining optimal psychological focus, in particular the ability to control ones thoughts. Lee and co-workers (1992) studied the students' thoughts during tennis instruction. Analysis revealed that those students who had skill technique thoughts were more likely to be successful during practice, and that negative self-evaluation thoughts may have influenced the students' inability to perform an effective technique.

A study about coaches' strategies to develop self-efficacy in players (Weinberg and Jackson, 1990) concluded that the most used strategies to enhance self-efficacy, as well as those strategies found most effective were: encouraging positive self-talk, modelling confidence oneself, using instruction and drills, using rewarding statements liberally, and using verbal persuasion.

Self-talk in competitive tennis was examined by Van Raalte and co-workers (1994). The analysis revealed that negative self-talk was associated with losing and that players who reported in believing in the utility of self-talk won more points than players who did not. These results suggest that self-talk influences competitive tennis outcomes.

ATTENTION, INFORMATION PROCESSING, ANTICIPATION AND VISUAL PATTERNS

Goulet and co-workers (1988) analysed the different visual patterns of expert and novice tennis players preparing to return tennis serve. They found that during the execution of the serve, fixation sequences of experts are linked to cues originating from the racquet and the arm holding the racquet, whereas novices organise their sequences by using a greater number of cues. During this phase, experts terminate their visual search on the racquet at the moment of impact, whereas beginners frequently prolong their processing by following the ball trajectory after impact.

Nougier and co-workers (1989) studied the information processing and attention with tennis players. Results showed that the more expert and the older the players were, the shorter was the reaction time. These results suggest that orienting of attention is submitted to growing up and learning, and younger players have an attentional deficiency when compared to adults since they have a more attentional focused beam and lower attentional flexibility.

Goulet and co-workers (1989) studied the information process for the serve and concluded that expert players appear to select valuable information during the placement of the ball and the initiation of upper body rotation and the arm/racquet complex motion. Goulet and co-workers (1992) evaluated the attentional demands of the processes leading to anticipation of the type of tennis serve. Results demonstrated that attentional effort is maximal before identification of the most important cues necessary for adequate performance.

In a study about the trainability of anticipatory skills for tennis (Singer and co-workers, 1994), it was concluded that expert tennis players focus on more meaningful and predictive cues that enhance quick and accurate decision making. In addition, highly skilled performers are better at resolving uncertainty concerning an opponent's actions from earlier cues than are beginners.

PARENT INVOLVEMENT AND MENTAL TRAINING

The development of the junior player has related to it the influence of parental involvement, and research has been completed to determine the exact influence of this involvement on the player (Crespo and Miley, 1999). Hoyle and Leff (1997) studied the role of parental involvement in youth sport participation and performance. Players who reported a high level of parental support reported significantly greater enjoyment of tennis, viewed tennis as a more important part of their lives and fell lower in state ranking than players who reported a lower level of parental support. The data provided no evidence that parental pressure is an important influence on the participation and performance of young tennis players.

DeFrancesco and Johnson (1990) studied the athlete and parent perceptions in Junior Tennis. Results determined that winning is very important to the players and over one-third of the parents. Only 5% of the players and 7% of the parents indicated that they become upset following loses in which players put forth considerable effort. 33% of the players indicated that their parents had caused them embarrassment during tennis matches. Because adults serve as role models to young athletes, educating parents about appropriate tennis-related behaviours is warranted.

Davis (1992) presented a model of mental training programme for junior tennis players that included goal setting, video analysis of play, imagery, relaxation and energising. It also covered coping with mistakes and losing, coping with stress, anxiety and anger, what to do between points, games and sets, attentional focus, post competition review, and analysis of game performance. The programme was considered to be successful and reinforced the belief that children over 11 years of age can cope with mental skills training.

THE IMPLICATIONS FOR TENNIS COACHING

Since the psychological demands of tennis are so apparent it is essential that the tennis player develop on and off-court mental skills to counter act the often-difficult aspects of the game in order to enhance performance in all areas. The goal of all training and coaching is to achieve the ideal performance on the tennis court as often as possible. The Ideal Performance State (IPS) has been labelled as the goal by which to achieve this consistent level of performance. Controlling the own IPS, which is individual to each player, and avoiding overarousal or underarousal is directly related the player's ability to utilise effective mental skills (Loehr, 1990). The purpose of mental training is to help players acquire mental skills. There are several general areas of mental skills that need to be developed. They include self-confidence, arousal control, attention control, visualisation and imagery control, motivational level, positive energy control and attitude control. It is essential that the tennis coach develops routines and drills to train players to develop each of these areas of importance: i.e. breathing, etc., (Weinberg, 1988).

As per attention and anticipation t he findings suggest that teaching the ability to select and extract meaning from certain anticipatory cues of an opponent's serving motion can greatly enhance the decision-making capabilities of the receiver. Coaches should not overlook the need to guide young players in determining appropriate anticipatory cues in various tennis situations (Crespo and Miley, 1998).

THE BIOMECHANICAL CHARACTERISTICS OF TENNIS

Biomechanics has been termed the "Science of Technique" and to a great extent that has been impact it has had in the field of tennis coaching. Biomechanics has allowed the tennis coach to develop a greater understanding of the principles behind 'good' technique. However the judgement of what constitutes good technique should be based on a number of important factors. These important factors have been reported to be past experience, current world trends, individual flair plus an understanding of the biomechanics of stroke production (Schonborn, 1999).

GROUNDSTROKES: FOREHAND AND BACKHAND

Groundstrokes have geared much of the attention of the biomechanical studies of tennis strokes. Elliott, Marsh and Overheu. (1989a) studied forehand drives of elite tennis players. Results showed that the strokes began with flexion of the knees and hips accelerating the body down towards the court. Deceleration of the body then applied stretch to the muscles, which resulted in the subsequent storage of elastic energy in muscles. This stored energy was then partially used to assist the lower limb drive in moving the player to the ball. A loop in all forehands, which produces a more fluent stroke and allows the racket to accelerate over a larger distance, characterised backswings. Rotation of the trunk and low limb drive increased racket velocity, which is higher just before impact than at impact. "Leading with the elbow" forehands produced higher racket and ball velocities than "Conventional" ones.

Elliott, Marsh, and Overheu (1989b) studied backhand drives of elite tennis players. Results showed that no significant differences were recorded in the three different backhands at the completion of the backswing phase. At impact, a smaller shoulder joint angle, a more acute shoulder alignment, a larger wrist angle and a racket inclined further forward was recorded for cross court backhands when compared to down the line ones. The running backhand reported a more vertical trunk at impact when compared to the two stationary strokes.

Elliott and Christmass (1993) studied the slice backhand drives of elite tennis players. Results showed that for the high slice backhand, the initial flexion of the knees was not so evident. Players, with respect to trunk rotation, prepare for slice and topspin backhands in a similar manner. At impact, the racket face was more open for the lower stroke than for the higher one. In the slice backhand, the ball is impacted closer to the body than in the topspin or flat backhand. The key movements during the backhand forward swing are forward rotation of the upper arm followed by extension of the forearm at the elbow.

Blackwell and Cole (1994) investigated the wrist kinematics in players performing the backhand stroke. Results showed that expert players hit the backhand with the wrist more extended than novice players, which may contribute to lateral tennis elbow in the latter group.

In a study about the role of the elbow in groundstrokes (Elliott, 1995) concluded that it plays an important role in the development of racket speed in the backhand groundstrokes. For both the topspin and backspin strokes movement at the elbow helped to generate considerably greater velocity.

The comparison between the one and two-handed backhands (Groppel, 1983) showed that the one-handed backhand is basically a multiple-segment motion in which the hips, trunk, arm, forearm, and hand and racket move in an extremely co-ordinated fashion. The two-handed backhand was observed to be a two-segment motion where hips rotate then the trunk and upper limbs rotate simultaneously.

SERVICE

Power is one of the major goals for the modern serve. Elliott (1988) pointed out that the parts of the body act as a system of chain links whereby the energy (or force) generated by one link (or part of the body) is transferred in a succession to the next link. The primary source of power for the advanced server is found in the leg action (knee bend-flexion and extension). It is with this action that the source of power is transferred through the link system.

As per the serve motion, Elliott and co-workers (1995) stated that the degree of body rotation, backward bend and arm extension during impact was positively related to success in serving. In order to produce some forward spin on the ball in the "flat" power serve players need to use an "up and out" action prior to and after impact. The author also points out that this appears to be a characteristic of more advanced players and may often be accompanied by the creation of forward rotation due to an upward racket trajectory.

These authors also studied the "foot-up" and "foot-back" styles used in the service action. The foot-up style results in higher impact positions and better "up and out" racket trajectory compared to the foot-back style. However players using the foot-back technique recorded higher horizontal forces and some may be more advantageous for rapid movement towards the net

VOLLEY AND APPROACH SHOTS

These strokes have also been analysed mechanically. Kernodle, Groppel, and Campbell (1982) studied the mechanics of the punch versus the drive volley for skilled competitors. The results showed that the skilled players were more accurate with the punch volley than they were with the drive volley. It was concluded that although more force can be created with the drive volley by increasing the range of motion of the upper limb and racquet head, accuracy is lost in creating more force.

In a study about the action of nine muscles during the execution of the volley (Van Gheluwe and Hebbelinck, 1986), results showed that during the acceleration phase there is less elbow flexion during a volley when compared to a forehand groundstroke. This is reasonable since during a volley there is less swing of the arm, therefore putting less demand on the elbow flexors. The volley requires less forceful muscle action than the forehand, still demands strong muscular effort in order to be executed properly, except for the elbow flexors.

A study about the volley swing (Roetert and Garrett 1987) concluded that advanced players produced a great amount of kinetic energy and greater segmental velocity measures while using shorter swing than the intermediate players. The advanced players also tended to use a continental grip while the intermediate players tended to use an eastern forehand grip.

Elliott, Overheu, and Marsh, (1988) compared volleys hit at the service line and closer to the net. Results showed that the racket was positioned behind the hitting-shoulder for volleys played at the service line by high level players, while in volleys played closer to the net the racket was relatively closer to the shoulder. Advanced players recorded greater wrist and tip of racket velocities when compared to the intermediate group. They moved their racket forward and downward after impact while the intermediate players moved their racket in a "dishing" action where the racket face opened and moved more in a downward trajectory.

Chow and co-workers (1999a; 1999b) studied the muscle activation and the movement characteristics in the volley. They concluded that muscle activity increased with increasing ball speed. Players did not tighten their grip and wrist until moments before ball impact. As per the movement, results showed that when the ball velocity was low, players initiated movement by leaning sideward, when the velocity was high movement was initiated by a vigorous push off of the contralateral foot.

Forehand approach shots of elite tennis players were studied by Elliott and Marsh (1990). Results showed that the mechanics of the topspin and backspin forehand approach shots are significantly different. A reduced backswing is needed compared to the regular forehand groundstroke. The backspin was characterised by a reduced backswing when compared to the one used in the topspin. Values show that it was important to keep moving towards the net in an approach shot especially in the topspin shot. The racket velocity was higher in the topspin shot than in the backspin one. These data support the common belief that the backswing approach shot "keeps low" while the topspin shot "rises" after bouncing.

THE IMPLICATIONS FOR TENNIS COACHING

In looking at the area of advanced stroke techniques, one has to appreciate that there are many variations of good technique. There is no correct answer as to which forehand is the best. What is without dispute is that all of these forehand strokes are extremely effective. In looking at what are correct stroke techniques at the advanced level, the player needs to be much more concerned with the effectiveness rather than the conformity or the cosmetics of the stroke. In attempting to improve the effectiveness of a stroke, today's player should look much more to the biomechanics of the stroke rather than just to the look of the stroke (Crespo and Miley, 1998).

The present literature clearly points the presence of power strokes in today's game. In order to generate enough power to be effective, it is stressed the need for the adequate co-ordination and timing of the kinetic chain for all strokes. Other important practical aspects include a well timed split step, and the need for short pre stroke preparation that allow for the maximisation of potential elastic energy in the musculoskeletal system together with the need of creating more angular momentum. The coach working with young players and children should be aware of the importance of developing a sound technique based in the adequate biomechanical principles. Topspin shots are very important to master since they provide more consistency and aggressiveness. However, today's game needs slice to ensure reach and variation. On the other hand, the two-handed backhand seems to be an easier shot to learn.

The modern attacking forehand has developed into a multi segment stroke with each part of the arm generating greater overall force since there is less emphasis on control and more on the ability to generate power. In order for the modern tennis player to succeed it is apparent that the player should try to develop within their own individual flair a more multi segment forehand technique. In the backhand groundstrokes the player will generate greater velocity by optimising elbow extension in the later part of the stroke. In the service the coach should be aware of the importance of the "up and out" technique for producing the more advanced flat serve with capable players. The two different styles of foot technique should also aim to match the strengths and weaknesses or the game strategy of the individual.

The coach needs to be aware that in the volley the position of the racket needs to be relative to the distance from the net, with the drive volley (with racket positioned behind the hitting shoulder) being used further from the net. In contrast the block volley (racket positioned closer to the shoulder) is used closer to the net. The coach also needs to be aware that advanced players should have a forward and downward action after impact in their volleys.

THE DEVELOPMENTS IN TENNIS TEACHING AND COACHING METHODS

It is clear from much of the previously mentioned literature that tennis is a game of some complexity that can be highly demanding and challenging. In order to tackle the very technical and complex nature of playing tennis the traditional emphasis of tennis teaching and coaching has been to teach the technical aspects of tennis in isolation. The apparent basis for this was to allow the student to be able to perform the necessary motor patterns. The outcome of this style of teaching was that the student began to be able to perform many of the technical aspects of tennis in this isolated practice situation. However, many students who can perform the specific skill in the practice situation cannot perform or utilise the skill in the game or match situation because they do not know how to do it. In fact they do not understand the tactical use of technique. This criticism has been one of the reasons for the evolution of what has been termed the "Game based approach" to coaching tennis (Crespo and Cooke, 1999).

GAME BASED APPROACH TO TEACHING TENNIS

Thorpe and co-workers (1986) who developed the original concept of the game based approach within the physical education setting suggest that game appreciation and the development of tactical awareness should precede development of the motor skills of the game. This is in stark contrast to how traditional tennis coaching has been described. Advocates of the game based approach point to a number of issues in teaching that relate to the overall enjoyment and longer-term development of the student in support of this approach. The game based approach has been suggested to develop greater interest and excitement within the student since the object of classes is not to perform isolated skill but the more enjoyable game setting (Griffin and co-workers, 1997). By understanding the tactical components of tennis students will be empowered to develop the necessary skills to fulfil the essential tactical constituent (Crespo and Cooke, 1999). In terms of the learning of the technique within the context of the game or match situation it is expected that the student who learns skills in an open skill situation will perform better than the student who learns the skill in closed isolation (Cooke, 1999).

FROM COMMAND TO DISCOVERY LEARNING

Traditional teaching and coaching has used the so-called "command style" together with the line formation, analytic or "part" methods and technical correction based on the coach's tips. On the other hand, the new approach to teaching methodology for tennis is centred in the use of discovery and problem solving styles, task assignment, global methods and correction based on buddy teaching, optimal challenge, positive reinforcement or "positive sandwich", effective questioning or facilitation procedures (Crespo, 1999). In one study the discovery method of teaching was compared to the command method in the learning of the forehand and backhand groundstroke. The results showed that the players who had learnt through the discovery method had significantly greater retention for both groundstrokes when later tested (Mariani, 1969). This clearly highlights the importance of the discovery versus command method in teaching tennis.

Another study by Claxton (1988) described and analysed systematically the coaching behaviours of more and less successful high school boy's tennis coaches during practice sessions. Results showed that coaches demonstrated more instructional behaviours than any other behaviour. Management and silence categories accounted for almost 75% of all intervals. The less successful coaches instructed more than did the more successful coaches. The less successful coaches used praise more. More successful coaches asked a significantly greater number of questions of their players than did the less successful coaches. This type of approach by the more successful coaches is indicative of a more discovery based approach.

Harries-Jenkins and Hughes (1993) analysed the behaviour of female tennis coaches observed during three practice sessions. The results showed that all coaches made significantly more organisational comments than skill comments to the players. Each coach exhibit more positive than negative tone in her comments.

Efran, Lesser, and Spiller (1994) studied the use of metaphors in tennis instruction. Players visualised themselves enclosed in a bubble, cocoon, or chrysalis that separated them from non-task stimuli. Results showed that when compared to players receiving regular instruction, players taught to use the metaphors improved significantly in terms of performance and ability to concentrate and were rated higher in enjoyment of the experience, motivation, and a display of appropriate behaviour.

OTHER TEACHING AND COACHING INITIATIVES

In their study of the sequence of teaching the tennis contents, Eason, Smith, and Plaisance (1989) concluded that there was negative transfer when the players who first learned the forehand were required to learn the backhand. The study recommended that "the traditional instruction model of teaching forehand ground strokes be re-examined. In tennis, teaching forehand and backhand concurrently may initially present a confusing task to the learner, but such a procedure may lead to higher ultimate performance".

Herbert, Landin, and Solomon (1996) studied the effect of practice organisation. The results indicated that the practice organisation (blocked or alternated) influenced beginner players' performance more than highly skilled players did.

Another study by Chun (1996) investigated peer tutoring or "buddy teaching" in tennis using a fellow player to observe the performance of the criterion skills was. Results were also successful when higher skilled players worked with players with lower motor capabilities. The higher skilled students were challenged to find ways to assist the less skilled. In turn the less skilled receive individual assistance and guidance, and ultimately find success.

Periodisation and planning the tennis training has been one of the major concerns of coaches and researchers in our sport. Tennis is one of the few sports without an off season for professionals. This fact increases the risk of injury, getting stale or even burn out. When performing at the highest levels of tennis, the focus of the training process should be on training in an efficient way (quality/not quantity). Different periodisation and season plans have been presented in the most recent tennis literature applied to the various age groups and levels of competition (Crespo and Miley, 1998; Schonborn, 1999).

THE IMPLICATIONS FOR TENNIS COACHING

The traditional way of teaching tennis was for the coach to focus on the technique or production of the strokes. It was more "coach centred". The coach used precise models of the strokes to "show" the player how to play. Once the player had mastered the "model" techniques, the coach then focused on the tactics (i.e. implementing the techniques in a game situation). In tennis teaching today, tennis is viewed as an open skill sport with each shot being hit different. The player never plays the same shot twice! Each shot require the player to go through the following process: Perception ⇒ Decision ⇒ Action ⇒ Feedback. This process is more "player centred". Current technique (the action) should be seen as a function of the biomechanical principles outlined and as a means to implement tactics more efficiently. A player's stroke technique should always depend on his tactical intention. Simply stated form (technique) should follow function (tactics) (Crespo and Miley, 1998).

CONCLUSION

This paper concludes that sports science has influenced on all fields of tennis coaching. Studies show that the Physiological, Psychological, Biomechanical and Methodological characteristics of tennis and tennis coaching addressed in the different research projects can have a direct application both in the training of players and coaches. Successful tennis coaching involves the integration of coaching knowledge and the scientific basis of sport. Although this influence of sports sciences applied to tennis is clear, it appears that more research and an improvement of communication between coaches and scientists are needed in order to achieve a full application of these findings in the practical context.

REFERENCES

Balaguer, I., & Atienza, F.L. (1994). *Principales motivos de los jóvenes para jugar al tenis*. Apunts d'Educació Física i Esport, 31, 285-299.

Balaguer, I., Duda, J.L. & Crespo, M. (1999). *Motivational climate and goal orientations as predictors of perceptions of improvement, satisfaction and coach ratings among tennis players*. Scandinavian J. Med. & Sci. in Sports, 9, 381-388.

Bergeron M, Maresh C, Kraemer W, Abraham A, Conroy B, Gabaree C. (1995) *Tennis: A physiological profile during match play*. International Journal of Sports Medicine 12: 474 – 479.

Blackwell, J. R.. & Cole, K.J. (1994). Wrist kinematics differ in expert and novice tennis players performing the backhand stroke. *J. Biomechanics*, 27, 5, 509-516.

Brabenec J. (1994) *Creating efficient training sessions*. ITF Coaches Review. 5, 1-3.

Cooke, K. (1999) *The importance of implicit learning in skill development*. ITF Coaches Review. 19, 7-8.

Crespo, M. (1999) *Teaching methodology for tennis*. ITF Coaches Review. 19, 3-4.

Crespo, M. & Cooke, K. (1999). The tactical approach to coaching tennis. *ITF Coaches Review.*19, 10-11.

Crespo, M. & Miley, D. (1998). *ITF Advanced Coaches Manual.* ITF. London.

Crespo, M. & Miley, D. (1999). *How to be a better tennis parent.* ITF. London.

Chandler J. (1998) *Conditioning for tennis: preventing injury and enhancing performance.* In: Science and Racket Sports, Lees A, Maynard I, Hughes M, Reilly T (Eds), E & FN Spon, London.

Christmass M, Richmond S, Cable N, Hartmann P. (1993) *A metabolic characterisation of singles tennis*. Journal of Sport Science 11: 543.

Chow, J.W., Carlton, L.G., Chae, W., Shim, J., Lim, J. & Kuenster, A.F. (1999). *Muscle activation during the tennis volley.* Med. & Sci. Sports & Ex., 31, 6, 846-854.

Chow, J.W., Carlton, L.G., Chae, W., Shim, J., Lim, J. & Kuenster, A.F. (1999). *Movement characteristics of the tennis volley.* Med. & Sci. Sports & Ex.. 31, 6, 855-863.

Chun, D. (1996). *Peer Tutoring in Tennis*. JOPERD. Vol.67(2):12-15.

Claxton, D.B. (1988). *A systematic observation of more and less successful high school tennis coaches*. Journal of Teaching in Physical Education, 7, 302-310.

Davis, K. (1992). *A mental training program for elite junior tennis players.* SportsCoach. 15 nº 3. , 34. July-September.

DeFrancesco, C. & Johnson, P. (1990). *Athlete and parent perceptions in Junior Tennis*, Journal of Sport Behaviour, vol. 20, nº 1, 29-36.

Dunlap, P. y Berne, L. (1991). *Addressing competitive stress in junior tennis players.* JOPHERD, 62, 1, January, 59-63.

Eason, R.L., Smith, T.L. & Plaisance, E. (1989). *Effects of Proactive Interference on Learning the Tennis Backhand Stroke.* Perceptual and Motor Skills, 68: 923-930.

Efran, J.S., Lesser, G.S. & Spiller, M.J. (1994) *Enhancing Tennis Coaching with Youths using a Metaphor Method.* The Sport Psychologist, 8: 349-359.

Elliot, B. (1988). *Biomechanics of the serve in tennis.* Sports Medicine, 6, 285-294.

Elliot B, Dawson B, Pyke F. (1985) *The energetics of singles tennis.* Journal of Human Movement Studies 11:11-20.

Elliott, B.C., & Christmass, M. (1993). *The slice backhand in tennis.* Sports Coach, July-September, 16-20.

Elliott, B.C., & Marsh, T. (1990). *The forehand approach shot in tennis: a coach's perspective.* Sports Coach, July-September, 11-15.

Elliott, B.C., Marsh, A. P. & Overheu, P.R. (1989). *A Biomechanical comparison of the Multisegment and single unit topspin forehand drives in tennis.* International Journal of Sport Biomechanics, 5, 350-364.

Elliott, B.C., Marsh, A. P. & Overheu, P.R. (1989). *The topspin backhand drive in tennis: A biomechanical analysis.* Journal of Human Movement Studies, 16,1-16.

Elliott, B.C., Marshall, R.N., Noffal, G..J. (1995). *Contributions of upper limb segment rotations during the power serve in tennis.* Journal of Applied Biomechanics, 11, 4, 433-443.

Elliott, B.C., Overheu, P.R. & Marsh, A. P. (1988). *The service line and net volley in tennis: a cinematographic analysis.* Austr. J. of Sc. & Med. in Sport, 20, 10-18.

Gould, D., Udry, E., Tuffey, S., & Loehr, J. (1996). *Burnout in competitive junior tennis players. I. A quantitative psychological assessment.* The Sport Psychologist, 10, 322-340.

Gould, D., Udry, E., Tuffey, S., & Loehr, J. (1996). *Burnout in competitive junior tennis players. II. A qualitative psychological assessment.* The Sport Psychologist, 10, 341-366.

Goulet, C., Bard, C. & Fleury, M. (1989). *Expertise differences in preparing to return a tennis serve: A visual information processing approach.* J. Sport and Exercise Psychology, 11, 382-398

Goulet, C., Bard, C. & Fleury, M. (1992). *Les exigences attentionnelles de la préparation au retour de service au tennis.* Can. J. Sport Sciences, 17, 98-103.

Goulet, C, Fleury, M., Bard, C, Yerlès, Michaud, & Lemire (1988). *Analyses des indices visuels prélevés en réception de service au tennis.* Canadian Journal of Sport Sciences, 13, 79-87.

Griffith, L.L., Mitchell, S.A. & Oslin, J.L. (1997). *Teaching sport concepts and skills. A tactical games approach.* Human Kinetics. Champaign, Ill.

Groppel J, Roetert E. P. (1992) *Applied physiology of tennis.* Sports Medicine 14: 260-268.

Groppel, J.L. (1983). *Teaching one and two handed backhand drives.* JOPERD. n. 38.23- 26.

Harries-Jenkins, E. & Hughes, M. (1993). *A computerised analysis of female coaching behaviour with male and female athletes.* In T. Reilly, M. Hughes & A.Lees (Eds.) Science and racket sports, 238-243.

Herbert, E.P., Landin, D. & Solmon, M.A. (1996). *Practice Schedule Effects on the Performance and Learning of Low- and High- Skilled Students: An Applied Study.* Research Quarterly for Exercise and Sport. Vol.67 (1): 52-58.

Hoyle, R.H. & Leff, S.S. (1997). *The Role of parental involvement in youth sport participation and performance*. Adolescence. Spring.

Jacobsen W. (1992) *Match analysis*. Coaching Excellence, L.T.A. Winter.

Kernodle, M., Groppel, J.L., & Campbell, K. (1982). *A kinematic analysis of the forehand drive volley*. In J.Groppel (Ed.) Proceedings of the Fourth International symposium on the effective teaching of racquet sports, Champaign, Il..

Lee, A.M., Landin, D.K. & Carter J.A. (1992). *Student Thoughts during tennis instruction*. Journal of Teaching in Physical Education, 11, 256-267.

Loehr, J.E.(1990). *The mental game*. Pelham Books. Lexington, Ma.

Mariani, T. (1969) *A comparison of the effectiveness of the command method and the task method of teaching the forehand and backhand tennis strokes*. The Research Quarterly, Vol. 41, No. 2: 171-174.

McCarthy P. (1997) *Influence of fatigue and dietary manipulation strategies on skilled tennis hitting performance*. Unp. Doctoral Th, Loughborough Univ. Engl.

Newton, M. & Duda, J.L. (1993). *Elite adolescent athletes' achievement goals and beliefs concerning success in tennis*. J. Sport & Exercise Psychology. 15, 437-448.

Nougier, Azemar, Stein, Ripoll. (1989). *Information processing and attention with expert tennis players according to their age and level of expertise*. Proceedings of the 7th World Congress on Sport Psychology, 237.

Pandelidis, D. (1997). *Is an 11-year-old tennis player indifferent to competition stress?*. Arch. Pediatr. March.

Ransom, K. & Weinberg, R.S. (1985). *Effect of situation criticality on performance of elite male and female tennis players*. Journal of Sport Behaviour, 8, 144-148.

Roetert, E. P. & Garrett, G.E. (1987). *A kinematic and kinetic analysis of the tennis volley in 12-15 year old children*. Proc. XI Int. Con. Biom, 267. Free University Press. Amsterdam.

Schonborn, R. (1999). *Competitive Tennis*. Meyer and Meyer Sport. Oxford.

Singer, R.S., Cauraugh, J.H., Chen, D., Steinberg, G.M., Frehlich, S.G. & Wang, L. (1994). *Training Mental Quickness in Beginning/Intermediate Tennis Players*. The Sport Psychologist, 8, 305-318.

Tilden, W.T. (1925). *Matchplay and the spin of the ball*. Kennikat Press. N.Y.

Thorpe, R., Bunker, D. & Almond, L. (1986). *Rethinking games teaching*. Univ. of Technology, Loughborough.

Van Gheluwe, B. & Hebbelinck, M. (1986). *Muscle actions and ground reaction forces in tennis*. International Journal of Sport Biomechanics, 2, 88-99.

Van Raalte, J.L., Brewer, B.W., Rivera, P.M. & Petitpas, A.J. (1994). *The relationship between observable self-talk and competitive Junior Tennis Players' Match Performances*. Journal of Sport and Exercise Psychology,16, 400-415.

Weinberg, R.S. (1988). *The mental advantage.*. Human Kinetics. Champaign Ill.

Weinberg, R.S. & Jackson, A. (1989). *The effects of psychological momentum on male and female tennis players revisited*. Journal of Sport Behaviour. 12, nº3, 167-179

Weinberg, R.S., & Jackson, A. (1990). *Building self-efficacy in tennis players: A coach's perspective*. Journal of Applied Sport Psychology, 2, 164-174.

Weinberg, R.S., Richardson, P.A., & Jackson, A.J. (1981). *Effect of situation criticality on tennis performance of male and female*. I.nt. Jour. Sport Psychology, 12, 253-259.

Wughalter, E.H. &Gondola, J.C. (1991). *Mood states of professional female tennis players*. Perceptual and Motor Skills, 73, 187-190.

Relation between selected psychological features and sports abilities in junior tennis

P. Unierzyski
University School of Physical Education, Tennis Department, Poznan, Poland

ABSTRACT: Nationally ranked male tennis players aged 11 (n= 42) and 14 (n=45) underwent psychological tests of temperament and achievement motivation. The tested features are described as permanent characteristics of the nervous system, strongly inborn and difficult to modify during training and education. These characteristics are considered among others, as a base for mental toughness, so the information about its level may be important to coaches and might be used, among others, as a part of talent identification procedures.

The results showed that the tested features, with the exception of achievement motivation, had no influence on the sport results of 11-year-old players. Among 14-year-old players strong influence on the sports level was found as far as the stimulation of the nervous system, the balance between stimulation and inhibition and achievement motivation are concerned. The results proved that the features increased their influence on results in tennis with age and that the possession of strong types of temperament and higher achievement motivation would be an advantage as far as sport development and results are concerned.

INTRODUCTION

The state of knowledge concerning psychological conditions of sports level in tennis indicates the existence of still substantial reserves. There is a particular scarcity of researches on psychological conditions that have the character of lasting properties, such as temperament or achievement motivation. There is also a lack of researches on how the above-mentioned factors change their importance with age. Separate application of psychological knowledge to coaching is the reason for many mistakes made by coaches as well as causes disappointment to those players who are not predestined to practise high-professional tennis by the basics of their personality.

Thus it seemed desirable to conduct researches whose main task would be to examine the influence of diverse factors on the sports results of young tennis players. It was suggested to examine chosen psychological properties, especially including so-called (Schönborn 1984) "driving properties of the nervous system" (mainly its reactivity and achievement motivation) which are known to have a permanent character i.e. they do not easily undergo education or training, however, they apparently determine sports level in tennis to a substantial degree.

AIMS OF WORK

The general aim of the researches was to establish to what degree the given features determined the sports level in the early period of tennis practice. The objective was to check the formation and a possible change in the importance of these features among people practising tennis at different age. The aims of the work were as follows:

- To determine whether temperamental features and achievement motivation influence sports results in tennis as already from the initial stage of tennis career.
- To check the importance level of these factors among players from different age groups and of a different sports level.
- To create scientific data for talent identification programmes.

MATERIAL AND METHODS

The group of subjects was made up of players aged 11 (n=42) and 14 (n=45), taking part in tournaments organised by the Polish Tennis Association. On the basis of their results in tournaments, the players were put by the Association on up-to-date national ranking. This laid the foundations so as to determine in the research the relation between sports results and the level of the tested psychological properties. In order to examine temperament, researchers applied a questionnaire constructed by J. Strelau. In order to measure achievement motivation an the questionnaire constructed by M. Widerszal-Bazyl (1978) was used.

The following proceedings were taken to examine the influence of temperamental features on the sports level of the tennis players aged 11 and 14:

- The comparison of significance of differences between groups of leading (top 10 on National Ranking) tennis players in relation to other players (position 10-40 on National Ranking);
- The analysis of the correlation between the independent variables under investigation and the sports level (position on the classification list).

The division into two age groups enabled researchers to search the changes in the possible influence of the independent variables occurring with age.

RESULTS

Table 1 shows average values, standard deviations and significance of differences between the groups of tennis players of a different sports level. Table 2 presents rank correlation between the tested variables and the sports level of boys aged 11 and 14.

It is noticeable that in both age categories the players of the higher sports level are generally characterised by greater stimulation of the nervous system and its smaller inhibition, which entails a greater value as far as their reciprocal relation is concerned.

The better players are also characterised by greater values in the range of achievement motivation. No differences were found as far as the mobility of the nervous system is concerned. However, the differences were not statistically significant. Researchers did not also find any statistically significant correlation between these features and the sports level.

The differences between two groups of 14-year-old players are, in comparison with younger players, generally greater. A correlation between the stimulation of the nervous system, the relation of stimulation to inhibition and to achievement motivation and the subjects' sports level was also found.

Table 1. Average values, standard deviations and significance of differences between groups of players aged 11 and 14, within the range of the psychological features under investigation.

Psychological Feature	**11**			**14**		
	Group A n=14	Group B n=28	Difference A - B	Group A n=15	Group B n=30	Difference A- B
Stimulation	61,9 ±9,2	58,9 ±12,1	3,0	69,4 ±9,0	65,7 ±8,0	3,8
Inhibition	60,9 ±14,5	63,5 ±11,5	-2,6	64,0 ±15,3	69,4 ± 8,4	-5,4
Balance Stimulation/ Inhibition	1,06 ±0,3	0,93 ±0,2	0,13	1,13 ±0,24	0,94 ±0,1	0,19**
Mobility	62,0 ±9,0	60,5 ±11,1	1,5	64,3 ±9,8	62,8 ±8,7	1,5
Achievement motivation	66,6 ±4,6	63,3 ±7.3	3,3	68,1 ±3,8	65,1 ±6,5	4,0

** Significant difference at level $p<0.01$
* Significant difference at level $p<0.05$

Table 2. Correlation between tested psychological properties and the sports level among boys 11 and 14 years of age

TESTED PSYCHOLOGICAL PROPERTIES	11 years n= 42	14 years n=45
1. Stimulation	0,15	0,29*
2. Inhibition	-0,07	-0,10
3. Balance - stimulation/inhibition	0,28	0,40**
4. Mobility	0,06	0,03
5. Achievement motivation	0,30*	0,32*

** Significant correlation at level $p<0,01$
* Significant correlation at level $p<0,05$

DISCUSSION

It seems obvious that the researches did not consider all of the psychological properties that could influence the sports level in the early period of tennis career. They tried to pay attention to those psychological features that were known to have a permanent character i.e. they do not considerably undergo rapid training.

The analysis of chosen psychological conditions of the sports level in tennis indicates that the influence of the psychological properties on young players' results is ambiguous. On one hand the players of a better sports level are characterised by higher values in majority of the psychological features. However, the differences and correlation (except for achievement motivation) are not statistically significant among 11-year-old boys. Some of the temperamental features (among others the power of stimulation processes) and achievement motivation among 14-year-old boys could be described as the properties which considerably determine good results in sport. Thus one cannot clearly ascertain that temperamental factors and achievement motivation considerably influence sports results as early as the initial period of a career in tennis. It is worth emphasising that the relation between low reactivity (i.e. high stimulation of the nervous system), achievement motivation and a sports level is greater in the older age group. This can indicate that the influence of the tested psychological features on the sports level in tennis increases with age.

RELATIONS BETWEEN TEMPERAMENTAL FEATURES AND YOUNG PLAYERS' PERFORMANCE AND DEVELOPMENT

Temperamental features are considered to be relatively permanent properties of an organism, which manifest themselves in all performed activities. Such features can be found at the energy level (the manner of energy storing and release) as well as in the characteristics of time reactions (rapidity and mobility). They are relatively stable and thus they have the character of certain permanent properties (Strelau 1974, Gracz 1989, Gracz, Sankowski 1995).

Reactivity is considered (after Strelau (1985)) to be the basic temperamental feature under investigation. It is defined as relatively permanent and characteristic of a given individual intensity or the power of reaction to stimuli. Reactivity is one of the most important variables determining a sportsman's behaviour in a stressful condition, including a tennis match. The reactivity is greater when an individual is characterised by greater sensitivity and even the smallest stimulus might cause disruption. The inverse of reactivity (i.e. low reactivity), in turn, involves considerable resistance that is low sensitivity to stimuli and significant efficiency of the nervous system.

As was stated above, the results indicate that the players who achieve better sports results could be described as less reactive. Such players could also be depicted as strong, little sensitive and very efficient (as far as the nervous system is concerned). The external stimulation in such individuals is suppressed by the physiological mechanism of reactivity, which manifests itself in the decrease in the intensity of reaction in comparison with the high-reactive persons. Strong stimulation, which is usually given in the situation of a tennis match, provokes optimum activation in low-reactive persons. This means that the individuals characterised by strong types of temperament should be more successful in sport i.e. able to face difficulties better, could be braver and more determined. Such players would be more effective in the stressful situation during an important tennis match, their reactions would be rapid, strong and durable, the level of movement automation and volitional features would not decrease (Gracz, Sankowski 1995, Orawiec, Dańczyk 1994, Czajkowski 1995). As the stress grows together with the increase in the level of the competitions in subsequent age categories, one can assume that the tennis players characterised by lower reactivity and a higher relation of the power of stimulation processes to inhibition, would be more resistant as far as the effectiveness of their play is concerned. Thus one can suppose that the importance and influence of this feature would increase together with age and growing requirements imposed on professional sportsmen.

Those players, who achieve worse sports results are characterised, in the light of the results, by higher reactivity. There is a concern that those players at older age, together with the increase of psychological stress, may not be up to situations during a match or might require special individualised treatment as they are characterised by increased emotional and sensory sensitivity and lowered resistance of the nervous system (Gracz, Sankowski 1995).

INFLUENCE OF ACHIEVEMENT MOTIVATION ON RESULTS IN TENNIS

The research into the achievement motivation of all 11- and 14-year-old boys indicated that there was a correlation between its higher level and the achieved sports results.

Achievement motivation is a significant element of human personality. It directs his activity and makes it more dynamic. This property, the "driving power of activity", should be understood after J. Atkinson and Feather (1966) as the joint function of the motive power (which is a permanent property of personality) and of what a given individual expects as the consequences of his own activities. This action is a product of two tendencies: to achieve success and to avoid failures. The people of greater achievement motivation prefer tasks and situations where they can influence the result and their actions are successful (Gracz, Sankowski 1995, Sankowski 1991). Such persons continue long-lasting insoluble tasks more effectively and reveal greater persistence (Atkinson, Feather 1966). Situations of this kind are dominant in sports activity. These are e.g. an even matches during Davis Cup tie where players feel great responsibility for the result. Thus this is obvious that those tennis players who achieve better results should be characterised by high achievement motivation.

The above-mentioned observations were confirmed in the researches on tennis players conducted by D.S.Butt, D.N.Cox (1992) who indicated a higher motivation level among class tennis players in relation to university players in the USA. Similar correlations were described by R.Schönborn (1984), E.Morjaen (1992), F.van Frayenhoffen and V.Mion (1995). On the other hand, the so-called negative motivation is characteristic of people with low achievement motivation, who are not confident and want to avoid a failure. This often leads to a lack of progress and in the situation of a match usually evokes excessive stimulation and lowers the quality of action.

High achievement motivation and the previously discussed low reactivity often coexist in a given individual and manifest themselves in, among others, the optimum level of stimulation in difficult situations and the realistic level of aspirations (Gracz, Sankowski 1995, Czajkowski 1995).

RECAPITULATION

On the basis of the analysis it is not possible to come out with a clear-cut statement that the temperamental factors and achievement motivation strongly influence the sports results achieved from the beginning of one's sports career. However, it does not mean that psychological properties have no significant influence on sports level in tennis. The importance of some psychological features should increase at the most advanced stages of a sports career.

The findings into selected conditions of the sports level at the early stage of tennis practice show that the importance of individual abilities and leading features can differ from the established model of a champion (Królak 1990) or the factors that would limit the sports level in tennis (Schönborn 1984, Unierzyski 1994, 1996). There are many psychological features that are significant if one wants to achieve good results among senior players but only a few of them have a considerable influence on the sports level of 14-year-old boys. This means that coaches must focus not only on the features determining current sports results but also be able to recognise the properties which can make the progress in sport feasible or difficult or which can determine the potential.

The talent identification based only on the analysis of sports results in the early period of sports practice is burdened with a serious error. The high results in the early periods of training do not determine a sports career (Savarik 1984) and at each stage of the career the sports level is determined by different factors (Unierzyski 1994, 1996b). Thus on one hand, coaches should consider the level of such features whose importance increases with age, on the other hand, they should consider such variables whose importance decreases with age. This statement concerns not only psychological features but also other factors that influence the progress in sport.

CONCLUSIONS

As a result of the research proceedings and the analysis of results, one can draw the following conclusions:

(1) On the average, the higher level of the tested psychological features was found among those young boys who were characterised by a higher sports level. However, the differences between the top players and other players were usually not statistically significant.

(2) High achievement motivation is an important factor affecting performance in tennis already at the early stages of career.

(3) In case of 14-year-old players there a significant correlation between the sports level and some temperamental properties (like low reactivity of nervous system) was found.

(4) The tendencies manifesting themselves in the increase of the importance of some other psychological properties with age can indicate the possibility of the increase in their influence on the sports results achieved at older age i.e. over 14.
(5) Because of the fact that many features that determine the sports level in professional tennis do not reveal their importance among very young players, the selection or talent identification in tennis based only on the analysis of sports results among 11-year-old boys could be burdened with an error. The process of talent identification has to be based not only on the sports results, but also on the analysis of possibly all properties, features and abilities (including psychological ones) that determine the sports level in professional tennis.
(6) As it is important, as a part of talent identification process, to recognise the level of permanent psychological properties already at the early stages of tennis career because they would strongly determine future results (progress in sport).

REFERENCES

Atkinson J. Feather F.(1966) *An introduction to motivation.* Van Nostrand. Princeton.

Butt D.S.,Cox D.N. (1992) Motivational Patterns in Davis Cup, University and recreational Tennis Players. *Int J.Sport Psych.*, 23:1-13.

Czajkowski Z. (1995) Znaczenie pobudzenia w działalności sportowej cz.1 i 2. *Sport Wyczynowy* nr 7-8, 9-10.

Dracz B. (1978) *Psycho-społeczne uwarunkowania tenisa.* Kraków AWF, Monografia nr.10.

Gracz J., Sankowski T.(1995) *Psychologia Sportu.* AWF POZNAŃ.

Fraayenhoven van F., Mion V. (1995) *Developing the Mentality of Tennis Players.* 9th ITF Worldwide coaches Workshop, Barcelona.

Gracz J. (1989). *Informacja dotycząca oceny wybranych psychologicznych kryteriów powodzenia w tenisie.* Maszynopis.

Królak A. (1990) Droga do sukcesów w tenisie. *Trening nr 2.*

Morjaen E. (1992) *Aspecten van mentale training.* LCK Brussel.

Orawiec A., Dańczyk R. (1994) Temperament a wybrane zdolności psychomotoryczne zawodników hokeja na lodzie. *Trening* nr 4.

Sankowski T. (1991) Motywacja osiągnięć i jej implikacje w działalności sportowej. *Roczniki Naukowe AWF w Poznaniu*, zeszyt 40.

Safarik V. (1985) Prispevek k vyberu talentovane mladeże v tenisu. *Teoria i Praxe Tel.Vych.* nr 4.

Schönborn R. (1984) Leistungslimitirende und Leistungabestimende Faktoren. In. Gabler H., Zein B. *Talentsuche und Talentforderung im Tennis* Ahresburg. Czwalina.

Schönborn R. (1993) Players' performance and development. ITF Coaches Review no 2.p 1.

Schönborn R. (1994) Modern Complex Training in Tennis. ITF Coaches Review no 4.pp 1-4

Strelau J. (1974) *Rola cech temperamentalnych w działaniu.* Ossolineum. Wrocław.

Strelau (1985) *Temperament, osobowość działanie.* PWN, Warszawa.

Unierzyski P. (1993) Tendencje rozwojowe tenisa ziemnego, (in): Drozdowski Z. (red.) *Tendencje rozwojowe sportu.* Monografie nr 300. AWF Poznań.

Unierzyski P. (1994) Relations Between Experience, Fitness, Morphological Factors and Performance Level with Reference to the Age. *ITF Coaches Review* no. 3 p 8.

Unierzyski P. (1996) *Morfologiczne, motoryczne i psychiczne uwarunkowania poziomu sportowego we wczesnym okresie uprawiania tenisa ziemnego*. Doctoral Dissertation. AWF Poznan.

Unierzyski P. (1996b) A retrospective analysis of junior grand slam winners. *ITF Coaches Review*, Issue 9.p 2.

Wiederaszl-Bazyl M. (1978) Kwestionariusz do mierzenia motywu osiągnięć. *Przegląd Psychologiczny* nr 2.

Kinematic analysis of the service stroke in tennis

C. Papadopoulos, M. Emmanouilidou
Aristoteles University of Thessaloniki, TEFFA, Serres, Greece
S. Prassas
Colorado State University, Fort Collins, USA

ABSTRACT: The purpose of this study was to investigate the service stroke of elite tennis players. During a 1998 international tennis tournament, service strokes of the first, second, third, and fourth female winners were videotaped with two 60 Hz videocameras. Subsequently, the service technique of each athlete was assessed utilizing an Ariel Performance Analysis System (APAS). Velocities of the racquet head, ball, shoulder, elbow, and wrist joints, and subjects' center of mass (CM) displacement were examined. Results showed that, with one exeption, the horizontal ball velocity was inversely related to the players' placement (38.11m/sec, 42.64 m/sec, 49.04 m/sec, and 46.89 m/sec for the 1st, 2nd, 3rd, and 4th winners, respectively) and only in two players (2nd and 3rd) the (resultant) maximum racquet velocity coincided with the ball contact. In addition, it was found that in all players, ball contact occurred when the body's CM was at the highest point. Lastly, it was found that the proximal-to-distal joint speeds sequentially increased in all players during the execution of the analyzed services, however in two players the timing of joint maximum speeds did not follow the motion sequencing described in the literature. It was concluded that the service technique was largely unique to each athlete and that certain characteristics of that technique did not necessarily relate, as expected, to the final placement of the athletes.

INTRODUCTION

Equipment/surface development, better training programs, and/or better skill technique(s) have changed the sport of tennis to the degree that today tennis is considered to be a dynamic and physically taxing activity. Indeed most scoring today occurs during the first few volleys after the service, or by the service alone (ace).

Research in tennis has been limited to the examination of string and racquet characteristics (Hatze, 1994), forehand stroke (Inoue, et al., 1997; Wang, et. al., 1999) and backhand stroke (Ray, 1995; Wang, et al. 1998; Groppel, 1986).

One- and two-handed backhand strokes have also been investigated and compared to determine which one is more likely to reduce elbow joint injuries (Groppel, 1986). A number of investigators have examined the service stroke to determine the contribution of body segments to racquet speed (Springs, et al, 1991; 1994; Elliot, et al., 1995). Results showed upper arm internal rotation to be the most important contributor to racquet head speed.

Although the importance of the service stroke has been recognized and has been investigated, the respective data collection was performed during non-competitive settings and the level of the players was not elite. Thus the purpose of this study was to investigate the service stroke of elite tennis player during competition.

METHODS

The first through the fourth placed tennis players competing in a 1999 international tennis tournament (Thessaloniki, Greece) were videotaped with two Panasonic 60 Hz videocameras. They were ranked 146th, 164th, 361th, and 127th respectively by the International Tennis Federation. The cameras were placed (approximately) perpendicular to and in line with the plane of motion of the ball.

The video data—beginning on the fifth frame prior to ball tossing and terminated on the fifth frame after impact—were analyzed utilizing an Ariel Performance Analysis System (APAS). Three-dimensional position data of 13 body points (ankles, knees, hips, shoulders, elbows, and wrists) and the head of the racquet were calculated by combining the video images of the two cameras utilizing the direct linear transformation (DLT) method (Abdel-Aziz and Karara, 1971). The raw data was digitally smoothed with a cut-off frequency of 5 Hz before being submitted to further analysis. Dempster' s (1955) data as presented by Plagenhoef (1971) were utilized to predict the segmental and total body anthropometric parameters necessary to solve the mechanical equations.

The following variables were examined:

(1) Resultant maximum shoulder ($Vmax_s$), elbow ($Vmax_e$), and wrist ($Vmax_w$) joint velocities.
(2) Resultant maximum racquet head velocity ($Vmax_{r.}$).
(3) Horizontal velocity of the ball (Vx_b).
(4) Initial, ($CM\mathbf{r}_y i$), lowest ($CM\mathbf{r}_y l$), and highest ($CM\mathbf{r}_y h$), center of mass vertical heights.
(5) Center of mass negative/downwards ($CM\mathbf{d}_y n$) and positive/upwards ($CM\mathbf{d}_y p$) vertical displacements.
(6) Time to impact.
(7) Time to maximum velocities.

RESULTS AND DISCUSSION

Table one presents maximum shoulder, elbow, wrist, and racquet resultant velocities and horizontal velocity of the ball at impact. As it is shown, joint velocities increased from the proximal (shoulder) to distal (wrist) joints in all subjects. Racquet head velocities were lower than the mean values reported previously (Elliot et al., 1986; Elliot et al.,

1995; van Gheluwe and Hebbelinck, 1985) ranging from 18.54 to 21.66m/sec. Horizontal ball velocities, however, were appropriate for the caliber of the players ranging from 38.11 to 49.04m/sec (approximately 90 to 110 miles/hour). The low framing rate (60 Hz) in this study may be a possible explanation for the low racquet head speeds.

Table 1 Joint, racquet, and ball velocities

Variable	First	Second	Third	Fourth
$Vmax_s$ (m/sec)	4.52	4.17	3.96	4.16
$Vmax_e$ (m/sec)	6.30	7.28	9.02	6.54
$Vmax_w$(m/sec)	8.99	9.36	9.67	9.63
$Vmax_{r.}$ (m/sec)	21.66	20.47	21.37	18.54
Vx_b (m/sec)	38.11	42.64	49.04	46.89

Temporal results are presented in Table two. The proximal-to-distal motion sequencing described in the literature (Elliot, et al., 1986; van Gheluwe and Hebbelinck, 1985; Plangenhoef, 1971) was observed in only the second and third placed players. As it is shown, the first and fourth placed players exhibited a different movement pattern, reaching maximum elbow joint velocity before reaching maximum shoulder joint velocity. Maximum racquet head speed coincided with impact in the second and third placed players, but was seen slightly afterwards in the first and fourth.

Table 2 Temporal results

Variable	First	Second	Third	Fourth
Time to $Vmax_s$ (sec)	0.837	0.770	0.737	0.954
Time to $Vmax_e$ (sec)	0.787	0.803	0.753	0.921
Time to $Vmax_w$(sec)	0.840	0.870	0.820	0.954
Time to $Vmax_{r.}$ (sec)	0.903	0.890	0.850	1.070
Time to impact (sec)	0.870	0.890	0.850	1.040

All athletes exhibited an initial reduction in center of mass height from the court surface followed by an increase (Table 3). Center of mass negative (downward) displacement ranged from 8.7 to 19.1 cm, followed by a positive (upward) vertical displacement ranging from 14.5 to 30.7 cm.. These results suggest that a moderate center of mass vertical motion may be advantageous when serving in tennis. It should be noted that, in all subjects, ball impact occurred when the center of mass was at its highest point.

able 3 Center of mass kinematic results

Variable	First	Second	Third	Fourth
CM$\mathbf{r}_y$i (cm)	93.0	92.6	102	100
CM$\mathbf{r}_y$l (cm)	80.9	73.5	93.3	83.3
CM$\mathbf{r}_y$h (cm)	95.4	104.2	117	112
CM$\mathbf{d}_y$n (cm)	12.1	19.1	8.7	16.7
CM$\mathbf{d}_y$p (cm)	14.5	30.7	23.6	29.3

CONCLUSION

The results of this study suggest that the service technique of elite level tennis players contain unique elements that may not comport to biomechanically optimum technique. Although the proximal-to-distal joint speeds sequentially increased during the execution of the analyzed services in all players, in two players, the timing of joint maximum speeds did not follow the motion sequencing described in the literature. Higher framing rate and larger number of subjects will enable a better understanding of the mechanics of the service stroke in tennis.

REFERENCES

Abdel-Aziz, Y.I., and Karara, H.M. (1971). Direct linear transformation from comparator coordinates into object space coordinates in close-range photogrammetry. In *Proceedings of the Symposium on Close-Range Photogrammetry*, pp. 1-18. Falls Church, VA: American Society of Photogrammetry.

Dempster, W.T. (1955). Wright-Patterson Air Force Base. *Space requirements of the seated operator*, pp. 55-159. WADC Technical Report, Dayton, Ohio.

Eliott, B. C., Marsh A., and Blanksdy, B. (1986). A three-dimensional cinematographic analysis of the tennis serves. *International Journal of Sport Biomechanics*, **2**(4), 260-270.

Eliott, B. C., Marshall, R.N., and Noffal, G. J. (1995). Contributions of upper limb segment rotation during the Power service in tennis. *Journal of Applied Biomechanics*, **11**, 433-442.

Gropel, J.K. (1986). The biomechanics of tennis: An Overview. *International Journal of Sport Biomechanics*, **2**, 141-155.

Hatze, H. (1994). Impact probability distribution, sweet spot, and the concept of an effective power region in tennis racquets. *Journal of Applied Biomechanics*, **10**, 43-50.

Inoue, Í., Iino Õ., and Kojima, Ô. (1997). Upper extremity torque during tennis forehand Volley. *Book of abstracts. XVI Congress of the International Society of Biomechanics* (pp.352)

Plagenhoef, S. (1971).*Patterns of human motion: A cinematographic analysis.* Englewood Cliffs, NJ: Prentice Hall.

Ray, J.B. (1995). The biomechanical analysis of the one-handed and two-handed backstroke in tennis. *Book of abstracts. XII International Symposium on Biomechanics in Sports* (pp.92-94). Hungarian University of Physical Education, Budapest, Hungary.

Sprigings, E., Marshall, R., Elliott, B., and Jennings, L. (1991). Determining the contributions that the anatomical rotations of the arm segments make to racquet head speed. *Book of abstracts. XII International Congress on Biomechanics* (pp.399-400).

Sprigings E., Marshall, R., Elliott, B., and Jennings, L. (1994). A 3-D kinematics method for determining the effectiveness of arm segment rotations in producing raquet-head speed. *Journal of Biomechanics*, **27**, 245-254.

Van Gheluwe, B. and Heddelinck, M. (1985). The kinematics of the service movement in tennis: A three-dimensional cinematographical approach. In *Book of abstracts. Biomechanics IX-B* (pp.521-526). University Park Press, Baltimore, USA

Wang, L., Wu, H.W., Su, F.C., and LoK, C. (1998). Kinematics of upper limb and trunk in tennis players using single handed backhand strokes. *Book of abstracts. XVI International Symposium on Biomechanics in Sports*, (pp.273-276). University of Konstanz, Konstanz, Germany.

Wang, L., Wu, H.W., Su, F.C., and Lo K.C. (1999). Kinematics of upper limb and trunk in tennis players using forehand stroke. *Book of abstracts. XVII International Symposium on Biomechanics in Sports* (pp.113-116). Edith Cowan University, Perth, Australia.

Dynamic characteristics of two techniques applied to the field tennis serve

L. Braga Neto, E.C. Bezerra, J.C. Serrão, e A.C. Amadio
Department of Human Movement Biodynamics, School of Physical Education and Sports, University of São Paulo, Brazil
E.P. Eche
Eche-Pommê Tennis Academy, São Paulo, Brazil

ABSTRACT: There is currently great interest in the tennis serve, a skill assumed to be a determinant of the game outcome. However, few studies have been performed to investigate this movement and to obtain further information about it for a better analysis and interpretation of its technique. In the present study, two of the most common techniques (the foot-back and foot-up techniques) were analyzed by dynamometry. The two techniques were found to have different standards, although neither one represented excessive overload of the locomotor system.

INTRODUCTION

Today, when we watch a tennis game we note a growing contribution of indefensible serves which are increasingly present in the game. Of course, this motivates the study of the serve since this is a technical element which strongly determines the performance of the player, who has total control on the execution of this closed skill, while the environment has little or no effect (Ashe, 1975; Elliott & Kilderry, 1983; King, 1981). Despite this advantage, the execution of the serve is difficult to muster since the arm that throws the ball must be raised slowly in order to place the ball in the ideal point of contact, while the arm that holds the racket must balance with a complex pattern in order to hit the ball with combined power and coordination. Not only do the arms perform different patterns of movement and rhythm, but they must also synchronize with the movements of the lower limbs and of the trunk (Elliott & Kilderry, 1983; Price, 1975).

According to Groppel (1986), this fact tends to continue to evolve, depending on many other variables such as evolution of physical conditioning and training methods applied to tennis, evolution of the technique, and evolution of the racket, among other conditioning and determining factors. According to Amadio & Stucke (1992), the specialized biomechanics literature is weakly described and reports little information for a better interpretation of the tennis serve.

Some studies have been conducted relating dynamometry and tennis using the ground reaction force (GRF). Nigg *et al.*(1981) stated that the force peaks can be altered by variables such as the velocity of movement. Thus, using dynamometry, which concerns all types of force as well as the distribution of pressure, we can interpret the dynamic behavior responses of human movement.

In the specific case of the tennis serve, Elliott & Wood (1983) have compared the foot-up and foot-back serve techniques using a force platform. The study reported that the foot-up technique (which consists of the approximation of the lower limbs in order to perform impulsion with an effective action of the latter) is characterized by a greater magnitude of vertical GRF in relation to its horizontal component due to the higher height of the racket-ball impact presented by those who follow this technique compared to the foot-back technique. In contrast, the foot-back serve strategy (which consists of maintaining a greater anteroposterior separation between the lower limbs) showed a great magnitude of the horizontal GRF since those who follow this technique presented a greater anteroposterior advance towards the net during the first step after the serve compared to those who follow the foot-up technique.

Van Gheluwe & Hebbelinck (1986) compared the three ground reaction forces recorded during the execution of the serve: transverse GRF, which is practically nonsignificant; horizontal GRF, in which the impulse generated by the body in the anteroposterior direction towards the net is low, being more evident in the initial phase of the serve, and vertical GRF, in which the greatest force magnitudes occur. The cited study thus shows that the forces executed through the body on the field surface are of relatively low magnitude, with the most outstanding being the vertical impulse, which does not exceed the relative value of 1/3 the body weight. Thus, it can be seen that the overload the athlete's body withstands during any phase of the serve is relatively low.

The objective of the present study was to compare the two serve techniques executed by the same athlete in order to differentiate the dynamic patters of these two forms of movement, and to compare the results obtained to those previously reported in the literature.

MATERIALS AND METHODS

The experimental sample consisted of a single volunteer male athlete weighing 98 kg, 1.83 m tall and aged 27 years. This athlete has a 20 year experience with tennis. Data collection consisted of 12 acquisitions for each serve technique. GRF was measured using a Kistler AG model 9287 A force platform based on piezoelectric transducers. The resting surface measures 600 x 900 x 150 mm, dimensions that satisfy the characteristics necessary to perform the movement. The platform determines the GRF components (F_x, F_y, F_z), the signals are then amplified with a Kistler 9865 B amplifier, and the data are acquired through an A/d (12 bits) converter, stored and processed with the Kistler Bioware software (Biomechanical Software Analysis System and Performance Module) using a PC. However, only the vertical (F_y) and anteroposterior (F_x) components were selected for movement analysis since they present relevant magnitudes in the study of movement, as described in the literature. The sampling frequency used was 1000 Hz

For data treatment we used a specific routine developed with the MATLAB software version 4.2, which filters the data based on the selected cut frequencies and later normalization as a function of total resting time, as described by Bezerra et al. (1998). On the basis of the routines implemented, this software automatically identifies and calculates the selected variables and exports the mean curves and standard deviations for each time base used in normalization.

The values of the mean curve for each serve technique were calculated on the basis of the twelve serve attempts executed for each technique on the platform. The coefficient of variation (CV) determined by Winter (1991) was used for statistical analysis of the variability of the serve pattern. Because of the complexity of the serve movement, the present study has some limitations, with restriction of the selected variables to those representative of behavioral dynamics.

ANALYSIS AND DISCUSSION OF THE RESULTS

On the basis of the experimental data collected, it was possible to construct Figures 1 and 2 which present the mean Fy and Fx curves and their respective standard deviations based on time for the two techniques. The time interval considered starts when the individual climbs on the force platform (beginning of the movement), continues through the phase of preparation for the serve, and ends at the beginning of the aerial phase (loss of foot contact with the platform), which occurs with the technical purpose of hitting the ball at the highest possible point.

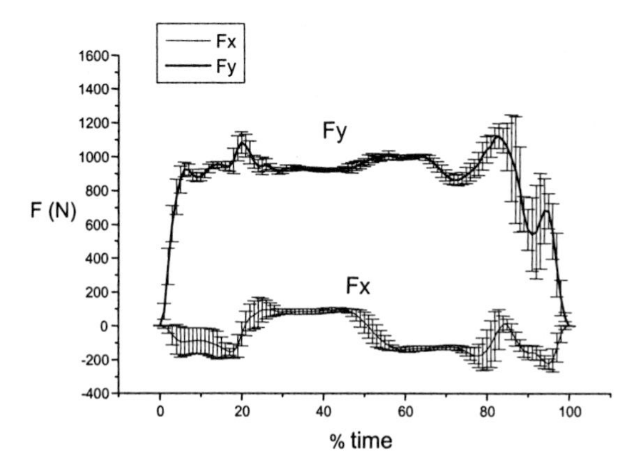

Fig. 1 Mean vertical ground reaction force (Fy) and horizontal ground reaction force (Fx) and the respective standard deviations for the foot-back serve technique (n=12).

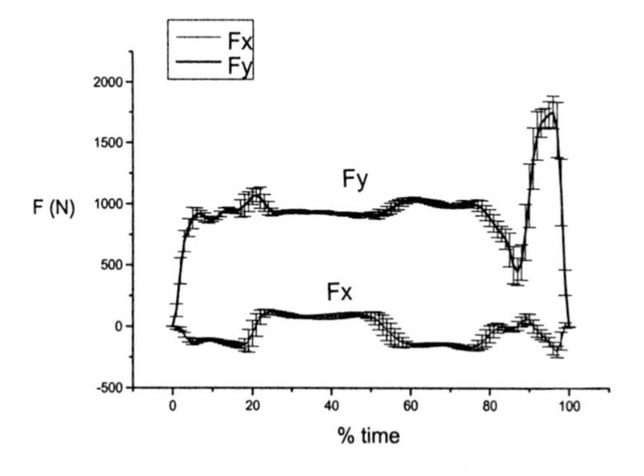

Fig. 2 Mean vertical ground reaction force (Fy) and horizontal ground reaction force (Fx) and the respective standard deviations for the foot- up serve technique (n=12).

The foot-back serve technique presented in its mean curve a maximum value of 1122 N (± 65) for the vertical component, representing 1.14 times the body weight of the executor. In contrast, for the foot-up technique, this value was 1753 N (± 133), representing 1.79 times the body weight of the executor. The relationship between the vertical GRF and the body weight of the executor obtained here does not agree with the results obtained by Van Gheluwe & Hebbelinck (1986), who reported that the vertical GRF is approximately 1.33 times the body weight. The values obtained here and illustrated in Figure 2 were higher than those reported in the literature. However, even so, these values do not represent significant overloads on the locomotor apparatus of the executor.

To determine the relationship between the dynamic patterns of the two serve techniques it became necessary to superimpose the mean curves for the vertical and horizontal GRF of both, thus facilitating the visualization of similarities and contrasts.

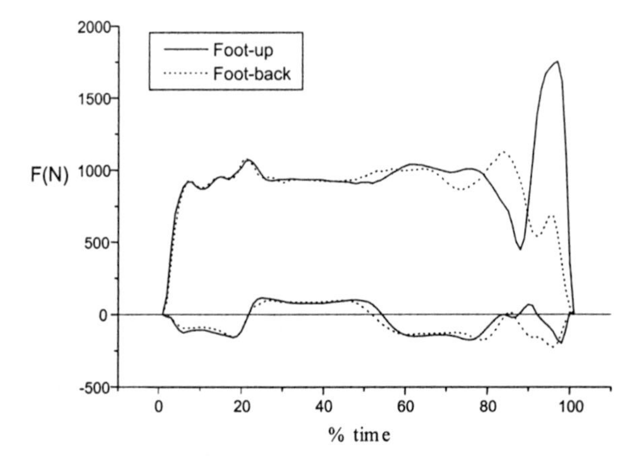

Fig. 3 Superimposition of the mean curves for the vertical and horizontal ground reaction forces of the two serve techniques (n=12).

Closely similar dynamic patterns can be seen for the two serve techniques by observing the superimposition of their mean Fx and Fy curves during the phase of preparation for the serve. This is due to the fact that this phase is executed in the same manner for the two techniques, with changes in movement pattern occurring from the time when the ball is thrown, as observed in the dynamic responses of GRF. A high-magnitude vertical GRF is quite evident in the foot-up technique compared to the foot-back technique, as previously reported by Elliott & Wood (1983). In the present study, the magnitude of vertical GRF in the foot-up technique corresponded to a value 1.56 times higher compared to the vertical GRF of the foot-back technique.

CONCLUSIONS

Thus, we confirm the conclusions of previous studies that have shown a greater magnitude of the vertical force during the foot-up serve strategy due to the greater impulsion during its execution. This magnitude, however, is not so great as to injure the locomotor apparatus because it does not exceed a value corresponding to double the body weight of the executor. The dynamic patterns observed in the two techniques analyzed during the phase of preparation for the serve are quite similar. Future studies should be conducted to record images of the serve with the synchronized use of a videocamera in order to confirm and quantitate the relationship between the two techniques by ordering the phases of the technical movement and the height of contact with the ball. Similarly, electromyography studies, which determine the patterns of muscle activation, could be used to obtain this serve pattern for the two techniques which, according to our results, presented distinct dynamic responses during the phase of preparation.

REFERENCES

Amadio, A.C., Stucke, H. (1992) Análise Biomecânica do Saque no Tênis de campo : comparação de dois métodos para a determinação da velocidade da bola. *IV Congresso Nacional de Biomecânica*, São Paulo, Brasil.

Ashe, A. (1975) Getting more firepower into the cannonball. *Tennis strokes and stratgies*, New York: Simon & Shuster.

Bezerra. E.C. (1998) Descrição dinâmica do padrão da marcha durante o transporte de carga através de mochilas. Monograph, EEFE-USP.

Elliott, B., Kilderry, R. (1983) *The art and science of tennis*, Philadelphia: Saunders.

Elliott, B., Wood, G. (1983) The biomechanics of the foot-up and foot-back tennis serve techniques. *The Australian Journal of Sports Sciences*, v.3(3), p.3-5.

Groppel, J.L. (1986) The biomechanics of Tennis: an overview. *International Journal of Sport Biomechanic*, v.2, p.141-155.

King,B.J. (1981) *Play better tennis with Billie Jean King,* Octopus Books Ltd, London.

Nigg, B., Demoth, J., Neukomm, P.A. (1981) Quantifying the load on the human body. Problems and some possible solutions. *Biomechanics*, VII, p.88-91.

Price, B. (1975) Body arc and serving power. *Tennis strokes and strategies,* New York: Simon & Schuster.

Van Gheluwe, B., Hebbelinck, M. (1986) Muscle actions and ground reaction forces in tennis. *International Journal of Sport Biomechanics*, v.2, p.88-99.

Winter, D.A. (1991) Biomechanics and motor control of human gait, *University of Waterloo Press, 2nd ed.,Ontario* p.35-37,.

Kinematics of trunk and upper extremity in tennis flat serve

L.H. Wang, C.C. Wu & F.C. Su
National Cheng Kung University, Tainan, Taiwan
K.C. Lo
King Shan Institute of Technology, Tainan, Taiwan
H.W. Wu
Foo Yin Institute of Technology, Kaohsiung, Taiwan

***ABSTRACT*:** The purpose of this study is to investigate the kinematics of trunk and upper extremity during flat serve. The video-based motion analysis system was used to quantitatively analyze the range of motion of the joints in upper extremity in four phases of serve (wind-up, cocking, acceleration and follow through). Eight Taiwan national tennis representatives were recruited in this study. The HiRES Expert Vision motion system with six cameras was used to collect the trajectories of the reflective markers at sampling rate of 240 Hz while the subject performed tennis serve. Ten trials were sampled for each subject. Each trial lasted for 5 seconds with 3 minutes rest between trials. Sixteen markers were placed on selected anatomic landmarks unilaterally to define the coordinate system of trunk, pelvis, upper arm, forearm and hand. The results showed that in the late phase of acceleration, the velocity of elbow extension and wrist flexion was 500 and 800 degrees/sec, respectively. The power of serve was mainly from the movement of trunk rotation and bending, and elbow extension.

INTRODUCTION

Serve is the most important technique in tennis. It is an active stroke without any interference from opponent to directly attack the opponent's weakness. There are three patterns in tennis serve: flat serve, spin serve and slice serve. Players of all levels perform the first serve by using the pattern of the flat serve which can help to develop a high-speed serve. Visual evaluation for serve is not scientific. The video-based motion analysis system can be used to quantitatively analyze the range of motion of the joints in upper limb in four phases of serve (wind-up, cocking, acceleration and follow through), Particularly through the individual movements of the upper extremity segments. The purpose of this study is to investigate the kinematics of trunk and upper extremity during flat serve. The result will be able to provide the basic guidelines of the tennis training and the tennis evaluation. Also, it is helpful for the physician and therapy to assist the diagnosis of sports injury and clinical treatment.

MATERIALS AND METHODS

In the kinematic model, the trunk, upper arm, forearm and hand were treated as a four-segment linkage system. For the spatial kinematic description, each segment was treated as a rigid body and each joint was assumed to be of the ball and socked type. Sixteen markers were placed on selected anatomic landmarks unilaterally to define the coordinate system of trunk, pelvis, upper arm, forearm and hand. The selected anatomic landmarks were processes xiphoideus, sternal notch, spinous process of the 7th cervical vertebra, acromin process, medical and lateral epicondyles of the elbow,
radical and ulnar styloid processes, knuckle II and knuckle V, anterior superior iliac pine, and posterior superior iliac pine. In addition, a set of triad markers was placed on the upper arm. The positions of the markers on the medial and lateral epicondyle during tennis single-handed backhand drive were calibrated using the local vectors with respected to the triangular frame on the upper arm in an anatomical neutral posture. This was done in order to avoid error resulting from skin movements.

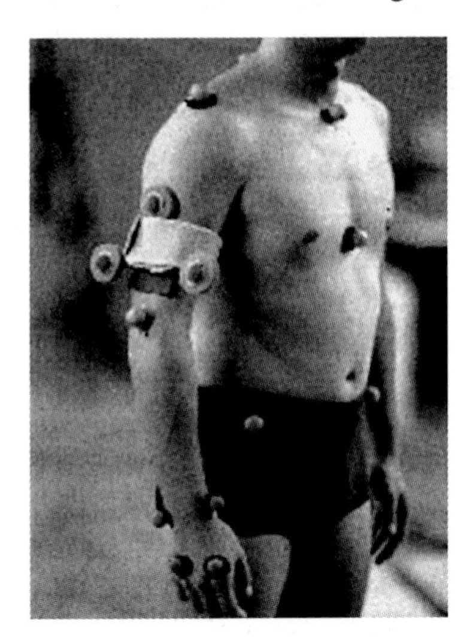

Fig. 1 Marker set

The rotation matrix used to describe the orientations of objects could be formulated based on these coordinate systems. The orientation of a distal segment coordinate system relative to a proximal segment coordinate system was used to describe the joint movement by

$$\boldsymbol{R}_j = \boldsymbol{R}_p^{\,T} \boldsymbol{R}_d$$

where $\boldsymbol{R}_j$ is the rotation matrix of joint movement in the global coordinate system and $\boldsymbol{R}_p$ and $\boldsymbol{R}_d$ are the rotation matrices of the proximal and distal segments, respectively.

To systemically describe the joint movements, the joint reference position was defined as that joint position that exists when the body is in the anatomical posture. The rotation of joint movement was modified as:

$$R = R_j \, ({}_oR_j)^T$$

where R is the rotation matrix of joint movement based on the anatomical posture and ${}_oR_j$ is the rotation matrix of the joint reference position.

Euler angles were used to describe the orientation of a distal segment coordinate system relative to a proximal segment coordinate system. The first rotation about the y-axis represents the flexion/extension angle (α). The second rotation about the x' axis represents the adduction/abduction angle (β). The third rotation about the z" axis represents segmental axial rotation (γ).

Eight male tennis players with right hand dominant were recruited in this study. Their mean age, height and weight were 26.9±7.4 years, 175.9±5.5 cm, 69.8±10.0 kg. A full three dimensional kinematic model of trunk and upper extremity was developed for studying tennis serve. Sixteen markers were placed on selected anatomic landmarks unilaterally to define the coordinate system of the trunk, pelvis, upper arm, forearm and hand (Fig. 1). The HiRES motion system with six cameras (Motion analysis Corp., Santa Rosa, CA, USA) was used to collect the position of the reflective markers at sampling rate of 240 Hz while the subject performed tennis serve. Ten trials were sampled for each subject. Each trial lasted for 5 seconds with 3 minutes rest between trials. Euler angles were used to describe the three-dimensional joint movements of trunk and upper extremity with respect to their neutral posture. The positions of the markers were smoothed using a generalized cross-validation spline smoothing (GCVSPL) routine (Woltring, 1986) at a cutoff frequency of 6 Hz. A custom program in MATLAB language was coded for the calculation of joint movements.

RESULTS AND DISCUSSION

Figure 2 shows that the movement of trunk was extension and slight right bending in the phase of wind-up (frame 1:200). The movement of shoulder and elbow was internal rotation and extension, respectively. The forearm was at 60° pronation position. In the phase of cocking (frame 200:370), the movement of trunk was continuing extension, right bending and right rotation. The movement of shoulder was external rotation, horizontal abduction and elevation. The movement of elbow, wrist and forearm was flexion, extension and supination. In the phase of acceleration (frame 370:408), the movement of trunk was flexion, left bending and left rotation. The movement of shoulder was first continuing the movement in the phase of cocking then maintained at that position. The movement of elbow, wrist and forearm in extension, flexion and pronation were the only peculiar character of the flat serve. In summary, the radial/ulnar deviation movement of wrist was small. These results were in agreement with the findings of Elliott et al., (1995). The main movements in the phase of acceleration were elbow extension and forearm pronation and this result was similar to those of Sprigins et al. (1994) and Jack et al. (1979). The shoulder movement was small.

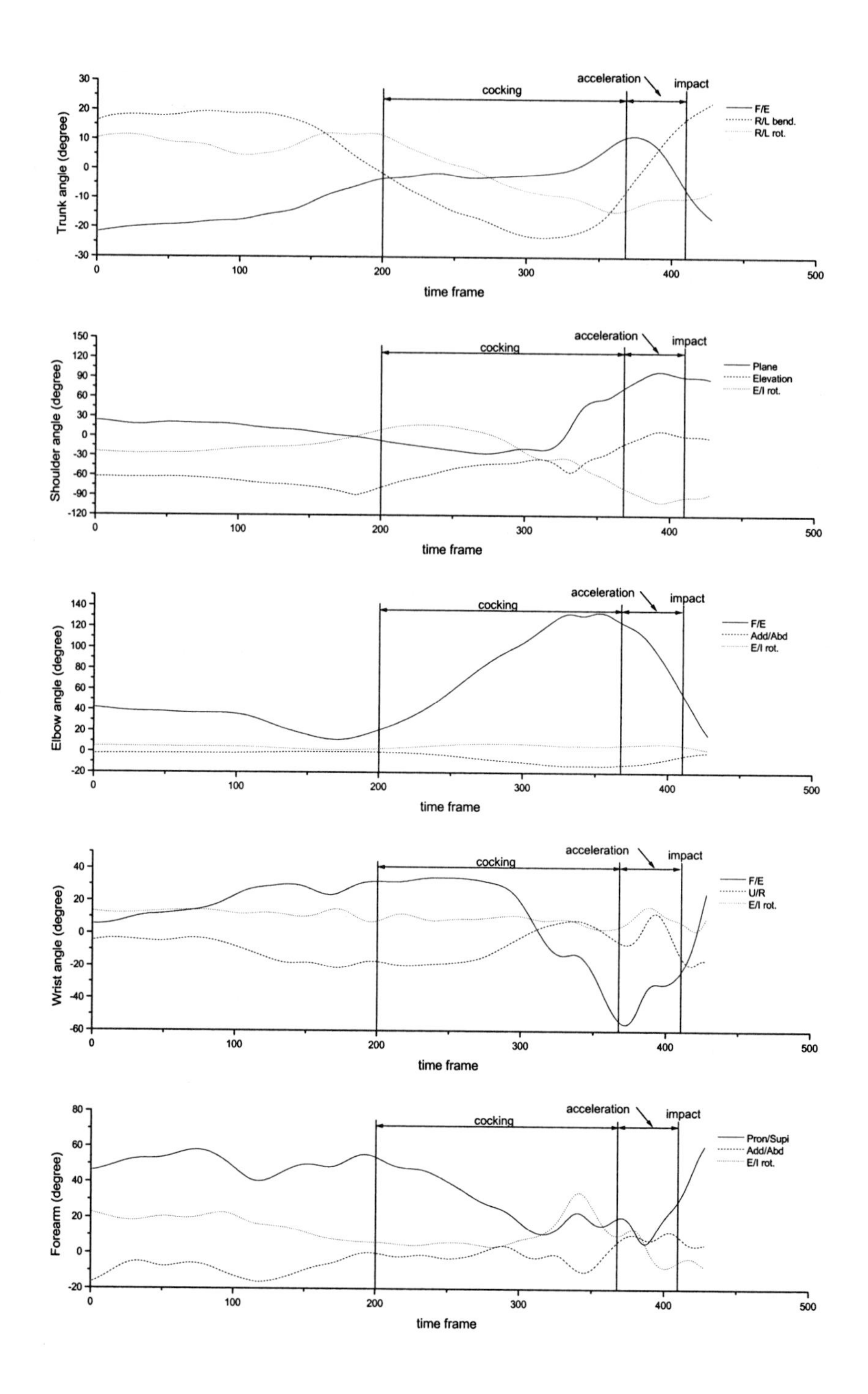

Fig 2 The angular velocity of upper extremity

The maximal angular velocity, occurred in the acceleration phase, was about 160°/sec in trunk bending, 180°/sec in trunk flexion, 300°/sec in shoulder abduction, 500°/sec in elbow extension, and 800°/sec in wrist flexion (Fig. 3). Also, the peak wrist and elbow angular velocities occurred at impact. The mean velocity of ball was about 120 km/hr. In addition, the change of joint angular velocity was much greater in acceleration phase than in the phases of wind-up and cocking. The results show that the power of serve was mainly from the movement of trunk rotation and bending, and elbow extension.

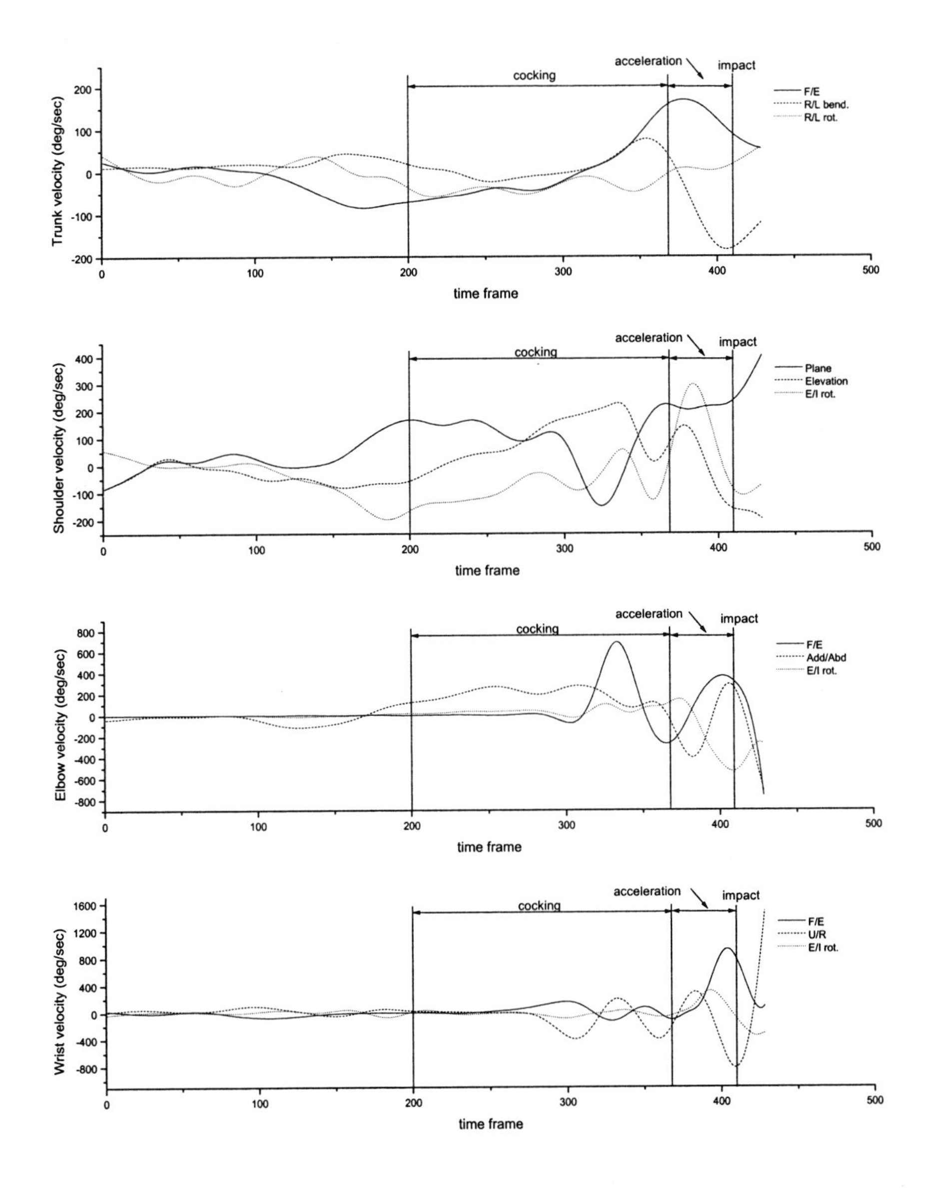

Fig. 3 The angular movement of upper extremity

CONCLUSIONS

Understanding of three-dimensional joint movements of upper extremity based on the biomechanics and anatomy during the tennis flat serve will allow the tennis coach and clinician to adjust the factors that produce injuries. The results show that the power of serve was mainly from the movement of trunk rotation and bending, and elbow extension. In the future, the more attentions we pay on considering the characteristics of ground reaction force of tennis serve the more information we will obtain for the study of the tennis serve technology.

ACKNOWLEDGEMENT

Supports from the National Science Council grants NSC88-2413-H-006-006-, Taiwan is gratefully acknowledged.

REFERENCES

An, K. N., Browne, A. O., Korinek, S., Tanaka, S. & Morrey, B. F. (1991) Three-dimensional kinematics of glenohumeral elevation. *Journal of Orthopaedic Research*, **9(1)**, 143-149.

An, K. N., Morrey, B. F.& Chao E.Y.S., (1984) Carrying angle of human elbow joint. *Journal of Orthopedic Research*, **1**, 369-378.

Balius, X., Turro, C., Carles, J., Jauregui, J., Escoda, J. & Prat, J. A. (1995) Improving the performance of a top ATP tennis player with a kinematical approach, and a 3-Dinteractive visualization of the serve. XVth Congress of the International Society of Biomechanics, 82-83, July, 2-6, Jyvaskyla.

Cohen, D. B., Mont, M. A., Campbell, K. R., Vogelstein, B. N. & Loewy, J. W. (1994) Upper extremity physical factors affecting tennis serve velocity. *American Journal of Sports Medicine*, **22(6)**, 746-750.

Elliott, B. (1983) Spin and the power serve in tennis. *Journal of Human Movement Studies*, 97-104.

Elliott, B. C., Marshall, R. N., & Noffal, G. J. (1995) Contributions of upper limb segment rotations during the power serve in tennis. *Journal of Applied Biomechanics*, **11**, 433-442.

Elliott, B., Marsh, T., & Blanksby, B. (1986) A three-dimensional cinematographic analysis of the tennis serve. *International Journal of Sport Biomechanics*, 260-271.

Ito, A., Tanabe, S. & Fuchimoto, T. (1995) Three dimensional kinematics analysis of the upper limb joint in tennis flat serving. XVth Congress of the International Society of Biomechanics, 424-25, July, 2-6, Jyvaskyla.

Jack, M., Adrian, M. & Yoneda, Y. (1979) Selected aspects of the overarm stroke in tennis, badminton, racquetball and squash. In J. Terauds (Ed.), *Science in Racquet Sports*, 69-80, Del Mar, CA: Academic.

Sprigings, E., Marshall, R., Elliott, B. & Jennings, L. (1994) A 3-D kinematic method for determining the effectiveness of arm segment rotations in producing racquet-head speed. *Journal of Biomechanics*, **27(3)**, 245-254.

Woltring H. J. (1986) A Fortran package for generalized cross-validation spline something and differential. Advance Engineering Software, **8**, 104-113.

Strategies for the return of 1st and 2nd serves

H. Kleinöder, J. Mester
German Sport University, Cologne, Institute for Theory and Practice of Training and Movement, Cologne, Germany

ABSTRACT: The available time budget and time management are the most important parameters regarding strategies for the return of 1st and 2nd serves. The mean initial velocities of 160 km/h ± 15 km/h (1st serve) and 117 km/h ± 10 km/h (2nd serve) result in a corresponding time budget of 841 ms ± 60 ms and 1010 ms ± 130 ms. There is, however, a considerable loss of service velocity on clay courts after ball ground contact in the service field, which corresponds to approx. 400 ms time budget for the return player until his point of impact. The improved perceptual conditions on the second half of ball flight suggest that the main regulation processes of the return movement occur after the reduction of the ball speed after ground contact.

Kinematic analysis (3-D, 200 frames/s) shows that the synchronisation of segment velocities of the striking arm follows the bio-mechanic model of the kinetic chain. The high 1st serve velocities cause enormous time pressure and demand precision-oriented co-ordination during low racket and segment velocities. This is also reflected in low changes of wrist angles approx. 100 ms before the point of impact and the failure to reach maximum racket velocity at impact. These decrease from 19.6 m/s ± 4.8 m/s to 16.7 m/s ± 4.9 m/s for the return on 1st serve and – significantly less – from 26.6 m/s ± 3.6 m/s to 25 m/s ± 5.1 m/s. The reduced velocity of the 2nd serve leads to an increased time budget (approx. 200 ms more), significant higher segment and racket velocities and more offensive returns indicating that the compromise between velocity and precision shifts to the velocity side. Another characteristic of world class players is the relatively low number of off centre points of racket impacts in contrast to players of less ability (Kleinöder 1998). This seems to be a very important factor with regard to the incidence of typical injuries to the striking arm.

INTRODUCTION

The return can be regarded as the most demanding technique in tennis because the players have to cope with very high time pressure caused by the high service velocities of up to and over 200 km/h. So the frame of analysis is the time budget which is available in this situation. Moreover, 3-D analysis of the return provides information about the micro-structure of co-ordination which is not visible to the human eyes. A comparison of the return with other tennis techniques helps to quantify the differing demands on co-ordination.

Focussing on the scientific approaches in this field, one can refer to various concepts. An often discussed model in bio-mechanics is the kinetic chain. This means that co-ordination occurs such that the movement of one segment begins as the velocity of the previous moving segment reaches its maximum, in a' 'staircase effect" (commonly called summation of velocities)..." (Elliott, 1991; see also: Braden/Bruns, 1977; Van Gheluwe/Hebbelinck, 1983; Elliott, 1989). It begins with the lower body segments up to the upper body segments and from the proximal to the distal joints of the hitting arm (in tennis). It is considered as a very important aspect of high racket velocities. It is to clarify, particularly for the return on 1st serve, whether this mechanism applies under the conditions of high time pressure.

The wrist as the last joint of the open chain plays an important role for the transfer of energy to the racket. In tennis, wrist and forearm actions produce high velocity shortly before the point of impact (Kleinöder and Mester, 1998). Again the question is what role these actions play under the special conditions of the return.

From the psychological viewpoint the following aspects of motor control can be identified. On the one hand, there is the concept of movement regulation with permanent feedback control during movement (closed loop). On the other hand, a programme oriented ballistic approach (open loop) especially for movements with a duration less than 200ms, such as box strokes, the bat-swing in baseball, typewriting or piano playing can be found (Schmidt, 1982). Gollhofer (1995) also characterises reactive movements in the stretch-shortening cycle (drop jump) as open loop ability. Thus a examination of spatial and time structure also contributes to a deeper understanding of movement organisation.

Finally, the level of co-ordination in tennis can be evaluated not only from the course oriented perspective but also from the result oriented point of view. The accurate placement on the court is strongly related to centrally impacting on the racket (Hennig and Milani, 1995). Off centre impacts lead to torsional loading on the racket resulting in an increased stress for the hitting arm. The following study will compare time budget and time management of world class players based on the above-mentioned scientific models. It is intended to work out some invariant elements of tennis technique as well as individual characteristics of the tennis return. Scientific results will be expanded by suggestions for the transfer into training practice.

EXPERIMENTAL CONDITIONS

Three-dimensional (3-D) high-speed-photography (NAC 400) with DLT calibration was used to investigate the existence of invariant elements of the return. Four internationally ranked players were filmed with 200 frames per second (fps) during competitive matches on clay court at the World Team Cup in Düsseldorf, Germany. From these 40 returns (20 forehand and 20 backhand) were subsequently analysed.

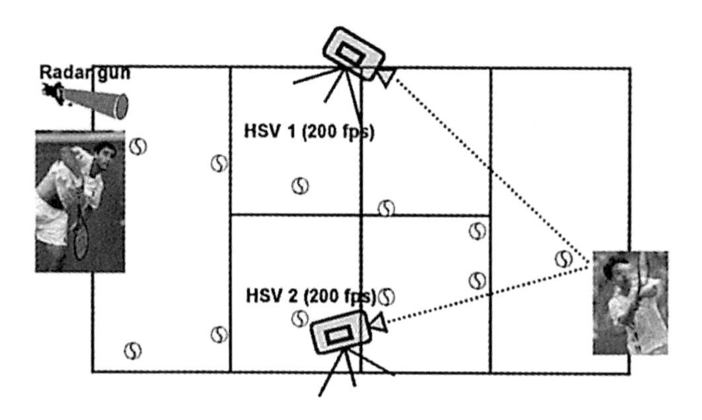

Fig. 1 Experimental conditions

ANALYSIS AND TREATMENT OF DATA

The frames of each selected ground stroke were stored in a PC and analysed with the motion analyse system BAS 2.0. Both 2-D images of the subjects were digitised using a HANAVAN (1964) model with 25 points and 2 pass points. The unknown 3-D co-ordinates of each of the subjects′ landmarks were determined and velocities, acceleration and angles were calculated. Mean, standard deviation as well as minima and maxima were determined. Statistically relevant differences were assessed using the Mann-Whitney U-test.

RESULTS

The mean initial velocity was found to be 160 km/h ± 15 km/h for the 1st serve and 117 km/h ± 10km/h for the second serve with noticable individual differences (maximum 187 km/h ± 1.41 km/h and 127 km/h ± 5.66 km/h). This results in a mean time budget of 841 ms ± 60 ms (1st serve) and 1010 ms ± 130 ms (2nd serve) from the impact of the service player to the impact of the return player. Approximately half of 841 ms elapses up to ball contact in the service field, the other half of the time budget is available for the comparably short distance until the return player hits the ball (Kleinöder 1997)., This suggests that the main adaptations of segment co-ordination take place during the second interval after the considerable speed reduction due to ball impact on the clay court (Kleinöder 1997).

The kinematic analysis of the 4 investigated players (n = 20 forehand and n = 20 backhand returns) reveals invariant elements and individual deviations from scientific models. The bio-mechanical concept of the kinetic chain is thus generally confirmed for the return (see table 1). The corresponding velocity-time-curves show that the synchronisation of segment velocities within the cinematic chain of the arm is apparent from the proximal to the distal segments, even under extremely high time pressure. This means that the shoulder achieves its maximum resultant velocity prior to that of the elbow, the wrist and the racket. This is an apparent sign of good co-ordination for all stroke techniques (Elliott 1991; Kleinöder/Mester 1998).

Table 1 Resultant segment velocities for forehand and backhand returns (maximum and impact in m/s) and the corresponding time intervals (ms) from maximum velocity to impact on the racket

	Racket	Wrist	Elbow	Shoulder
v_{max} (m/s) on 1st serve	19.6 ± 4.8	7.6 ± 1.5	5.1 ± 1.4	3.5 ± 1.4
v_{impact} (m/s) on 1st serve	16.7 ± 4.9	6.2 ± 1.8	3.3 ± 1.3	2 ± 0.8
v_{max} (m/s) on 2nd serve	26.6 ± 3.6	10.7 ± 2.5	5.4 ± 1.5	3.6 ± 0.9
v_{impact} (m/s) on 2nd serve	25 ± 5.1	9.1 ± 2.5	4.2 ± 2.2	2.4 ± 1.2
time interval (ms) v_{max} to v_{impact} 1st serve	20 ± 0	50 ± 20	60 ± 30	60 ± 40
time interval (ms) v_{max} to v_{impact} 2nd serve	20 ± 20	20 ± 10	40 ± 30	60 ± 40

Relatively large band widths in the standard deviation can be explained by the changing conditions during the competition as well as by physical differences of the individual players. These express the high variability of the bio-mechanic system to the ball flight. In contrast to the ground strokes and the service, the world class players fail to reach maximum racket velocity at impact (Kleinöder 1997). It decreases from 19.6 m/s ± 4.8 m/s to 16.7 m/s ± 4.9 m/s for the return on 1st serve and – significantly less – from 26.6 m/s ± 3.6 m/s to 25 m/s ± 5.1 m/s (2nd serve).

The reduced velocity of the 2nd serve (117 km/h ± 10 km/h) leads to an increased time budget (approx. 200 ms or more), to significant higher segment and racket velocities and to more offensive returns (see table 1), indicating that the compromise between velocity and precision shifts to the velocity side. Nevertheless world class players have their own individual characteristics of performance.. e.g. Sergi Bruguera returns constantly with low segment and racket velocities (11.68 m/s ± 3.49 m/s) as an answer to high service velocities (1st serve) from his opponent (172 km/h ± 13 km/s), suggesting that he prefers a more blocked return. In contrast, other players return with significantly higher racket velocities between 16 m/s and 20 m/s after 1st serve.

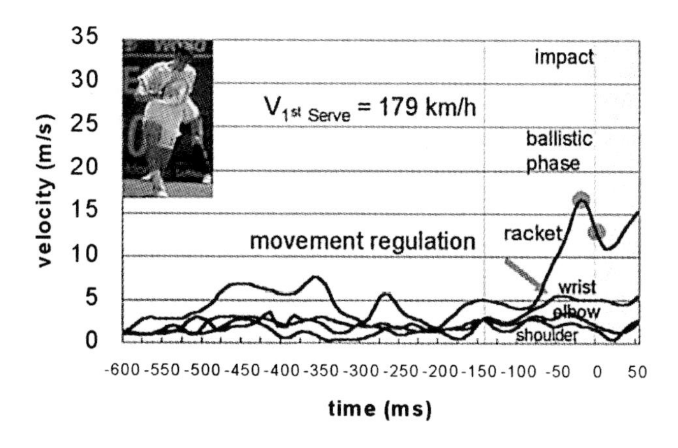

Fig. 2 Exemplary return of Sergi Bruguera

For a better comparison, the individual length of the velocity-time-curves was standardised to 0.6 s counting from the point of impact. The mathematically calculated start point of the racket forward swing divides the velocity-time-curves into two sections. The first can be classified as the interval of movement regulation (closed loop), the second can be interpreted as the ballistic phase with high racket acceleration. Typical of the first interval are the continuous adaptations of the segments of the hitting arm to the approaching ball. The slightly changing velocities of the segments of the arm stress its regulatory character. Regulation processes seem to be possible very close to the point of impact (approx. 100 ms before impact). With the vastly increasing racket velocity on beginning of the forward swing, it is probably impossible to have further control by conscious information processing. This very small time interval is described as the ballistic part without internal feedback except for reflex-based corrections (Schmidt 1988). Both described units are not strictly divided. The connecting corridor can – physiologically explained – be seen in the last adjustments of racket movement.
This is similar to other tennis techniques and the return on 2nd serve. The larger time budget allows significant higher segment velocities which result in higher racket velocities.

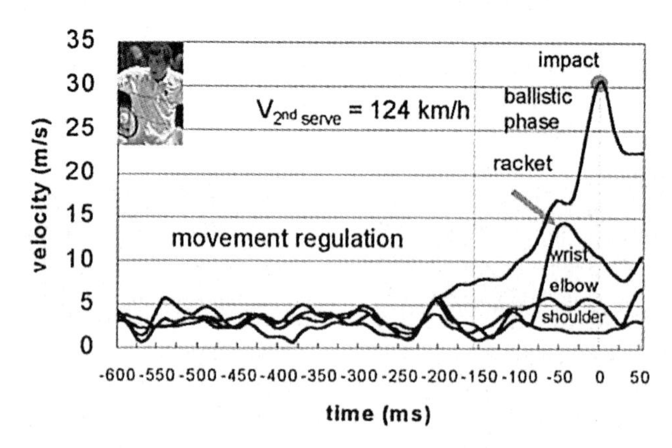

Fig 3 Exemplary return of Pete Sampras

The analysis of the wrist angle gives additional information about wrist and forearm actions. Because of the improved identification of the segments during competition, this was used as a model between elbow, wrist and racket head movement. This construction comes near to the anatomic wrist angle and delivers comparative values for all players. In contrast to the ground strokes there are only small changes of wrist angle shortly before the point of impact (Kleinöder, 1998). This shows the exemplary comparison between Sergi Bruguera´s forehand groundstrokes and forehand returns.

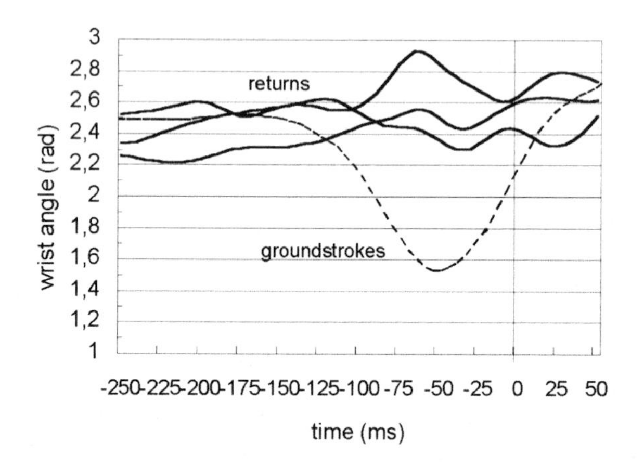

Fig 4 Wrist angles of Sergi Bruguera at the return in comparison to the groundstrokes (mean of n = 8 strokes)

From the result-oriented point of view, players of lower ability often have problems with the demands of regulating the return (Kleinöder 1997). This is reflected in the wider point of impact areas on their rackets. Thus leisure time players in contrast to world class players often hit the ball off centre (Kleinöder and Mester 1998), as can be seen in figure 4.

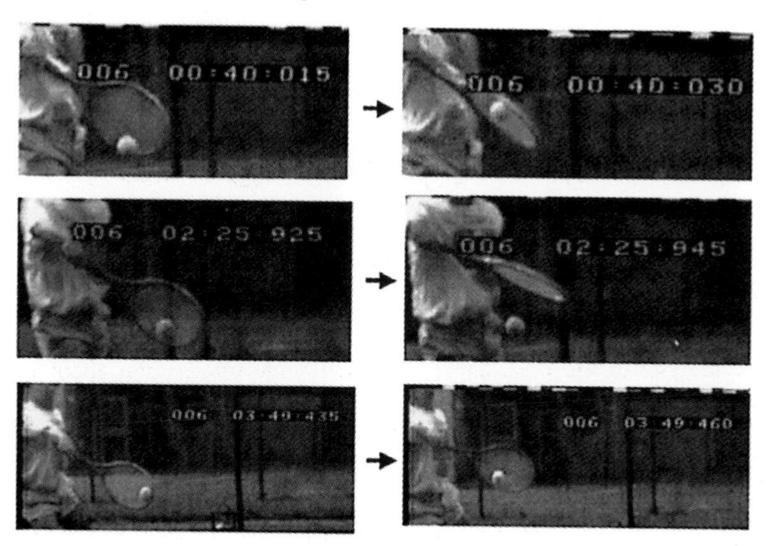

Fig. 5: Off centre points of impact at the return on 1st serve

CONCLUSIONS

The general scientific models as well as the individual interpretation of technique are discussed using time budget and 3-D kinematic analysis of world class players′ forehand and backhand returns. The scientific models generally lead to a better understanding of the invariant elements and their individual adaptations in top class tennis.

The investigations indicate that the bio-mechanic model of the kinetic chain can truly serve as an basic invariant of tennis strokes even with a very limited time budget. World class players are able to find an individual compromise between precision-orientation after the 1st serve and velocity production after the 2nd serve. Wrist and forearm actions are very important in the last segment of the open kinetic chain (Elliott, 1991; Kleinöder 1997). For the return on 1st serve these actions are responsible for precision of racket swing and stability. Smaller variations in the racket impact area are typical of world class players, in contrast to players of lower ability (Hennig and Milani 1995; Kleinöder 1997). Precise regulation is a prerequisite for achieving central impact on the racket, this being extremely important in minimising the risk of typical injuries to the striking arm.

For the return on 2nd serve, wrist and forearm actions produce racket velocity by the use of stretch shortening cycle similar to the ground strokes (Kleinöder and Mester, 1998). Thus reflex activities and stored elastic energy are used within the last segment of the kinetic chain supporting economic and fast playing at the same time (Elliott, 1991).

According to these results, tennis coaches should integrate special training of wrist and forearm actions as standard practice. Special strength training is recommended to develop the relatively small muscle group in the forearm.

REFERENCES

Braden, V., Bruns, B. (1977). *Vic Braden's tennis for the future*. Boston.

Elliott, B.C. (1988). Bio-mechanics of the serve in tennis. *Sports Medicine*, **6**, 285-294.

Elliott, B.C., Marsh, T. (1989). A bio-mechanic comparison of the topspin and backspin forehand approach shots in tennis. *Journal of Sport Sciences*, **7**, 215-227.

Elliott, B.C. (1991). The role of bio-mechanics in power strokes. *Tennis the Australian way manual*, **1**, 1-8.

Hanavan, E.P. (1964). *Mathematical model of the human body*. Ohio: Wright Patterson Air Force Base.

Hennig, E., Milani, Th. (1995). Die Auswirkungen der Haltekraft am Tennisschläger auf Ballgeschwindigkeit und die Vibrationsbelastungen des Unterarms. *Deutsche Zeitschrift für Sportmedizin*, **46**, 169-174.

Kleinöder, H. (1997). Quantitative Analysen von Schlagtechniken im Tennis: Intra- und interindividuelle Studien von Spielern unterschiedlichen Leistungsniveaus. Unpublished doctoral dissertation, German Sport University Cologne.

Kleinöder, H., Neumaier, A., Loch, M., Mester, J. (1995). Cinematographic analysis of the service movement in tennis. In Tennis: Sports Medicine and Science (edited by H. Krahl, H.-G. Pieper, B. Kibler, P. Renström), pp. 16-21. Düsseldorf: Walter Rau Verlag.

Kleinöder, H., Mester, J. (1998). Belastungen des Schlagarmes durch Handgelenk- und Unterarmaktionen im Tennis. *Deutsche Zeitschrift für Sportmedizin (Sonderheft)*, **1**, 217-220.

Roetert E.P., Brody, H., Dillman, C.J., Groppel, J.L., Schultheis, J.M. (1995). The Biomechanics of tennis elbow. An integrated approach. *Clinic Sports Medicine,* **1**, 47-57.

Schmidt, R.A. (1982). *Motor Control and Learning. A behavioral emphasis*. Champaign, Illinois: Human Kinetics.

Van Gheluwe, B., Hebbelinck, M. (1983). The kinematics of the service movement in tennis: A three-dimensional cinematographical approach. In *Bio-mechanics IX-B* (edited by D.A. Winter et al.), pp. 521-526, Champaign, Illinois: Human Kinetics.

The development of an emotion control programme for adolescent tennis players

P.J. Dent
School of Hebes, Enfield Campus, Middlesex University, UK
R. Masters
School of Sport and Exercise Sciences, University of Birmingham, UK

***ABSTRACT*:** Coping with stress and emotional turmoil is an integral part of sport competition. A major objective must therefore be the development of cognitive skills to cope with acute and chronic stressors by reducing self-reflection and controlling disadvantageous types of cognitions. The aim of this pilot investigation was to assess the suitability of the Emotional Control Questionnaire (ECQ; Roger & Najarian, 1989) and the Coping Styles Questionnaire (CSQ; Roger, Jarvis & Najarian, 1993) in the sports domain and to evaluate if a sports specific version of Rogers & Masters ' (1997) emotion control training programme would provide a tool for coping with stress and emotional turmoil in adolescent tennis players. Taken together the Emotional Control Questionnaire and the Coping Styles Questionnaire comprise six empirically discriminable maladaptive coping mechanisms seen in most populations (rehearsal, emotional inhibition, aggression control, benign control, avoidance and emotional coping) and two adaptive mechanisms (rational coping and detachment). There was an increase in the total use of adaptive mechanisms and a decrease in the total use of maladaptive mechanisms. The decrease in the use of rehearsal was accompanied by an increase in the control of aggression. The results confirmed that the training programme was effective in modifying behaviour in adolescent tennis players. Suggestions were made for further work to investigate age, gender and ability in a longitudinal study with a modified sports specific form of the ECQ and CSQ. This should also include an examination of self-presentation as a maladaptive coping mechanism.

INTRODUCTION

Coping with stress and emotional turmoil is an integral part of sport competition. A major, but often overlooked, objective in dealing with stress in sport competition should be the development of cognitive skills to cope with acute and chronic stressors by reducing self-reflection and controlling disadvantageous types of cognitions (Krohne & Hindel, 1988; Orlick, 1980; Orlick & Partington, 1986; Smith, 1980).

Coping consists of learned behavioural responses that successfully lower arousal by neutralising or minimising the importance of a dangerous or unpleasant condition (Lazarus & Folkman, 1984). Thus, coping is a conscious process that allows a person to master, reduce or tolerate stressful demands. Failure to cope with and respond constructively to acute stress may lead to ineffective cognitive processes, energy reduction, performance failures and other debilitating outcomes (Smith, 1986). Experiencing an unpleasant event often creates emotional turmoil, and may undermine efficient sport skill execution. Coping with acute stress requires regaining composure, establishing proper mental set (i.e., the psychological readiness to respond to subsequent stimuli) and maintaining optimal arousal and concentration (Anshel, 1990).

The demands that competitive tennis places upon a player make it one of the most stressful sports (Harwood, 1999). Few sports place such pressure on the "self" within such a punishing scoring system. Harwood (1999) states that emotional responses exhibited by players include outbursts of anger and frustration - emotions, which are uncontrolled prior to the next point. Negative physical reactions and lack of a positive physical presence in response to points lost or a poor period of play, result in low competitive image control and withdrawal of effort. "Tanking", in order to save face is common, blaming lack of trying, not personal ability, for the failure to win.

COPING RESEARCH WITHIN SPORT

Compared to the general psychology literature, little research on coping within the sports domain existed prior to Krohne & Kindel's (1988) study investigating 36 top German table tennis players. The authors used a variety of anxiety assessments to examine the relations between general and sport specific trait anxiety, self-regulatory techniques, emotional reactions to stress situations, coping situations and athletic performance. The authors concluded that top table tennis players reduce threat and anxiety in stressful situations by using avoidant as opposed to vigilant coping and by exhibiting less worry during competition. The importance of this work was to demonstrate how coping and stress interact to influence athletic performance in top athletes.

Madden, Kirkby & McDonald (1989) and Madden, Summers & Brown (1990) used a sports specific version of the Ways of Coping Checklist (WCC; Folkman & Lazarus, 1980, 1988) demonstrating that older athletes and subjects experiencing more stress employed more coping strategies such as social support and problem focused coping along with increased effort and resolve. This work was followed up by Grove & Prapavessis (1995a,b) who concluded that efforts to increase batting self efficacy in Australian baseball players may also have promoted problem-focused (adaptive) coping when an athlete faced a batting slump.

Crocker (1992) examined how competitive athletes cope with stress. Male and female athletes (N = 237), representing a variety of sports, completed a 68-item modified version of the WCC, revealing that a number of coping strategies were employed that could be classified into eight separate dimensions: active coping; problem-focused coping; seeking social support; positive reappraisal; self-control; wishful thinking; self-blame and detachment. Cocker (1992) suggested that active coping and problem focused-coping are highly adaptive whilst wishful thinking and detatchment could be maladaptive. The author also concluded that the WCC lacked internal factor consistency and was limited by the need to perform an exploratory factor analysis each time.

Gould, Eklund & Jackson (1993a) utilised in-depth qualitative interviews to examine coping strategies used by US Olympic wrestlers. Content analysis of the findings revealed that the wrestlers employed four major categories of coping, the largest of which were thought-control strategies. These consisted of blocking distractions, perspective taking, positive thinking, coping thoughts and prayer. The second category, task-focused strategies, included narrow more immediate focus and concentrating on goals. The final two categories of coping were behavioural-based strategies (consisting of changing or controlling one's environment and following set routines) and emotional control strategies such as arousal control and visualisation. In accordance with Folkman and Lazarus (1985) and Compas (1987), it was concluded that "coping efforts reflect an ever-changing complex process involving multiple strategies, often used simultaneously and in combination with one another" (Gould et al., 1993a). In a second study, Gould et al. (1993b) interviewed 17 current or former US national champion figure skaters to examine the relationship between the use of coping strategies and the particular sources of stress. Of the various categories identified rational thinking and self-talk, isolation and deflection, and ignoring the stressor were evident. It was concluded that the skaters employed a variety of strategies, which differed depending on the source of stress. The authors stated that this notion clearly supports Lazarus and Folkman's (1984) transaction model of coping. Finally it was concluded that Carver, Scheier & Weintraub's (1989) COPE subscales reflect the coping strategies which emerged through these content analyses better that the subscales of other coping inventories.

The Athletic Coping Skills Inventory-28 (ACSI-28) was administered by Smith & Christensen (1995) to 104 minor league baseball players revealing that the psychological skills measured (especially confidence and achievement motivation) were significantly related to hitting and pitching performance. As ACSI-28 measures other psychological resources related to coping, it may not be "a very pure measure of coping strategies" (Hardy, Jones & Gould, 1996). Anshel and his associates (Anshel 1990; Anshel, Gregory & Kaczmarek, 1990; Anshel, Brown & Brown, 1993) have investigated whether an educational-intervention programme, COPE, is effective in helping athletes deal with stress and improve performance. The aims of the programme are to help the athlete control emotions (C), organise input (O), plan responses (P) and execute skills (E) in an efficient and effective manner. Initial research has been promising, however more research with larger samples is needed (Hardy, Jones & Gould, 1996).

EMOTIONAL CONTROL WITHIN SPORT

Little research exists on control of emotions within the sporting arena. Early research on emotion control arose from suggestions that the impact of stress may be moderated by personality. The effects of emotional rumination (rehearsal) had been referred to in early research (see for example Cameron & Meichenbaum, 1982), but a review of the literature disclosed no psychometrically satisfactory instruments available for measuring it. This led to the development of the Emotion Control Questionnaire (ECQ; Roger & Nesshover, 1987; Roger & Najarian, 1989), which comprises four empirically discriminable scales entitled rehearsal, emotional inhibition, aggression control and benign control. Rehearsal measures the tendency to ruminate on emotionally upsetting events, and in conjunction with emotional inhibition has been shown to play a significant role in delayed physiological recovery following stress (Roger, 1988; Roger & Jamieson, 1988). Aggression control and benign control were found to be moderately correlated - benign control correlating substantially with established measures of impulsiveness (Roger & Nesshover, 1987).

Coping styles have also been described as moderators of stress responses (Monat & Lazarus, 1991). The Coping Styles Questionnaire (CSQ; Roger et al., 1993) comprises four scales entitled rational coping, detachment, emotional coping and avoidance. Taken together the Emotional Control Questionnaire and the Coping Styles Questionnaire comprise six empirically discriminable maladaptive coping mechanisms seen in most populations (rehearsal, emotional inhibition, aggression control, benign control, avoidance and emotional coping) and two adaptive mechanisms (rational coping and detachment).

On the back of these factors an emotion control training programme has been developed, validated and extended in a number of different contexts including populations as diverse as the North Yorkshire Constabulary (Roger & Hudson, 1995) and inmates at HMP Wakefield (Roger & Masters, 1997).

Personality and coping are described as pivotal in determining whether or not an event is perceived to be stressful and how the individual responds to it (Roger & Masters, 1997). Much of the emotion control training programme is designed to encourage participants to increase the number of coping mechanisms available to them. This is achieved through regaining control of attention, which becomes distracted by various stressors. Participants are encouraged to select various strategies to deal with evoked stressful situations and to avoid the tendency to ruminate.

The aim of this pilot investigation was to assess the suitability of the Emotional Control Questionnaire (Roger & Nesshover, 1987; Roger & Najarian, 1989) and the Coping Styles Questionnaire (Roger, Jarvis & Najarian, 1993; Roger, 1995) in the sports domain and to evaluate if a sports specific version of the emotion control training programme would provide a suitable tool in coping with stress and emotional turmoil in adolescent tennis players.

METHOD

PARTICIPANTS

Participants were 16 adolescent tennis players invited to attend the programme by Hertfordshire Lawn Tennis Association. Participants were lost from the sample during the course of the study, through illness or conflicting commitments. Full sets of questionnaire data were obtained from 10 players. The sample consisted of two female and eight male players, all of whom had obtained a County standard or better (LTA Rating 2.2 - 5.1). Mean age was 15.2 years (range 13 - 18 years).

PROCEDURE

The study employed a pre- and post- training design. The training was structured around eight 60 minute group sessions, separated by a period of 1 week. The sessions alternated between on and off court. The basis of the training programme has been described in detail elsewhere (see Roger & Hudson, 1995; Roger & Masters, 1997), and was adopted for use in the present context with tennis specific examples used to highlight certain key points. The main features included a definition of stress relevant to the demands faced by adolescent tennis players, preoccupation with emotional upset, discussion of identified stressors highlighting controllable versus non-controllable factors and emotional versus problem focused solutions to generated scenarios.

RESULTS

For the manipulation check, mean scores from the questionnaires obtained at the beginning and the end of the study were compared. The mean scores on each of the scales are shown in Table 1.

Table 1 Mean (standard deviation) scores on the ECQ and CSQ scales at first (before training) and second (after training) administrations.

SCALES	1ST ADMINISTRATION	2ND ADMINISTRATION
ECQ		
Rehearsal	6.30 (2.22)	5.00 (2.50)
Emotional inhibition	7.40 (4.06)	6.40 (4.56)
Aggression control	5.90 (3.96)	7.33 (3.28)
Benign control	6.50 (3.75)	6.22 (2.95)
CSQ		
Rational coping	17.50 (6.15)	19.11 (4.34)
Detached coping	12.80 (4.54)	14.50 (4.32)
Emotional coping	15.30 (5.96)	13.33 (6.08)
Avoidance	13.60 (2.99)	14.44 (3.78)

DISCUSSION

The results reported of this pilot study suggest that the training programme had an effect on the attitude of the players who participated. Rehearsal scores on the Emotion Control Questionnaire were higher on first administration than on the second administration. This suggests players were spending less time dwelling on stressful events such as bad calls and gamesmanship. These events can be highly detrimental with the focus of attention remaining on past events rather than searching for relevant attentional cues. Emotional inhibition was decreased indicating a greater tendency to exhibit emotion rather than "bottle it up". This has to be balanced with not revealing too much information to your opponent in terms of your feelings, whilst releasing tensions in a controlled manner. Aggression control was increased on the second administration suggesting a higher degree of control over aggression. The demands of the game are such that reacting to a situation in a hostile or aggressive manner can be detrimental to decision making yet holding in feelings of aggression can be just as detrimental. The benign control scores marginally decreased, suggesting the tendency to act before thinking, equivalent to impulsiveness, though the difference was too small to draw any conclusions from. This may have arisen from the players trying not to dwell on the past (rehearsal) resulting in the desire to get on with the next point This obviously, can cause problems with poor decision making arising from not considering all options.

Increases were exhibited on both of the adaptive coping mechanisms measured on the Coping Strategies Questionnaire with players exhibiting an increased tendency to seek rational and reasonable ways of dealing with problems. This increase was also reflected in the ability to detach or disengage oneself from the problem. The results suggest that the players had increased their ability to see the problem as something independent of themselves, reflecting a greater ability to detach oneself from the problem and seek a rational solution. There was also a decrease in emotional coping, which reflects the tendency to become overwhelmed with emotion. This can result in confusion and distress and a sense of helplessness.

An increase was noted in avoidance coping implying that the players were trying simply to ignore a problem in the hope that it would go away. This may be a misinterpretation of the need to use a coping strategy with the players believing that avoidance is actually a means of dealing with it. This is a maladaptive strategy as it is extremely difficult to avoid problems on court (Harwood, 1999) and can lead to the missing of vital relevant cues. Self-presentation was noted as another possible coping mechanism with players avoiding the situation or issue by "tanking" to preserve their self-image. The use of "tanking" as a coping mechanism was more prevalent in the older players suggesting that self-presentation does not become an issue until later in an adolescents' career, perhaps when the results, leading to thought of self-image, have more perceived worth. Factors that increase perceived likelihood of poor personal performance lower the athlete's ability to convey a desired image to their audience (James & Collins, 1997). The authors also identified social evaluation and self-presentation as a general source of stress in its own right.

This pilot work raises the question of the likely long-term benefits of the training and asks whether the increased use of adaptive coping mechanisms will impact on players' actual performances. This will only become clearer with the use of longitudinal studies in which there is reinforcement of adaptive coping strategies and rationalisation of the use of maladaptive mechanisms.

CONCLUSIONS

Despite the problems posed by intervention studies amongst adolescent tennis players, the results suggest the training programme was effective in reducing maladaptive coping strategies whilst increasing the use of adaptive strategies. Further work will examine the relationship between age, gender and ability with a much larger sample size as well as investigating the benefits of using a smaller pool of items (104 at present) to allow younger children to complete the questionnaires. Some of the items were not relevant to adolescents ("I hate being stuck behind a slow driver") and could be removed, therefore propagating the development of a more sports specific coping styles and emotional control questionnaire. The ECQ is constructed with true or false statements, which generated a certain amount of indecision. Combining the ECQ and CSQ scales to read as "always, often, sometimes or never" may add more clarity. With the development and validation of such a questionnaire and the use of longitudinal studies it may be possible to investigate how coping behaviours, resources and dispositions are developed, as suggested by Hardy, Jones and Gould (1996). Investigations need to focus on how each subscale interacts, so moving towards a more transactional model of coping. This could also include an investigation into the use of self-presentation as a coping mechanism in adolescent athletes as recommended by James and Collins (1997). The challenge is to develop an emotional control training programme that deals with the changing emotional make up of adolescent athletes where image may be everything.

REFERENCES

Anshel M.H. (1990) Toward a validation of a model for coping with acute stress in sport. *International Journal of Sport Psychology*, **21**, 58-83.

Anshel M.H., Gregory W.L. & Kaczmarek M. (1990) Effectiveness of stress training program in coping with criticisms in sport: A test of the COPE mode. *Journal of Sport Behaviour*, **13**, 194-218.

Anshel M.H., Brown J.M. & Brown D.F. (1993) Effectiveness of a program for coping with acute stress on motor performance, affect and muscular tension. *Australian Journal of Science and Medicine in Sport*, **25**, 7-16.

Cameron R. & Meichenbaum D. (1982). The nature of effective coping and the treatment of stress related problems: A cognitive behavioural perspective. In L. Goldberger & S. Bernitz (Eds), *Handbook of Stress*. New York: Free Press.

Carver C.S, Scheier M.F. & Weintraub J.K. (1989) Assessing coping strategies: A theoretically based approach. *Journal of Personality and Social Psychology*, **56,** 267-283.

Compas B.E. (1987) Coping with stress during childhood and adolescence. *Psychological Bulletin*, **101**, 393-403.

Crocker P.R.E. (1992) Managing stress by competitive athletes: Ways of coping. *International Journal of Sport Psychology*, **23**, 161-175.

Folkman S. & Lazarus R.S.(1980) An analysis of coping in a middle-aged community sample. *Journal of Health and Social Behaviour*, **21**, 219-239.

Folkamn S. & Lazarus R.S. (1985) If it changes it must be a process: A study of emotions and coping during three stages of a college examination. *Journal of Personality and Social Psychology*, **48**, 150-170.

Folkman S. & Lazarus R.S. (1988b) *Manual for the Ways of Coping Questionnaire,* Consulting Psychologists Press, Palo Alto, CA.

Gould D., Ecklund R.C. & Jackson S.A. (1993a) Coping strategies used by U.S. Olympic wrestlers. *Research Quarterly for Exercise and Sport*, **64**, 83-93.

Gould D., Finch L.M. & Jackson S.A. (1993b) Coping strategies used by national champion figure skaters. *Research Quarterly for Exercise and Sport*, **64**, 453-468.

Grove R.J. & Prapavessis H. (1995a) Correlates of batting slumps in baseball. In: *Understanding Psychological Preparation for Sport,* pp. 216.Chichester, John Wiley & Sons.

Grove R.J. & Prapavessis H. (1995b) Self-handicapping tendencies and slump-related coping in sport. In:*Understanding Psychological Preparation for Sport,* pp. 216. Chichester, John Wiley & Sons.

Hardy L., Jones, G. & Gould, D. (1996) *Understanding Psychological Preparation for Sport.* Chichester, John Wiley & Sons.

Harwood C. (1999) Understanding the mental performance factor. *Coach to Coach*, **5**, 19-21.

James B. and Collins D. (1997) Self-presentational sources of competitive stress during performance. *Journal of Sport and Exercise Psychology*. **19**, 17-35.

Krohne H.W. & Hindel, C. (1988) Trait anxiety, state anxiety and coping behaviour as predictors of athletic performance. *Anxiety Research*, **1**, 225-234.

Lazarus, R.S. & Folkman, S. (1984) *Stress Appraisal and Coping*, Springer, New York.

Madden C.C., Kirkby R.J. & McDonald D. (1989) Coping styles of competitive middle distance runners. *Quarterly Journal of Experimental Psychology*, **38a**, 659-670.

Madden C.C., Summers J.J. & Brown D.F. (1990) The influence of perceived stress On coping with competitive basketball. *International Journal of Sport Psychology,* **21**, 21-35.

Monat A & Lazuarus R.S. (1991) *Stress and Coping*, 3rd ed. New York; Columbia University Press.

Orlick T.D. (1980) In *Pursuit of Excellence.* Champaign, IL, Human Kinetics.

Orlick T.D. & Partington J. (1986) Psyched: Inner Views of Winning. Ottawa, Canada: Coaching Association of Canada.

Roger D. (1995) Emotion control, coping strategies and adaptive behaviour. *Stress and Emotion*, **15**, 255 -264.

Roger D. (1988) The role of emotion control in human stress response. Paper presented at the Annual Conference of the British Psychological Society, Leeds University, April.

Roger D & Hudson C.J. (1995) The role of emotion control and emotional rumination in stress management training. *International Journal of Stress Management*, **2**, 119-132.

Roger D & Jamieson J. (1988) Individual differences in delayed heart-rate recovery Following stress: The role of extraversion, neuroticism and emotional control. *Personality and Individual Differences*, **9**, 721-726.

Roger D., Jarvis G & Najarian B. (1993) Detatchment and Coping: The construction and validation of a new scale for measuring coping strategies. *Personality and Individual Differences,* **15,** 619-626.

Roger D & Masters R.S.W. (1997) The development and evaluation of an emotion control training programme for sex offenders. *Legal and Criminal Psychology*, **2,** 51-64.

Roger D. & Najrian B. (1989) The construction and validation of a new scale for measuring emotion control. Personality and Individual Differences, **10**, 845-853.

Roger D & Nesshover W. (1987) The construction and validation of a scale for measuring emotional control. *Personality and Individual Differences,* **8,** 527-534.

Smith R.E. (1980) A cognitive-affective approach to stress management training for fathletes. In C.H. Nadeau, W.R. Halliwell, K.M. Newell, and C.G. Roberts (eds), *Psychology of Motor Behaviour in Sport - 1979*, Human Kinetics, Champaign, IL pp. 54-72.

Smith R.E (1986) Toward a cognitive affective model of athletic burnout, *Journal of Sport Psychology*, 8, 36-50.

Smith R.E. & Christensen D.S. (1995) Psychological skills a predictors of performance and survival in professional baseball. *Journal of Sport and Exercise Psychology*, **17**, 399-415.

Professional tennis players in the zone

J.A. Young
Tennis International Consultants Pty Ltd, Melbourne, Australia

ABSTRACT: This paper presents research into the nature and conditions of the optimal experiential and performance state, termed the zone or flow. In a study conducted in conjunction with Tennis Australia, thirty-one Australian professional female tennis players related their experiences of a time that stood out from average, one involving total absorption and which was rewarding in and of itself. Participants also responded to the Flow State Scale (Jackson, 1995) and Experience Questionnaire (Privette, 1984) and nominated factors which they thought facilitated, interrupted and prevented the zone from occurring. The qualitative analyses of participants' narratives of zone experiences revealed a high degree of correspondence with the eight dimensions posited by flow theory, namely challenge-skill balance, concentration, action-awareness merging, clear goals and feedback, loss of self-consciousness, paradox of control, transformation of time and autotelic experience. Factors influencing the zone generally corresponded with the conditions proposed by flow theory. Practical strategies for attaining and maintaining the zone are highlighted.

INTRODUCTION

After Pete Sampras won the Wimbledon Singles Championship in 1998, and Pat Rafter won his first US Open Singles Championship in 1997, both players described his own performance as one in which he played in "the zone". Such a description denotes a highly focused, effortless and almost flawless performance when everything seems to "click". The ball appears the size of a football, time seems to slow down and there is ample time to prepare for and execute shots at will. It is a time when a player performs to the best of his/her ability. As such, the zone represents the pinnacle of achievement for a player.

While there is consensus that the zone describes a special moment in tennis, there is little agreement as to the origin of the term. Shainberg (1989) claims that its origin is unknown, whereas Loehr (1995) claims the term can be traced to the 1975 Wimbledon final between Arthur Ashe and Jimmy Connors. Describing the match, Ashe recalled, "It suddenly occurred to me that I had picked the Wimbledon weeks to go through the zone. I was totally relaxed in the finals, never nervous for a moment … I believed I could win" (Ashe cited in Loehr, 1995, p.36).

An explanation of the zone or "flow" state (the terms zone and flow are used interchangeably and synonymously in the sport psychology literature) can be gleaned from flow theory (Csikszentmihalyi, 1975, 1990). In brief, flow theory proposes that a phenomenological structure of eight dimensions describes the zone experience for individuals across occupations, cultures and demographic groups. These dimensions are listed by Csikszentmihalyi (1990) as: balance between challenges and skills, clear goals and feedback, action and awareness merged, concentration on task, sense of potential control, loss of self-consciousness, altered sense of time and autotelic (self-rewarding) experience. These conditions are deemed to constitute the conditions necessary for the occurrence and continuation of the zone.

While recent sport psychology literature is replete with references to the zone (e.g., Goldberg, 1998, Murphy, 1996), little research has been conducted on this highly desirable and much sought after state. To address this deficiency, a qualitative study was conducted by the author in conjunction with Tennis Australia, the governing body of tennis in Australia. The purpose of this first-known study of the zone in tennis was to delineate the phenomenon's nature and conditions. In doing so, it was anticipated that strategies could be developed to guide players in achieving and maintaining this optimal state.

METHOD

PARTICIPANTS

Thirty-one Australian professional female tennis players competing in the Tennis Australia Challenger and Satellite Circuits and the Australian Open Championships participated in the study. Participants had played tennis and competed professionally for a mean number of 12.2 years (*SD* = 3.8) and 2.2 years (*SD* = 1.9) respectively, and had a mean age of 22.7 years (*SD* = 2.8) at the time of the study. The best-performed participant had a world ranking of 80 in singles and 32 in doubles.

MATERIALS

Participants completed a self-report instrument consisting of qualitative and quantitative measures. The qualitative measures consisted of a series of questions asking participants to relate a zone experience and nominate factors that they thought facilitated, interrupted and prevented the zone from occurring. The quantitative measures consisted of the Flow State Scale (Jackson, 1995), items from the Experience Questionnaire (Privette, 1984) and ratings of challenges, skills and frequencies of zone occurrences in training and competition.

DATA COLLECTION

Data were collected at eight tournament sites (return rate = 82%). Participants could complete the questionnaires at their convenience during the series of tournaments.

RESULTS

Thirty-one (100%) participants reported a description of a zone experience. Three narratives representative of the clear and concise descriptions of the tennis zone were:

> I was really focused on what I was doing and there are no other thoughts in my mind. Nothing to distract me. Everything was flowing and seems easy. I am giving everything that I have and it feels great. I feel I can do anything. I am in total control;

I'd probably describe it as running on auto, just playing on instinct. There was no self-doubt. I felt invincible. I knew I could play any shot at any given time. My concentration and focus was so spot on it felt like I has so much time to hit the ball where I wanted to; and

I am fully focused on the ball. My timing is sweet and I am feeling strong and fit. I am in full concentration, which is automatic, and nothing can distract me... The only thing that stands in my match is this match and each point. As I win more points, my adrenalin is pumping. I feel on top of the world.

In further findings, the zone occurred more frequently in training than in competition (t [31] = 6.31, $p < .05$). Challenges and skills were closely matched at a high level with a mean challenge rating of 8.16 (*SD* = 1.84) and a mean skill rating of 8.45 (*SD* = 1.17). The zone was associated with an optimal yet not necessarily a victorious performance.

ZONE DIMENSIONS

Two hundred and forty five raw data themes (including 215 independent themes) were extracted from participants' narratives and deductively content analysed for a fit with one of the theoretical dimensions (listed above). A summary of the results of this analysis is presented in Table 1.

Table 1 Dimensions of the zone and miscellaneous category, percentage* of tennis players citing themes within each dimension, and percentage of all raw data themes represented by each dimension.

Zone dimension	Percentage of tennis players	Percentage of all raw data themes
Concentration	71	22
Paradox of control	68	23
Action-awareness merging	68	16
Clear goals and feedback	52	12
Challenge/skills	36	5
Loss of self-consciousness	29	5
Autotelic experience	24	16
Transformation of time	16	2
Miscellaneous	29	5

Note. * Participants may contribute more than one theme in each narrative.

As illustrated in Table 1, 95% of raw data themes were classified into one of the theoretical dimensions. The dimension of concentration ranks highest in terms of the percentage of participants citing a theme, while the dimension of paradox of control ranks highest in terms of the percentage of raw data themes.

FACTORS INFLUENCING THE ZONE

Participants nominated 410 factors (including 360 independent themes) that they considered facilitated, prevented and disrupted the zone. For clarity in the presentation of these factors, a series of inductive content analyses were conducted. A summary of the results of these analyses is presented in Table 2.

Table 2 General Factors Facilitating, Preventing and Disturbing Zone Experiences of Australian Professional Female Tennis Players.

Facilitate the zone	Disrupt the zone	Prevent the zone
Preparation (physical and mental) (1)	Preparation problems (6)	Preparation problems (2)
Positive mood (2)	Non-optimal mood (4)	Non-optimal mood (3)
Experience and control of arousal (3)	Inappropriate experience And control of arousal (5)	Inappropriate experience and control of arousal (4)
Motivation (4)	Problems with motivation (3)	Problems with motivation (5)
Focus (5)	Inappropriate focus (2)	Inappropriate focus (1)
Situational/environmental Conditions (6)	Situational/environmental Conditions (1)	Situational/environmental conditions (6)
Positive feedback (7)	Negative feedback (7)	Negative feedback (7)

Note. Numbers in parenthesis reflect the relative importance of each general factor for each category.

Table 2 lists 21 general dimension of factors that were identified as influencing the zone process. The general dimensions of factors describing a similar notion are listed in abbreviated form across the rows in the three columns. The Table illustrates the close correspondence between the general dimension of factors facilitating, preventing and disrupting the zone. The general dimensions representing factors that prevent and disrupt the zone are identical and contrast with the seven general dimensions of factors facilitating the zone.

DISCUSSION

Tennis players' descriptions of the zone corresponded with flow theory's posited structure of the zone consisting of eight inter-related dimensions, although there were differences in the salience of individual theoretical dimensions to the experience. While tennis players described the zone in different words, the phenomenon depicted was a similar one characterised by total involvement, a unity of mind and body and a sense of personal fulfilment at an optimal level of performance. Without exception, the quality of the zone experience was one of the highest levels of sport experience. Tennis players described such experiences as "out of this world", “unforgettable” or simply, “indescribable”.

In terms of the conditions affecting the zone, the study found the zone was a sensitive and volatile state influenced by an array of personal and situational factors. These factors generally corresponded to those proposed by flow theory. While tennis players considered they had control over many of the factors, this momentary experience was lost if the player reflected on it during its occurrence.

The study’s identification of factors influencing the zone process suggests that a program to guide tennis players to attain and maintain the zone should be designed to:

(1) Optimise those factors perceived by tennis players to facilitate the onset of the zone.
(2) Minimise or negate those factors perceived by tennis players to prevent or disrupt the zone.
(3) Assist players to re-interpret or restructure factors that they perceive to hinder or disrupt the zone.

In general, factors affecting the zone suggest tennis players be guided to:

(1) Set goals.
(2) Prepare (mentally and physically) fully for the challenge.
(3) Immerse themselves in the task to the exclusion of attending to disruptive situational factors.
(4) Maintain a positive attitude.
(5) Respond to feedback.
(6) Maintain an appropriate response to arousal.

Moving beyond these “blanket” guidelines, any program designed to guide tennis players to attain the zone needs to be tailored to the individual and address the individual’s awareness and interpretation of salient factors. It is the individual’s thoughts and feelings in response to factors, rather than the objective factors per se, which need to be addressed.

For example, maintenance of an appropriate response to arousal is considered important for the majority of tennis players to attain the zone. For some individuals, this means feeling relaxed and calm prior to and during competition. Alternatively, some tennis players need to feel excited and energised to play their best. Similarly, considerations of being physically fit and mentally prepared involves an individualised approach. Considerations may relate to an individual’s level of fitness, practice regime, pre-match routine and warm-up, acquisition of sound technique or mental training. In a further example, maintenance of appropriate focus may involve considerations of a number of specific task cues, including those watching the ball or adopting an aggressive strategy for playing vital points throughout the match.

In a similar vein, considerations of eliminating, negating or re-interpreting situational and environmental factors are important if the zone state is to be maintained. For an individual, disruptive extraneous factors may include those of poor weather and playing facilities, unlucky draws in tournaments, family problems and bad umpiring decisions. Tennis players can be guided to focus on the task at hand to the exclusion of these irrelevant and disruptive factors. Alternatively, tennis players can be trained to re-interpret extraneous factors. For example, for the tennis players who report windy conditions as a disruptive factor, these individuals can be guided to re-interpret this factor as the opportunity to use the weather conditions to their advantage to unsettle the game of their opponent.

CONCLUSIONS

The study's findings highlight that the zone is a highly valued and extremely meaningful state. In gaining an understanding of the nature and conditions of the phenomenon, the study suggests that the zone is no longer to be considered "out-of-reach" or a chance event. The opportunity exists for others to join Pete Sampras and Pat Rafter in finding the zone.

REFERENCES

Csikzentmihalyi, M. (1975). Play and intrinsic rewards. *Journal of Humanistic Psychology, 15,* 41-63.

Csikzentmihalyi, M. (1990). *Flow: The psychology of optimal experience.* New York: Harper & Rowe.

Goldberg, A. S. (1998). *Sports slump busting.* Champaign, Il: Human Kinetics.

Jackson, S. A. (1995). *Flow State Scale.* Brisbane, Australia: University of Queensland.

Loehr, J. E. (1995, July). Six keys to getting and staying in the zone. *Tennis*, p.36.

Murphy, S. (1996). *The achievement zone.* New York: Berkley.

Privette, G. (1984). *Experience Questionnaire.* Pensacola, FL: The University of West Florida.

Shainberg, L. (1989). Finding the zone. *New York Times Magazine*, pp. 34-6, 38-9.

Development of a new machine for tennis players to strengthen the lower limb adductor muscles

R. Tomosue
Yasuda Women's University, Hiroshima, Japan
K. Yoshinari
Shirayuri College, Tokyo, Japan
M. Yamane
Institute of Sports Medicine and Sciences, Aichi, Japan
M. Suda
Satellite Co. Ltd, Tokyo, Japan

ABSTRACT: The purpose of this study was to develop a training machine to strengthen the adductor musculature of tennis players and to determine the effect of using this lower limb training machine. The machine consists of a steel frame and two disks. The tennis player sets a foot on each of the disks, and then proceeds to rotate the disks by internal rotation of the lower limbs. Three male and 2 female tennis players participated in the testing of this new machine in the present study. The recorded electromyogram demonstrated that the muscle adductor magnus was used to rotate the disks to the inside. The electromyographic patterns were found To be similar to those of a forehand ground stroke. In squat training, however, the muscle adductor magnus did not act like a forehand ground stroke. After 5-weeks of using this machine, the muscular strength of the medial rotation was increased significantly. These results indicate that this new machine can be used to make an important contribution to tennis performance.

INTRODUCTION

The adductor muscles - the inside muscles of the thigh - are important for tennis players. They not only provide power for changing directions, but they also play an important part in rotating the trunk for the ground stroke. The majority of elite players emphasize trunk rotation and use an open stance in forehand stroke production in preference to stepping into the ball (Schonborn, 1992). Although the importance of training of the lower limb adductor muscles has been pointed out, there has not been a conclusive method for strengthening this muscle group. The purpose of this study was to develop a training machine to strengthen the adductor musculature of tennis players and to determine the effect of training using this lower limb training machine.

METHODS

THE MECHANISM OF THE LOWER LIMB TRAINING MACHINE

The machine consists of a steel frame (73cm × 35cm) and two disks (r=17.5cm) (Fig. 1). Each disk is connected to two tension springs by wire. The tension springs work to stop the rotation of the disk (Fig. 2).

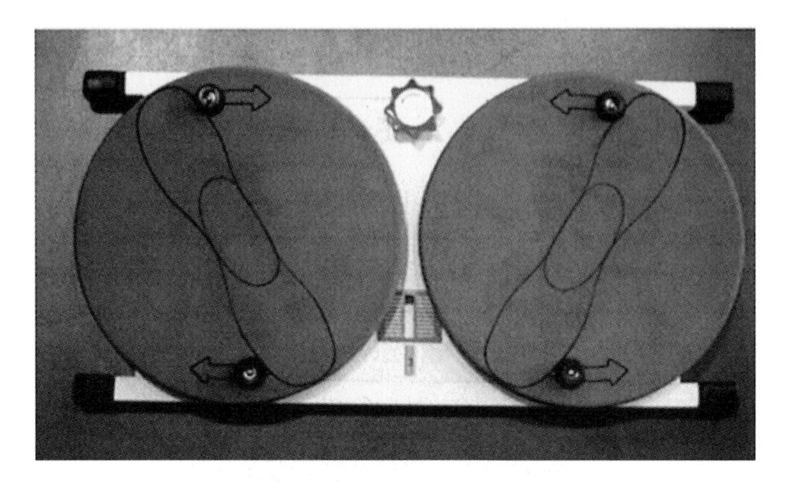

Fig. 1 Exterior view of the lower limb training machine.

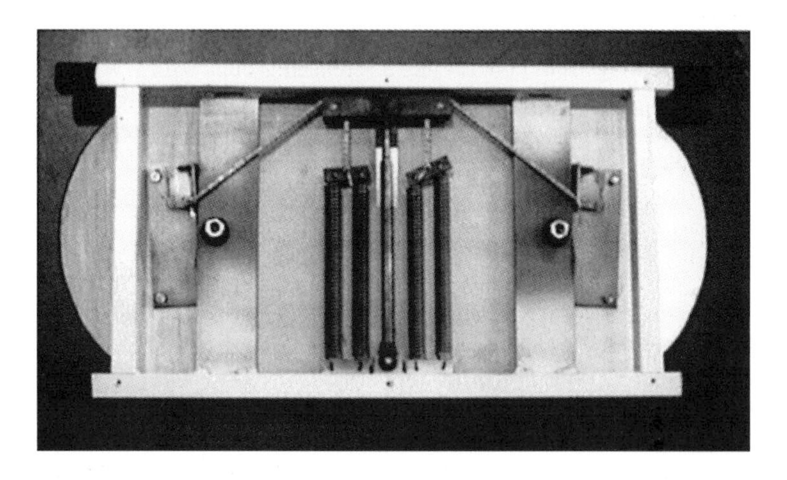

Fig. 2 Internal view of the lower limb training machine.

The tennis player sets a foot on each of the disks, and then proceeds to rotate the disks by internal rotation of the lower limb (Fig. 3). The torque of the disk rotation is increased or decreased via the torque knob. The torque knob can set the torque within a range of 5 to 20Nm.

Fig. 3 Posture during lower limb training.

EMG ACTIVITY

The subjects were 3 male and 2 female tennis players. The EMG activities of 4 muscles that are used to rotate the trunk were recorded using bipolar surface electrodes (Muscle tester ME 3000 professional, MEGA Electronics Ltd.). The M.rectus femoris, M.vastus lateralis, M.vastus medialis and M.adductor mugnus were chosen for the rotator muscle group. The EMG activity of the selected muscles was assessed for action involved in the tennis forehand stroke, squat training and medial rotation, while the subjects were using this new machine. For the recording of EMG, the scale of the torque knob was set on 7, which is almost equal to 11Nm.

THE EFFECTS OF THE TRAINING

We carried out the training experiment to determine the effect of the new training machine. Before and after training, the isokinetic peak torque of the medial rotation of the lower limb was measured by using an isokinetic dynamometer (Cybex II $^{+}$, Lumex Inc. USA). The test speed was 60 degree/sec. The subjects exercised three times per week for the 5 weeks of the experiment. Training consisted of 3 sets of 20 repetitions of medial rotation of the lower limb. The scale of the torque knob was set at 40% isokinetic peak torque. When the subjects could complete the sets easily, the resistance was increased, until it had been increased substantially by the end of the trial.

RESULTS AND DISCUSSION

Using a dynamometer, we checked the accuracy of the torque settings of the 2 machines to each other, and the relationship of the machine's torque scale to the actual torque value. The results are shown in Figure 4 and show that the torque scale of the two machines were similar, and that there was a close linear relationship between the scale of the torque knob and the real torque.

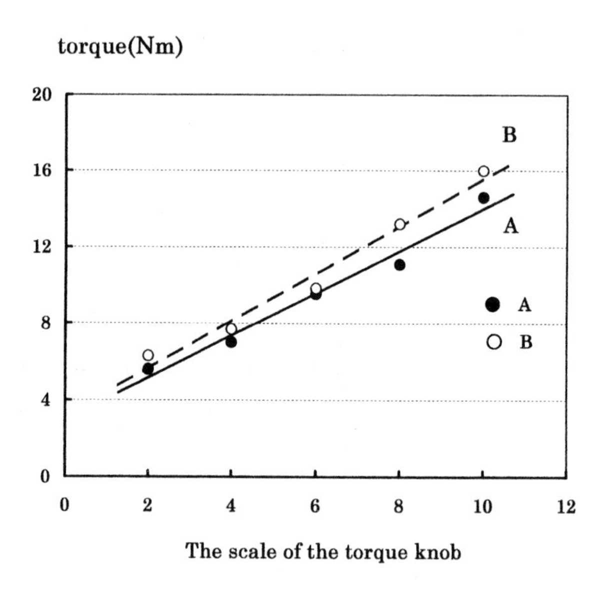

Fig. 4 Relationship between the scale of the torque knob and the real torque for the 2 machines (A&B).

The results of the EMGs are shown in Figure 5, 6 and 7. The recorded electromyogram demonstrated that the muscle adductor magnus was used to rotate the disks to the inside. The electromyographic patterns were found to be similar to those of a forehand ground stroke. In squat training, however, the muscle adductor magnus did not act like it would for a forehand ground stroke. These results show that movement using this new machine involves the rotator muscle group. This type of movement is difficult to perform other than when done intentionally. It seems that training with this new machine effectively strengthens the M. adductor magnus, in contrast to squat training using free weights.

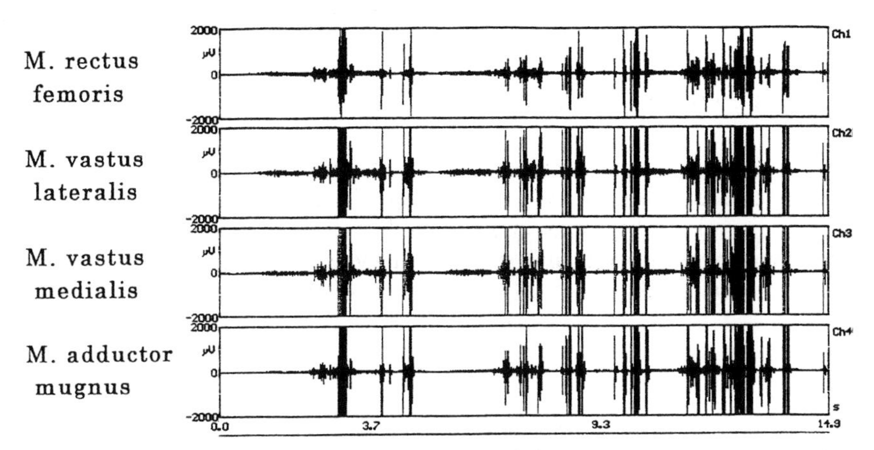

Fig. 5 Typical EMG pattern during lower limb training.

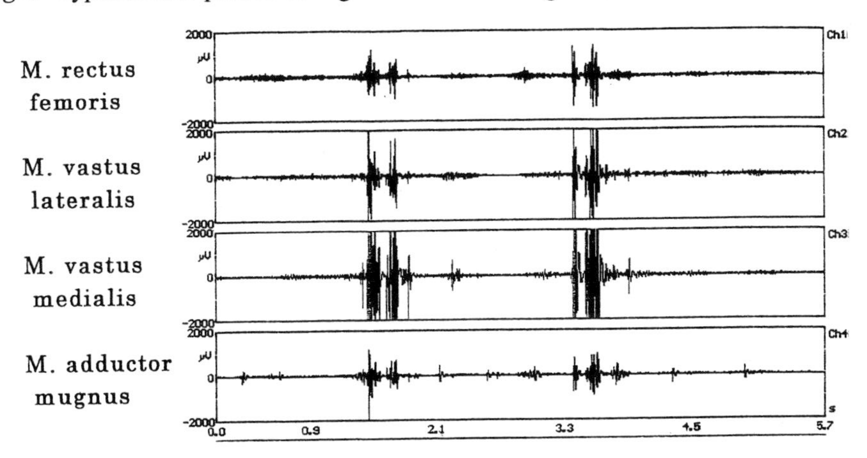

Fig. 6 Typical EMG pattern during tennis forehand stroke.

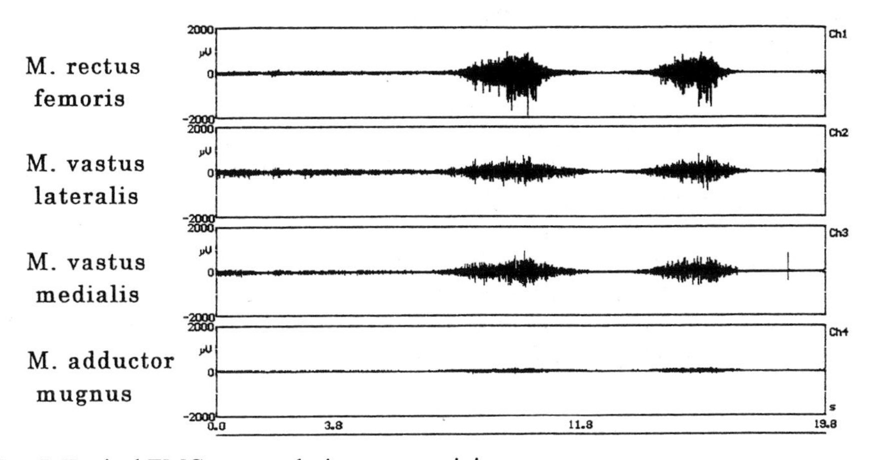

Fig. 7 Typical EMG pattern during squat training.

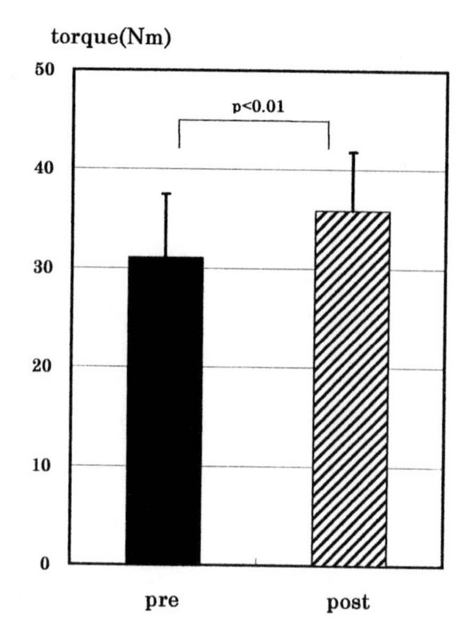

Fig. 8 The mean change of maximum torque between pre- and post-training.

The results of training using this new machine are shown in Figure 8. The changes of torque between pre- and post-training for the torque of the medial rotation significantly increased from 31 to 36Nm. The changes in the torque of the medial rotation were expected and indicate that this lower limb training machine can be used to make an important contribution to tennis performance.

CONCLUSION

To strengthen the adductor musculature of tennis players, lower limb training machine was developed. The electromyographic patterns were found to be similar to those of a forehand ground stroke. After 5-weeks of using this machine, the muscular strength of the medial rotation was increased significantly. These results indicate that this new machine can be used to make an important contribution to tennis performance.

REFERENCES

Schonborn, R. (1992) Paper presented 6th I.T.F. East Asian Coaches Workshop, Indnasia.

Yamane, M., Tomosue, R., Wakayama, A., Matsui, H., Yoshinari, K. & Masashi, Suda. (1998) The development of the hip joint training machine for alpine skiers. Proceedings of International Meeting of Sports Science Commemorating the1998 Winter Olympics in Nagano. pp.201-204.

Endurance in tennis – a complex approach

P. Unierzyski
University School of Physical Education, Tennis Department, Poznan, Poland
E. Szczepanowska
University School of Physical Education, Department of Sport Physiology, Poznan, Poland

ABSTRACT: Endurance is one of most important factors affecting performance in tennis. The classical understanding this ability is associated generally with aerobic endurance and cyclic types of sport activities like long distance running, bicycling, swimming etc. but is too narrow and does not meet the specific needs of modern game of tennis. Many of the various factors affecting tennis-specific endurance are only partly correlated with work physiology and physiological capacity. Hence the definition of tennis specific endurance must be different from the classical one. It should have referred mainly to the efficiency during a long-lasting, repetitive performance. Therefore the endurance of a tennis player must be different from a marathon runner or swimmer and should reflect complex character of the game. Tennis-specific endurance should be understood as the ability to maintain the quality of performance (e.g. shot precision, starting speed, co-ordination etc) on the highest possible level during given period of time. Training methods should be modified, respect the real requirements of the game and individual needs of a player.

INTRODUCTION

The game of tennis belongs to the sport disciplines of acyclic activities with a non-standardised movement structure. During a match players have to react quickly to constantly changing stimuli like: the distance, speed, height and rotation of coming balls, weather conditions, opponents' actions etc. Because of the relatively small weight of the equipment this kind of activity is dominated by dynamic muscle contractions. A tennis match consists of a series of rallies, most of them lasting maximum between 5 to 10 seconds (75,9% of rallies end within 5 seconds), followed by a break (20-30 or 90 seconds). The whole match usually lasts around one and a half - two hours but may last up to 6 hours (Lupo 1988, Collette 1991, Crespo 1993, Unierzyski 1993).

The game at junior level does not differ much in this aspect from the professional tennis. The average duration of rallies for under 14's is between 8 seconds for boys and 9.3 seconds for girls (Grivas, Bashalis, and Mantis 1992).

Statistical analysis of the game and observations made from the physiological point of view as well as motor characteristics of tennis (many sprints, accelerations, jumps, rapid stops and changes of running direction) (Unierzyski 1993) show that the game of tennis can be characterised as a short, intense exercise of interval type (Lupo et. al. 1988; Weber 1987). ATP and phosphocreatine are the main sources of energy for work during a match. Taking these facts into account tennis can be defined from a physiological point of view as an interval and speed-explosive power oriented game. The intervals between rallies, reaching about 75% of the total match time, are usually sufficient for full recovery. At the same time it can be stated that achieving a top performance level in tennis also depends on the development of optimal physical capacity and endurance.

ENDURANCE IN TENNIS

Within classical definition endurance is understood as the ability to perform hard or long-lasting physical activities involving large muscle groups, without the fast increase in fatigue and changes in inner environment of the organism. This definition also includes the strain tolerance and the ability of recovery. (Astrand 1987, Kozłowski, Nazar 1999). It should be asked if this definition really suits to demands of tennis?

Some authors (Weber 1987, Unierzyski 1992, 1993) showed that "optimal tennis endurance" means not necessary possessing aerobic endurance on the highest possible level. The endurance of a tennis player must be different from a marathon runner or swimmer and should reflect the specific character of the game. This ability has a very complex character and occurs in every rally during every set, in every round of a consecutive tournament. Poor shot execution in a long match can be caused by various reasons, e.g. a missed volley may result not only from bad technique, but also from a variety of other reasons (Schönborn 1994, Unierzyski 1996): insufficient explosive power or starting speed, bad co-ordination or agility, inadequate morphological build of a player, low stimulation or high inhibition of the nervous system, wrong tactical decision (anticipation, sport intelligence) or low level of achievement motivation. In various factors affecting performance in tennis the component of endurance appears in a long match as an integral part of a motor ability, psychological feature or technique. It means that we can talk about several kind of endurance, for example endurance in starting speed, in concentration etc.

If a tennis match lasts long in time and a player has to hit a fast serve hundred times, perform many jumps and sprints or stay focused for 5 hours facing an unfriendly crowd he must have a tennis-specific endurance on high level in order to maintain his performance capability. Many of various different factors affecting tennis specific endurance are only partly correlated with work physiology and physiological capacity. Hence the definition of tennis specific endurance must be different from the classical one. It should not be analysed from a purely physiological point of view. It should have more complex character and refer mainly to the efficiency of a long-lasting, repetitive performance. A tennis player with sufficient endurance level will be able to maintain his playing quality at the optimal level during a match, tournament or series of tournaments. So in our opinion tennis-specific endurance should be understood as the ability to maintain the quality of repetitive physical effort on the highest possible level during given period of time.

The tennis-specific endurance is affected by:

(1) Physical capacity.
- aerobic capacity
- anaerobic alactacid capacity
- anaerobic lactacid capacity

(2) Central Nervous System.
(3) General and specific co-ordination
(4) Level of conditional motor abilities.
(5) Mental attributes (motivation, intelligence, concentration etc.).
(6) Tennis technique and tactics.

If any of the above-mentioned factors affecting endurance decrease their level during any moment of a match or tournament it will cause tiredness and lower performance. For example a 5% decrease in starting speed (e.g. from 5.7 m/s to 5.2 m/s in a distance of 5 meters) makes the level of this ability insufficient for effective performance at a professional level.

Tiredness in tennis from a complex point of few can be described as decrease of the effectiveness of play and e.g. playing off balance, lower level of stroke precision, quality of footwork, worse concentration, anger or even bad decision making and as a result - lower performance level.

ENDURANCE REQUIREMENTS FOR TENNIS PLAYERS

The level of co-ordination, tennis technique and tactics, speed, agility and dynamic power

It is obvious that if the level of these abilities is higher the performance level of tennis player will be on higher level. At the beginning of a match, after the warm up all of them are usually at the individually maximal level. But together with tiredness of the nervous system and muscles the performance level decreases. If it decreases from a higher point it will reach the bottom later. A player having a kind of reserve of speed, co-ordination or technique at his disposal will perform in a more economical way. He will use less energy to play with a certain speed, he will be able to intensify the game when it is necessary and he may stay on a court longer without dropping his performance level.

Psychological requirements

A successful tennis player is able to stay focused and maintain the internal drive at a high level during the whole match. The base for it is a low reactivity of the nervous system and high achievement motivation. The majority of good players can be described as mentally strong, less sensitive with good stamina of nervous system. The long and strong stimulation (like a tough tennis match) causes an optimal activation among these persons. It means that players with the strong types of temperament are usually more able to face difficulties and fight better. They will be more effective in a stress situation during a long match and their reactions will be fast and precise and the will to win will not decrease (Sankowski 1981; Gracz, Sankowski 1991; Orawiec, Dańczyk 1994; Czajkowski 1995). This means that he/she has a good psychological base for the tennis specific endurance.

Also a high level of achievement motivation has the positive influence on tennis specific endurance. A player having this feature on a high level will continue long, unsolved tasks more effectively and they will show better persistence in problem solving situations (Atkinson J., Feather F. 1966) Motivation also plays a great role in overcoming subjective symptoms of fatigue (Kozłowski, Nazar 1999). The degree of motivation can effect performance during long matches and may enlarge the tolerance of fatigue changes (e.g. lactate). A person with low reactivity usually has high achievement motivation, too. As a result he is able to stimulate himself in the optimal way in tough situations, stay focused and keeping his emotions under control better is able to maintain the performance on optimal level for a long period of time.

Physiological requirements

A tennis player needs a high level of resistance to tiredness, a tolerance to changes of homeostasis, and above all a quick regeneration of ATP and phosphocreatine (CP) As a result of these factors he will be able to run at the same speed in the third hour of a match and hit all shots with the same precision as at the beginning. The base for tennis specific endurance is general endurance. Because general endurance is a base for other motor abilities it should be an important element of the training process, even in the first stage of a career. According to present knowledge (Weber 1987, Unierzyski 1992) a player should have the maximal oxygen uptake (VO_2 max) on the level around 60ml/kg/min. A further increase in (VO_2 max) does not affect much of the performance level in tennis. However, the most adequate oxygen consumption at the aerobic ventilatory threshold (VT) is around 35ml/kg/min for juniors and about 40 ml/kg/min for adults. These values are related to our studies and have not been confirmed by other researches up till now.

As it has already been mentioned tennis can be characterised from the metabolic point of view as the dominantly anaerobic-alactacid transformation of energy during short rallies. After each rally the energy is provided over aerobic transformation and CP and ATP can be rebuilt through oxidation. So a great difference ("space") between the oxygen consumption of the VO_2 max and AT levels is also very advantageous.

However it can not be forgotten that a game of tennis requires fast, dynamic movements so it is important to keep right proportions between speed and endurance during a training process. Many tennis careers were broken because of too high participation of the classical endurance training (e.g. long-distance runs), which caused the increase in ST fibres and, in the result, decreased the level of speed. So looking at the physiological part of endurance training it can be said: train endurance but not to run a marathon!

CONCLUSIONS

(1) Tennis-specific endurance should be understood as a complex ability to maintain the quality of performance on the highest possible level during the whole match or tournament.
(2) Specific endurance is a base for high level of speed, power and other psychomotory abilities necessary during tennis matches and tournaments.
(3) The great role in aerobic endurance plays the level of anaerobic ventilatory threshold - more important than maximal oxygen uptake.
(4) Training methods for tennis should be modified and respect the complex character of the game, individual predisposition of a given player and his personal game style.

REFERENCES

Astrand P-O, Rodahl K. (1987) *Textbook of Work Psychology*. McGraw. New York.

Atkinson J. Feather F.(1966) *An introduction to motivation.* Van Nostrand. Princeton.

Colette D.(1992) *Problem Solving as a pedagogical toll.* 3rd National Tennis Seminar. Melbourne, Tennis Australia. 1-13.

Crespo M. (1993) *Duration de los puntos en un partido de tennis.* ETA Symposium Rome. ETA.

Czajkowski J.(1995) Znaczenie pobudzenia w działalności sportowej cz.1 i 2. *Sport Wyczynowy* nr 7-8, 9-10.

Gracz J. Sankowski T. (1995) *Psychologia sportu.* AWF Poznań.

Grivas N., Bashalis S., Mantis K. (1992) *A statistical analysis of the 1992 European Championships (under 14).* ETA Coaches Symposium. Agios Nikolaos.

Kozłowski S., Nazar K.(red.) (1999) Wprowadzenie do fizjologii klinicznej. PZWL Warszawa.

Lupo et. al. (1988) Wybrane fizjologiczne i biochemiczne wskaźniki charakteryzujące zdolność wysiłkową tenisistów. *Sport Wyczynowy* 10.

Orawiec A., Dańczyk R. (1994) Temperament a wybrane zdolności psychomotoryczne zawodników hokeja na lodzie. *Trening* nr 4.

Sankowski T. (1981) Zróżnicowanie zespołów sportowych pod względem typu układu nerwowego i odporności na zmęczenie. *Wychowanie Fizyczne i Sport* nr 1.

Schönborn R. (1984) Leistungslimitierende und Leistungabestimende Faktoren. in. Gabler H., Zein B. *Talentsuche und Talentforderung im Tennis.* Ahrensburg, Czwalina.

Schönborn R. (1994) Modern Complex Training in Tennis. *ITF Coaches Review* no 4. pp 1-4.

Unierzyski P. (1992) *Model interval training for tennis.* ETA Coaches Symposium. Agios Nicolaos.

Unierzyski P. (1993) Tendencje rozwojowe tenisa ziemnego. w: Drozdowski Z. (red.) *Tendencje rozwojowe sportu.* Monografie nr 300. AWF Poznań

Weber K. (1987) Der Tennissport aus Internistisch-S2portmedizinischer Sicht. H.Richarz, St.Augustin.

Battery of tests for prediction and evaluation of tennis players

E.M. Isnidarsi, A.C. Gonçalves
Department of Physical Education, University of Ribeirão Preto, Brazil

ABSTRACT: Evaluation and selection of talents is a subject frequently discussed at various sport sciences congresses. This research traces a profile of the physical qualities and basic technical abilities indispensable for a tennis player to progress to higher levels. Renowned protocols of physical evaluation already exist in the literature. Model assessments of strength, agility, speed, flexibility and tennis skills were selected from a bibliographical search of published field-tested projects.

This project with its standardized methods of measurement, and data collection can be used to guide the physical trainer. He or she will have a practical data analysis tool to present and visualize results in graphs depicting the evolution of each physical quality and technique.

The data can then be used for subsequent comparisons. With a comprehensive evaluation, it is possible to analyze and diagnose progress with more precision. In order to prescribe the intensity of the training program it is necessary to have this longitudinal evaluation of the athletes' progress.

The focus of this work is the fusion of the existing methodologies with computer science technology. The principle of simplicity is essential in order to be able to reproduce the project model on any tennis court. This efficient and effective method can encourage development of new and existing grass roots tennis programs in schools worldwide using very little resources. Because this method is practical, schools, need no more than a court, a track and a backboard.

METHODS AND MATERIALS

EVALUATION

This project investigated seven (N=7) male tennis players varying in age from 8 to 13 years. The initial evaluation is especially important to determine the athletic condition of the player. After analysis of the findings, the initial threshold and individual program of training can be established.

To detect changes in performance, emphasis is given to precise measurement and recording of testing results. To standardize the amount of recovery time between exercises, the complete battery of tests are given to no more than two people at a time.

A "Starter system" was developed as a component of this project, to measure and strive to improve the speed of visual reaction, essential to a tennis player's success. The evaluation protocol also includes the following tests, some of which were administered three times as a part of each evaluation:

(1) Vertical jumps (VJ).
(2) Muscular Abdominal Resistance (MAR).
(3) Reaction Speed (RS).
(4) Agility (AG).
(5) Dynamic Balance (DB).
(6) Precision in Service (PS).
(7) Test of Mille (VO^2 max).
(8) Flexibility of Wells (FLEX).
(9) Horizontal Jumps (HJ).
(10) Broer-Miller for Tennis (BMT).
(11) Dyer for Tennis (DT).

The following materials were used to administer the battery of tests and to record and analyze the findings of each tennis player's evaluation:
(1) Computer with Microsoft Excel spreadsheet.
(2) Anthropometrics analyses were recorded using Physical Fitness for Windows.
(3) Physical Nutri for Windows was used to measure gain and loss of fat mass.
(4) Polar Heart rate Monitor-model Edge.
(5) Chronographer Timex - model Ironman.
(6) Wells Box for measuring flexibility.
(7) Traffic cones.
(8) Tape measure.
(9) Clipboard with evaluation form.
(10) Tennis balls.
(11) Tennis court.
(12) Track.
(13) Backboard.

In the original research, the combination of training and evaluation every six months accomplished a longitudinal study upon which the progress of the players could be measured and established the basis for the subsequent training regime.

TRAINING

The circuit time training model is used during the seventy minute sessions scheduled four times a week. Training focuses on tennis skills including jumps, turns, balance, technical shots, steps, explosion, agility, coordination and speed visual reaction. as well as strengthening of the extremities, and aerobic and anaerobic-alactic resistance.

The trainer continually monitors the plan, checking for overload and adjusting the thresholds accordingly. The athlete receives feedback as he is taught new tasks and adjusts to the progression of his physical abilities and organic adaptation.

Fig. 1 The student jumping in evaluation of Vertical Jump (VP)

Fig. 2 Evaluation of Vertical Jump (VJ)

Fig. 3 The trainer using the "Start System" for evaluation in Agility test (AG)

Fig. 4 Evaluation of Dynamic Balance adapted to court.

Fig. 5 Evaluation of Flexibility (Flex)

Fig. 6 Test of Mille to VO^2 max

Fig. 7 Antropometric evaluation

RESULTS

Table 1 Results of evaluations done at the beginning and end of ninety days of training.

	(VJ) *(Cm)*	*Ram* *(1 min)*	*(PS)** *(20 balls)*	*AG** *(Kmh.)*	*DB** *(9m)*	*RS** *(Cm.)*	*FLEX** *(Cm.)*	*DT* *(14 balls)*	*(HJ)* *(Cm.)*	*BMT** *(30 Sec.)*	*Mille** *(min.)*
Before	1.44 ± 0.05	43.71 ± 1.56	3.89 ± 0.60	9.36 ± 0.28	0.87 ± 0.09	16.71 ± 0.02	0.45 ± 0.12	0.54 ± 0.04	0.38 ± 0.11	7.67 ± 5.48	12.9 ± 0.50
After	1.53 ± 0.07	43.71 ± 2.61	7.39 ± 1.50	7.64 ± 0.25	1.05 ± 0.11	11.07 ± 1.71	0.30 ± 0.10	0.55 ± 0.03	0.31 ± 0.04	9.26 ± 7.03	15.3 ± 0.51
(%)	5.58 ± 2.07	-0.04 ± 6.83	66.35 ± 22.24	-18.75 ± 0.70	25.44 ± 14.93	56.59 ± 30.7	-52.7 ± 19.3	2.89 ± 8.24	7.73 ± 22.57	24.52 ± 8.46	18.3 ± 1.96

* indicate statistically significant differences for a P <0.05

Gonçalves (1997)

DISCUSSION

Currently Brazil has many tennis players, clubs and training centers. The Brazilian Confederation of Tennis reports that last year the number of championships grew by 30% .

The International Tennis Federation's training program for tennis coaches has had considerable success with more than 600 technical coaches trained and 24 courses given around Brazil in one year. Unfortunately, there has not been the same effort to develop physical trainers for tennis. Only supplying technical training to players with 30 to 60 minute sessions will not increase the number of players or improve their position in the ranking. Beginning tennis players particularly need and deserve a more comprehensive coaching team and program to develop their skills and maximize their potential.

The development of a complementary physical training program with a multidisciplinary team approach is going to be critical in the near future. The players will benefit by having more than one specialist, and this will make it possible to have a healthy generation of tennis players.

It is essential to have a comprehensive training program based on the accepted and efficient methods of evaluation included in this project while recognizing the need for low costs. This can be an important asset to the game of tennis, especially for the beginning player, and the coaching team.

CONCLUSION

In the studied group, the evaluation scores of participants at the end of the first training period showed overall improvement.. The bioenergetics testing including the Mille (M) and Agility (AG) declined.. The findings of testing that measure specific tennis abilities including the Precision in service (PS), Speed Reaction (SR), Dynamic balance (DB) and Broer-Miller Test (BMT), verified a significant improvement in the subjects score results. The agility and Mille scores could be improved with more time. Ongoing research will be needed to continue to develop this concept further, assuring its maximum effectiveness, and to make it widely available.

REFERENCES

Lea & Fediger. (1986) *Guidelines for Graded Exercise Testing and Exercise prescription,* American College of Sports Medicine. 3° rd, Ed. Philadelphia.

Astrand P. O. (1960) *Aerobic Work Capacity in Men and Women With special Reference to Age*, Act Physical. Scand. (Suppl.), 49:169.

Borg G. (1982) *Physiological Bases of Perceived Exertion.* Med. Science Sports. Exerc., 14:377.

Caldas P.R & Rocha.S.O. (1978) *Qualidades Físicas em Evidência Durante o Evento tênis*, Caderno técnico didático MEC.

Carnaval P.E. (1995) *Medidas e Avaliação em Ciências Dos Esportes*, Editora Sprint, Rio de Janeiro.

Caldeira S.; Matsudo V.; & LorenziniC.T. (1978) *Características do Tempo de Ação do Ttenista de Alto Nível e sua Relação Com o Metabolismo Anaeróbio.* Revista Brasileira de Ciências do Esporte, setembro. volume 5.

Carnaval P. E. (1995) *Medidas e Avaliação em Ciências dos Esportes*, Editora Sprint, Rio de Janeiro.

Groppel J.E., Loehr J.,.Quinn, A. Melville, Y. *eds.*(1992) *Science of Coaching Tennis.* Human Kinetic Champaign.III.

Kirmair C. A.. (1987) *Mexa-se é Fundamental*, Revista Match Point, volume II, n 5, ed Nesc, São Paulo.

Martens R. (1987) *Coaches Guide to Sport Psychology* Human Kinetics Publication Champaign. III.

Mathews D. K. (1986) *Medida e Avaliação em Educação Física*, RJ, Ed. Guanabara.

Matsudo V. K. (1983) *Testes em Ciências do Esporte*, 2° edição, São Paulo.

Naumko A. (1996) *Confrontação das Cargas de Treinamento e de Competição.* in "Encontro Internacional de Desportos Competitivos Rússia-Brasil", Curitiba, julho.

Petrovina A. (1996) *Qualidades Físicas do Tenista Profissional.* in "Encontro Internacional de Desportos Competitivos Rússia-Brasil", Curitiba, julho.

Runninger J. (1978).*Como Olhar Melhor a Bola.* Revista Tênis Esporte, volume II número 4, maio.

Skorodoumova A. (1996) *Heart Rate During the Match*, in "Encontro Internacional de Desportos Competitivos Rússia-Brasil", Curitiba, july.

Ulianov B. (1996) *Alguns Parâmetros da Ação dos Tenistas em Diferentes Tipos de Quadras de.* in "Encontro Internacional de Desportos Competitivos Rússia-Brasil", Curitiba, julho.

Longitudinal observation of physical and motor preconditions in tennis

J. Zháněl, F. Vaverka
Faculty of Physical Culture, Palacky University, Olomouc, Czech Republic
M. Černošek
TK Plus Prostějov, Prostějov, Czech Republic

ABSTRACT: The paper deals with the results of a three-year longitudinal observation of the level of physical and motor preconditions in women tennis players with the aid of a standard testing battery. On the basis of studies conducted on two young Czech women tennis players, an assessment of their developmental possibilities is presented with the help of time series and developmental trends analysis. Results are presented for comparison in numeric and graphic forms, indicating possible utilisation of current diagnostic techniques in the regulation and objective sports training of young tennis players focused on preconditions pertaining to physical and motor performance. The results inferred a certain degree of analogy in the character of developmental trends for some variables concurrent to the development of sporting performance in the observed players.

INTRODUCTION

An effective training process relies on meticulous and constant observation and analysis of substantial components of a sporting performance. Anthropometric, biomechanics, medical and specific testing methods were deployed in order to obtain necessary data. One of the progressive statistical methods enabling supervision of development of observed characters during the training process in the long run, is time series analysis. The presented paper aims to highlight the scope of this method in finding solutions to problems relating to individual assessment of the effective development of sportsmen during training.

OBJECTIVE

Assessment of general regularity during sports training is largely based on the acquisition and evaluation of selected data, coupled with a subsequent generalisation of the obtained results. This so-called group analysis has been criticised since the 70s, both from a practical point of view (individual values seldom correspond to the average values) as well as from a theoretical stand point (variations in performance need to be understood as complexly determined and can be explained only on the basis of knowledge of individual dimensions concerning the observed performance characteristics). The time series analysis thus appears to be a suitable diagnostic method in the training process. This method can be applied in three spheres: description of processes evolving the observed changes, verification of the effectiveness of interventions, and understanding the implications of the resultant effects (Lames, 1996). To study the process of individual components of the sporting performance is of particular interest from the point of view of sports training.

With special reference to the youth age category, where the performance development is often not well defined, the training process is focused mainly on development of sports-specific skills while adequate nurturing of the general condition – co-ordination skills, remain neglected (in terms of age and level of physical preconditions). Knowledge of individual developmental trend is a good starting point for gaining control of the training process, or for a timely selection of sportsmen desirous of entering a top-level sports activity. As already stated above, it is often difficult (especially in the youth) to evaluate measurement results and test physical and motor preconditions in groups. A more reliable result is obtained through an evaluation of individual development trend with respect to assessment of accelerated, or retarded performance (Mair, 1997). The possibility of comparison of individual results with those of the entire population, or with an age-corresponding group, renders the evaluation of an individual to be more precise. It is difficult to ascribe an exact definition to the validity of the observed data , i.e. assessment of relation between the level of physical and motor preconditions for sports performance. A more frequent means of evaluating sports performance – not referring to the hierarchy, which is a very problematic domain - is the so called 'expert judgement of performance development', i.e. professional assessment by a trainer for example. This is particularly true of the youth age group.

Thus, if we comprehend regular observation and data assessment as defining sports performance level to be a means of regulating and controlling sports training (see Fig. 1 – Blahuš, 1996), it is likely that evaluation of individual developmental trends in men and women tennis players can provide relevant information for further correction and optimal focusing of the training process, thereby assisting in estimating the performance development of a given young player.

The presented paper demonstrates an individual approach to analysis of physical and motor preconditions, based on a sample study of two identically aged women tennis players who entered with similar performance levels but whose performances exhibited a marked difference over a period lasting one year. We were keen to ascertain whether this difference in performance corresponded with developmental trends in the observed physical and motor features.

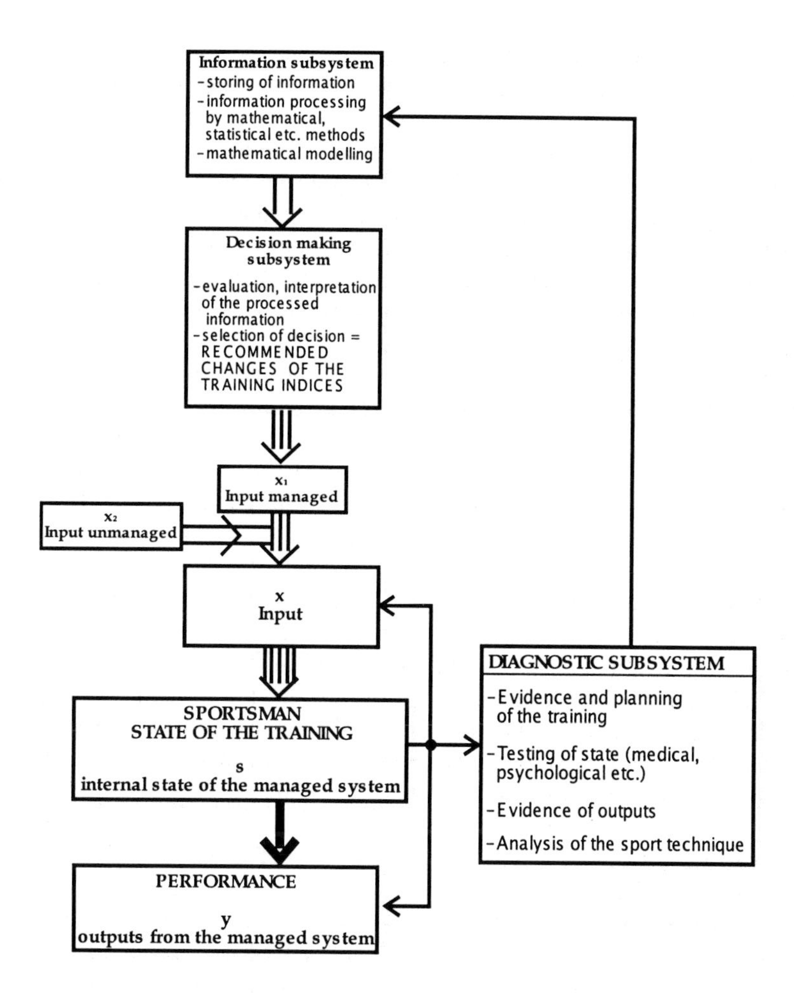

Fig. 1 Components of the system of the training process management and their operation (adapted by Blahuš, 1996).

METHOD

A comprehensive research on documented information (Arnot & Gaines,1990; Bös, 1986; Gabler, 1985; Roth & Thiel, 1987; Wohlmann, 1996, Kornexl & Müller, 1987), led to the creation of a test battery comprising nine diagnostic components for physical and conditioning-coordination preconditions for the tennis players in collaboration with top trainers from TK plus Prostjov. The battery encompasses three fields: physical, conditional and coordination.

The battery was applied between 1996 and 1999 on approx. 200 male tennis players and 150 women tennis players in various tennis clubs and also in the youth representation group of the Czech tennis association. Partial results were presented in scientific seminars and conferences (Zhán l 1998, 1999a, 1999b).

As a result of long term observation of two young Czech women tennis players of nearly the same age (the girls were tested in time periods: T 11.4 – 14.6 years;

C 11.0 – 14.2 years), developmental curves for selected indicators were projected. Both tennis players rank among the top Czech (and also European) women players, whose performance supervised by experts (personal trainers) was adjudged as follows: in 1996 the performance by player "T" was higher, during the 1997 and 1998 seasons, the performance of both players ranked equal, in 1999 player "C" improved remarkably while player "T" showed a certain deterioration.

Individual development trends were compared with those in the general population or with a group of women tennis players of similar age. In the presented paper we focused primarily on the analysis of physical preconditions (body height, weight, BMI) and conditional aspects (hand grip strength – left and right, speed – running speed test over cca 55 m with changes in direction, and endurance – short-time running endurance test over 400 m approx. with variation in direction). Data was processed on a computer program EXCEL 97 (construction of time series, calculation of regression equation, calculation of reliability coefficient R). Graphic representation of the curves of flow-lines of the trend was selected with respect to the highest attainable level of reliability coefficient R.

RESULTS

The results of longitudinal observation of individual physical and motor parameters are stated in a numeric form in table 1. Individual results or inter-individual differences are difficult to read from the numeric data when following classical descriptive methods. Measured individual data are subject to variation, the causes being hardly discernible and therefore remain unexplained.

Table 1 Individual performances in measured tests by tennis player "C" and "T" (longitudinal monitoring).

		I-96	II-97	XI-97	V-98	XI-98	IV-99
Age (years)	C	11.0	12.1	12.8	13.4	13.8	14.2
	T	11.4	12.5	13.2	13.8	14.2	14.6
Height (cm)	C	155	163	167	170	170	170
	T	146	155	157	158	158	159
Weight (kg)	C	40	47	54	58	60	62
	T	36	42	48	51	54	55
BMI	C	16.7	17.7	19.4	20.1	20.8	21.5
	T	16.9	17.5	19.5	20.4	21.6	21.9
Grip strength right (N)	C	19.6	29.0	29.8	31.8	33.3	34,2
	T	20.0	24.4	25.3	29.2	33.4	31.7
Grip strength left (N)	C	18.6	24.9	28.0	27.4	28.5	28.1
	T	15.2	19.2	20.8	23.5	26.7	28.1
Speed (sec)	C		18.0	15.6	15.8	14.8	14.1
	T		15.6	14.5	14.2	14.0	14.2
Endurance (num)	C	43	43	47	47	47	48
	T	48	49	48	48	50	50

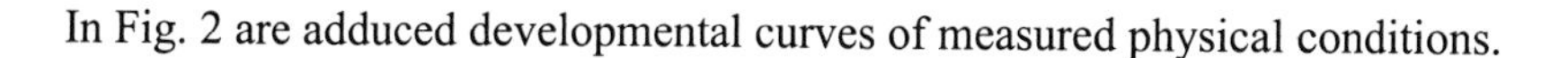

In Fig. 2 are adduced developmental curves of measured physical conditions.

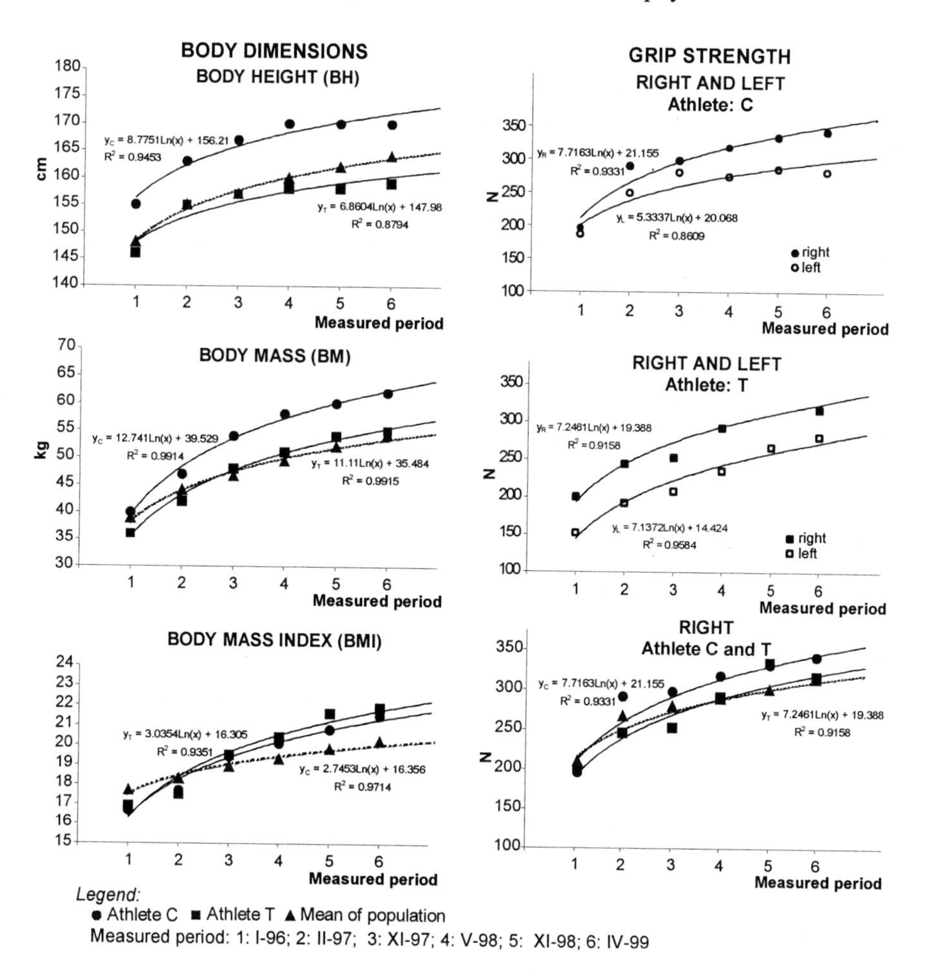

Fig. 2 Graphic expression of development trends for measured variables – body dimensions, grip strength.

BODY HEIGHT

Developmental trends pertaining to height in both players show distinctive differences (8-12 cm). While the height of player "T" is consistently below the population average, the height of player "C" is permanently above standard average. The height in both players stabilised during the course last year's measurements (age 13.5 – 14.5) and the ontogenetic evolution seems to have ceased.

BODY WEIGHT

Developmental trends in terms of weight in both players exhibit a differing tendency. Player "C" from the beginning had an above average weight (which corresponds with the better-than-average height) and is catching up to the weight standard of adult players. Player "T", on the other hand, at the outset of the recorded period was below average of the population and at around 13 years of age comes to a gradual accrual of the weight above the average of a given population (consequent to cessation in height gain).

BMI

Developmental tendencies in height and weight are also reflected in the developmental trends concerning BMI. In Player "T" BMI standards have shown accentuation much above the standard for the population, and from the age of 13, it was above that of Player "C". The value of BMI in both players at the last measure, i.e. at the age of 14.6 for Player "T" (BMI = 21.9) and at 14.2 years for Player "C" (BMI = 21.5) is relatively high considering the age.

STRENGTH

The grip strength of the right hand in Player "C" was higher than in Player "T" for the entire period and since the age of 12 it was higher above the level of general population. The strength in Player "T" oscillated at first below the average level and since the age of 13.8 it is above the average population level. From the very outset, strength in the left hand showed a higher, gradually increasing level for Player "C" (since the age of 12 remarkably above the level exhibited in women tennis players of that age), from 12.8 years age we observed cessation of power development. The development trend in the left hand for Player "T" followed a completely different course. From an initial below standard strength level it followed a rapid increase, rising to the level of Player "C" and above the level of women tennis players of the same age.

In Fig. 3 are presented developmental curves of measured running speed and endurance abilities.

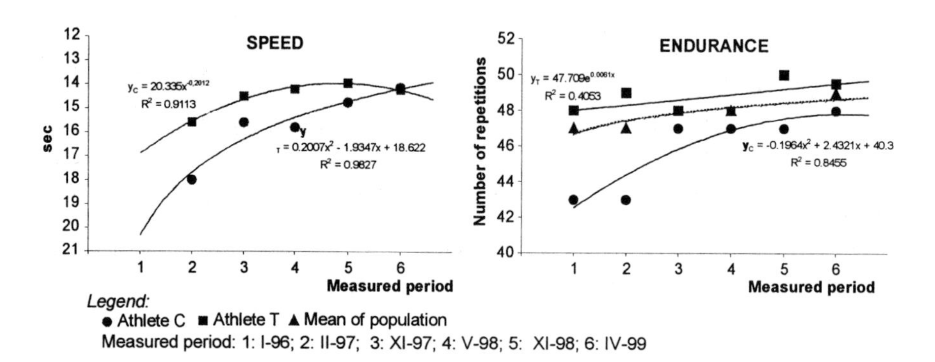

Fig. 3 Graphic expression of development trends for measured variables – running speed, running endurance.

SPEED

Comparison of speed development trends reveals different characteristics in both players. From the beginning, in Player "T" we can observe a higher and gradually decreasing trend of speed growth which stabilised during the last three measurements. Development trend in Player "C" has a remarkably accelerating character with gradual and continual increase in speed, culminating in Player "C" attaining better results than Player "T" in the last test run.

ENDURANCE

Developmental trends pertaining to short term endurance level in both players are similar to the speed. From the beginning of the observed period, Player "T" showed a higher gradually increasing endurance level, while at the end of the period she remained constant. The development trend for endurance in Player "C" showed an increasing character, beginning below the level of Player "T" and below the level of young women tennis players of the same age, hence approaching the average of tennis players of similar age.

DISCUSSION

In the introductory part we have put forward the idea of comparison of development trends of measured characters with the performance development of young women tennis players. The results showed certain remarkable relations between these variables. The above standard body height and weight of Player "C" when compared with that in the general population indicates development acceleration in the field of physical dimensions, which is a certain advantage in this age category. In comparison to Player "C", the height of Player "T" is under the average level while the body weight is slightly above it. This disproportion in the development of physical features is reflected in the BMI, which is disadvantageous for Player "T" at the age of 14 when compared with the top world tennis players. In comparison to similar values for 15 top women players in 1999 (average values: body height 174.2 ± 7.6cm; body weight 63 ± 7.6kg; BMI: 20.8 ± 1.0; variation in BMI: 19.2 – 22.1), we can see that Player "C" with BMI = 21.9 is closer to the indices of the best player than Player T with BMI of 21.9. The stated differentiation between Players "C" and "T" is dependent on body height (C: 170cm; T: 159 cm). Player "C" has better preconditions for reduction of BMI as a positive trend signalling further development of the athlete.

Analysis of strength preconditions (Fig. 2) progresses unambiguously better in Player "C" where the strength level is remarkably above average when compared with those of the general population and in relation to Player "T", has a more progressive growth tendency in the preferred extremity.

Development speed trend (Fig. 3) relates best to the performance of both players. Player "C" is at the age of 11 at a lower level than Player "T" whose performance, however, begins to stagnate from the age of 13 while Player "C" exhibits a constant improvement. Progress in endurance development (Fig. 3) indicates that endurance is probably not a significant determinant of sporting performance during the observed age period, in relation to physical dimensions, strength and speed.

CONCLUSIONS

(1) Longitudinal observation of two young women tennis players showed individual differences in development of measured variables.
(2) Analysis of development trends of some variables indicates dependence of physical dimensions, strength and speed skills on sporting performance.
(3) The research demonstrated usefulness of individual analyses of development trends in young tennis players.

REFERENCES

Arnot, R. & Gaines, Ch. (1990) *Sporttalent.* Orac Verlag, Wienn.

Blahuš, P. (1996) *K systémovému pojetí statistických metod v metodologii empirického výzkumu chování.* Univerzita Karlova, Praha.

Bös, K. (1986) *Handbuch sportmotorischer Tests.* Verlag für Psychologie, Göttingen..

Gabler, H. (1988) *Individuelle Voraussetzungen der sportlichen Leistung und Leistungsentwicklung.* Verlag Hofmann, Schorndorf.

Kornexl, E. & Müller, E. (1987) Zum speziellen motorischen Eigenschaftsniveau des jugentlichen Tennisspielers. In: *Spektrum der Sportwissenschaften* (Ed. by E. Kornexl). Östereichischer Bundesverlag, Wien.

Lames, M. (1996) Zeitreihenanalyse: Anwendung in der Trainingswissenschaft. In: *Zeitreihenanalyse und "multiple statistische Verfahren"* in der Trainingswissenschaft (Ed. by J. Krug). Bundesinstitut für Sportwissenschaft; Berichte und Materialien des Bundesinstitut für Sportwissenschaft; Bd 1996, 4. Köln, Sport und Buch Strauss.

Mair, H. (1997) Talentförderung im Tennis. Tennissport, **3**, 12-13.

Roth, P. & Thiel, E. (1987) *Der Sport-Talent-Test.* Goldmann Verlag, München.

Wohlmann, R. (1996) *Leistungsdiagnostik im Tennis.* Czwalina Verlag, Hamburg.

Zháněl, J., Balaš, J., Trčka, D. & Shejbal, J. (1998) *Diagnostika motorických předpokladů v tenise.* Seminar of Association of Czech tennis trainers. Prostějov.

TRIGGER – OSTEOPRACTIC: a novel approach for the treatment of tennis related injuries

W. Bauermeister
TRIGGOsan Zentrum für Schmerz – und Sport – Osteopraktik
Unnuetzstr. 17A D-81825 Germany
Hans-Henning Fries
Deutsche British Petroleum AG, Hamburg, Germany

ABSTRACT: TRIGGER – OSTEOPRACTIC ® regards the Trigger - Area as the major cause of acute and chronic recurrent problems of the musculo skeletal system. Overuse injuries activate latent Trigger – areas leading to pain and functional deficits of the muscle, tendon, tendon sheet and insertion, bursa, joint and intervertebral disk. In the beginning stages of the Trigger – Area development the symptoms can be controlled by conventional approaches. Over time, the symptoms mostly at the shoulder, low back, knee and foot, become chronic resulting in time out from training and tournaments. For a definitive resolution of the problems, the diagnosis of Trigger – areas leads to the cause of the symptoms. By treating the Trigger - Areas with the TRIGGOsan Probe through manual pressure or Trigger Shockwave Therapy, positive results can be achieved, mostly beginning with the first treatment. Trigger – Osteopraktik can be utilized for the diagnosis and treatment of acute and chronic problems of the musculo – skeletal system as well as for injury prevention.

INTRODUCTION

TRIGGER – OSTEOPRACTIC® (Bauermeister, 1997) is a newly developed diagnostic and therapeutic approach which recognizes the trigger as the causative pathology for most musculo - skeletal problems. A trigger is an area of permanently contracted and swollen muscle fibers within the muscle. These Trigger Areas (TA´s) refer pain to remote parts of the body and they have deleterious effects on the muscle, tendon, bone, joint, cartilage and intervertebral disk. Skeletal muscle is the largest organ of the body and with a total number of 400 muscles, it accounts for at least 40% of the body weight. Still the muscle receives little attention in medical school teaching and medical text books. Most sports medicine text books do not even mention muscular TA´s, instead they focus on the tendon, bursa, bone, joint and nerve. It is intended that this be changed and that attention be directed to muscle and its TA´s as the underlying pathology of chronic and acute musculo – skeletal problems.

The aim of this paper is to enable the reader to

- Define a Trigger Area (TA)
- Recognize the role of TA´s as *the* cause of musculo – skeletal problems
- Understand how overuse injuries contribute to the development of TA´s
- Follow the chain of pathologic changes reaching from the TA´s to the joint cartilage and the intervertebral disk
- Understand how to screen muscles for TA´s
- Understand how to diagnose TA´s directly through controlled pressure application with the TRIGGOsan Probe
- Understand how TA´s are dissolved through controlled pressure application and through Trigger Shockwave Therapy® (TST)
- Understand the importance of early TA – diagnosis for injury prevention

TRIGGER – OSTEOPRACTIC AND TENNIS

Statistics show that professional Tennis players suffer mostly from overuse injuries (OI) of the shoulder, low back, knee and foot (USTA, 1998). In spite of seemingly appropriate medical therapy, these problems can become chronic causing significant time out from practice and tournaments. This too was the case for several members of the German first and second national tennis league as well as professional junior players. They had failed to respond to conventional treatments and could not fully participate in the training and in tournaments. The players were diagnosed and treated according to the principles of the TRIGGER – OSTEOPRACTIC® (Bauermeister, 1999). Their problems were caused by muscular trigger – areas (TA´s) and were subsequently solved by eliminating the TA´s.

TENNIS PLAYERS AND THEIR PROBLEM - AREAS

The *shoulder problems* reported were pain in and around the shoulder blade, the shoulder joint area, pain along the upper arm and elbow, numbness and tingling sensations in one or more fingers. There were no neurological findings or other underlying specific causes.

The players with *back problems* had pain located in the lumbar and sacral area, and some had radiation into the hip, the hamstrings, along the side of the thigh and into the calf and foot, sometimes with numbness or tingling sensations.

Players with *knee problems* had pain located on the inner side, the outer side, around, below or above the kneecap.

Foot problems presented as pain around the Achilles tendon, the sole, on the inner, the upper and the outer aspect of the foot.

Table 1 lists the problem areas, the number of cases (n), the duration since the first occurrence of the problem and the number of treatments required for the resolution. In all cases it was possible to trace the problem back to TA´s and to successfully treat them with TO.

Table 1 Problem areas of tennis players treated

PROBLEM – AREA	n	Months duration of symptoms	Treatments until lasting relief
Shoulder	10	6 – 24	3 - 6
Low Back	15	3 – 40	1 - 8
Knee	12	2 – 24	1 - 6
Foot	10	2 – 36	2 - 8

THE NATURE OF OVERUSE INJURIES

Tennis with it's prolonged repetitive movements, such as serving thousands of times. the pivots, twists, sudden stops and starts and the pounding of the feet on hard surfaces will inevitably cause repetitive microtrauma of the muscles. It is postulated that microtrauma will cause Trigger Areas (TA´s) to develop (Travell et al., 1983) which are the cause of most musculo – skeletal problems of tennis players as well as of the general population. Since most professional players already start their tennis career at a very young age, the process of Trigger development spreads over many years. Mostly the young players do not experience lasting health problems until the late teens or early twenties. By that time the level of acquired small injuries has reached a critical point, that additional minor injuries will suddenly lead to significant chronic problems of the musculo – skeletal system due to TA´s.

THE TRIGGER AREA

PATHOLOGY

TA´s are by definition permanently contracted muscle fibers within the muscle. There is a failure of calcium ions to sequestrate. (Merskey et al., 1994) There appears to be a lack of oxygen in that area which is necessary for normal muscle function. (Brückle et al.1990) The exact mechanism of TA development is still unknown. In my opinion they are likely to be hereditary, and their pathologic growth in athletes is a result of repetitive microtrauma.

MICOSCOPIC STUDIES

Several muscle biopsy studies report about significant changes in the microscopic appearance of the muscle fibers. Even though there is no sufficient statistical evidence for the validity of the findings, there appears to be regions (Fig. 1)of shortened(1) and enlarged sarcomeres surrounded by elongated(2) ones as if they are being pulled apart.

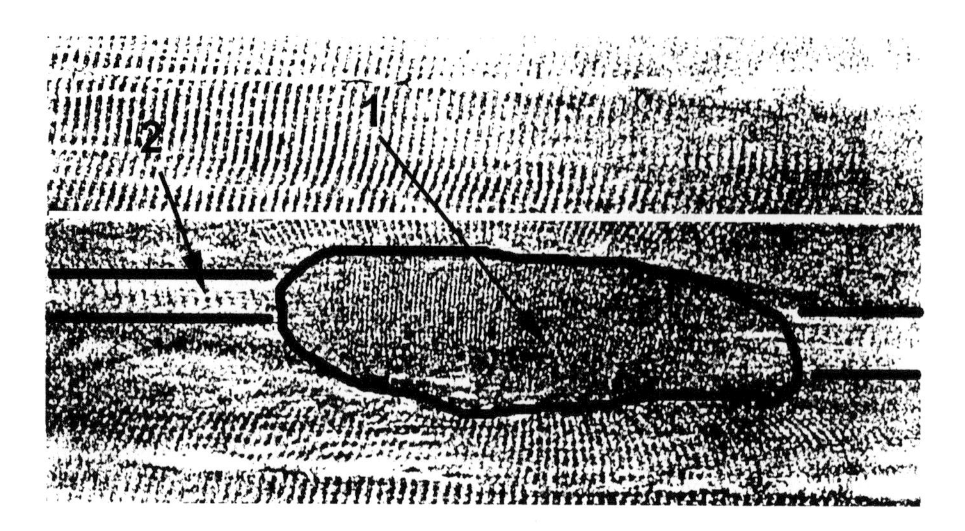

Fig. 1 Microscopic view of a trigger (Simons et al.1976)

TRIGGER –AREA EFFECTS

The *TA will shorten the entire muscle*, making it tight, limiting the joint range of motion (ROM) and/or the smoothness (Ease Of Motion EOM) while going through the range. This will reduce the number of functional muscle fibers causing measurable weakness and impaired coordination. TA´s make the muscle more susceptible to OI´s setting the stage for muscle stretch, tear and rupture.
The shortened muscle will cause overload on its tendon and the covering tendon sheet resulting in *tendonitis* and *tendosynovitis*. At the point of insertion it will cause *insertional tendonitis*. If the insertion is around a joint it will mimic *joint pain*. If the tendon is cushioned by a bursa it will cause bursitis.
The shortened muscle will cause overload on certain areas of the joint leading to joint degeneration, meniscus lesion, cartilage lesion, effusion, arthritis and joint pain.
The shortened muscle will cause *straightening* of the spine, vertebral *misalignment, disk disease, disk herniation, degeneration* of the vertebral body and the joints.
TA´s may cause *pain, numbness, tingling and cramping* at a significant distance from the actual TA (Travell et al., 1983). Treating the symptomatic area will not cure the problem instead it will cover up the symptoms and the TA will continue to grow in size and will become increasingly active.

DIAGNOSTIC CRITERIA FOR TRIGGER - AREAS

Each TA will have a structural effect on the muscle as listed above, which represents the *Structural - Trigger – Area (STA)*. A second type called Referral – Trigger – Area (RTA) will refer effects in remote areas of the body, such as pain, numbness, tingling and cramping. RTA´s are firm and tender "palpable" areas within the muscle which will reproduce the specific pain or other sensations through pressure application. The goal is to find the TA´s which will i.e. elicit the patient's exact pain pattern. Fig. 2 (left) shows pain referral patterns of the deep back muscles (Multifidi) and (right) of the deep hip muscle (Gluteus minimus).

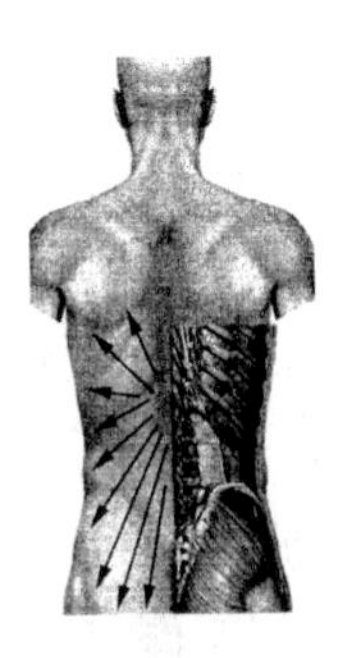

Fig. 2 Pressure on TA causes Referred Pain

STA´s are firm and tender "palpable" areas within the muscle which are responsible for the hardening and shortening of the muscle. Dissolving the TA´s will result in immediate softening and lengthening of the muscle.

DIAGNOSTIC PROBLEMS

The majority of TA´s lie too deep to be found through finger pressure. Several muscles may cover the TA and can only be found after relaxation of the superficial layers. The finger pressure approach, as recommended by (Travell et al., 1983), will leave most TA´s undiscovered. Several researchers had tried to overcome this problem by injecting saline solution into the deep muscle layers, suggesting that the observed referred pain pattern could be attributed to pressure activation of TA´s (Travell et al., 1983). This has produced trigger charts, which are used in daily practice, as a reference for the supposed localization of TA´s. In my experience and in the experience of many other co-workers, these localizations were seldom found to represent TA´s. Any therapy, especially injection of local anesthetics into these areas, mostly did not produce the desired results, probably because these suggested areas did not harbor the responsible TA´s.

RECOGNITION PROBLEMS

Even though TA related syndromes are recognized as a specific cause for various pain problems (Travell et al., 1983), TA´s so far have not played an important role in sports medicine. Various reasons can be made responsible for that: TA´s cannot be visualized i.e. by x-ray or Ultrasound. Until now there was no reliable standardized diagnostic approach. The treatment outcome was unpredictable and success was limited.

IMPROVEMENTS IN DIAGNOSIS AND THERAPIE THROUGH TRIGGER - OSTEOPRACTIC

The crucial point in developing Trigger – Medicine was, as a first step, to change the diagnostic approach. Through the development of the TRIGGOsan Probe (Fig. 3) it is now possible to diagnose previously unknown TA´s. It became clear that it often requires large forces to get to the TA´s. Pressure of a magnitude, which could not be achieved through finger palpation. The next step was to realize, that TA´s often are too large to be injected with local anesthetics without risking overdosing. Applying high pressure with the TRIGGOsan Probe (Fig.3) proved to develop into a very simple and successful therapeutic approach.

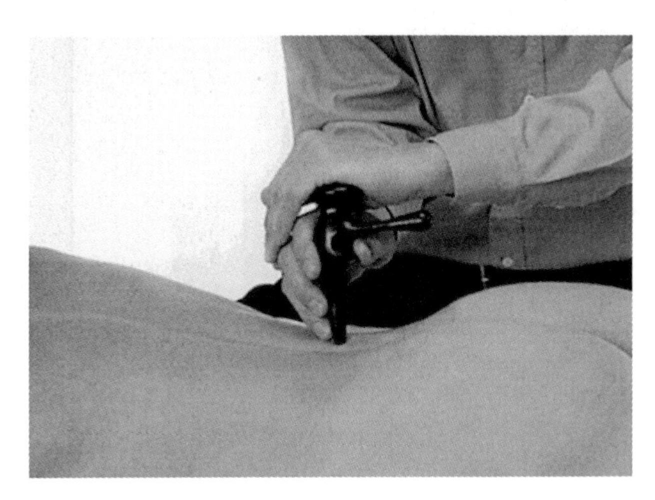

Fig. 3 TRIGGOsan Probe® applied to the low back

MODIFIED DIAGNOSTIC PROCEDURE

Through a screening test for Range of Motion (ROM), Ease of Motion (EOM), surface EMG and strength testing the trigger infested muscles or muscle groups can be identified.

STA´s can be easily detected as tight ropes or nodules within the muscle either through palpation, or with the TRIGGOsan Probe to reach into deeper structures. RTA´s require sufficient pressure to elicit referred pain, numbness, tingling, heaviness or tightness. It is the goal to find those TA´s which reproduce exactly the patient's complaint.

The pressure required varies with muscle size, depth and sensitivity of the TA´s. The pressure values can be assessed using a digital load sensor (Fig. 4), which can record the effort to achieve the required pressure at the tip of the Probe. The data can be send directly to a computer for documentation and analysis.

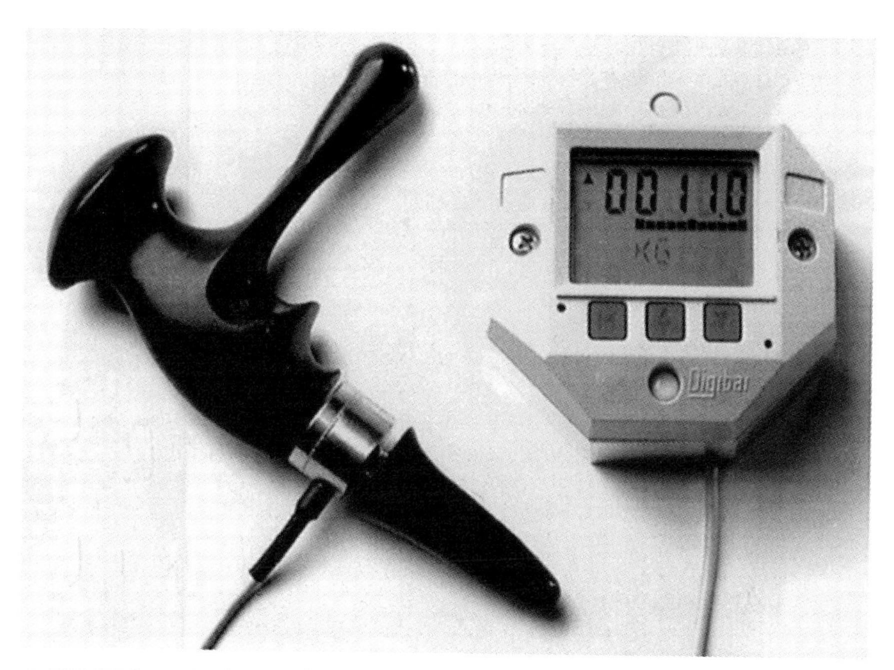

Fig. 4 TRIGGOsan Probe with Digital load sensor for trigger diagnosis

Table 2 Load required to trigger referred pain from RTA´s

Muscle	Load required in kg
Supraspinatus	12 – 35
Infraspinatus	50 – 26
Subscapularis	60 – 24
Multifidi	20 – 40
Longissimus	18 – 35
Iliocostalis	15 – 28
Quadratus lumborum	15 – 32
Gluteus medius	12 – 38
Gluteus minimus	15 – 40
Piriformis	11 – 27
Tensor fasc. latae	12 – 30
Hamstrings	12 – 25
Quadriceps	15 – 36
Triceps surae	11 – 24
Tibialis anterior	12 – 24
Peroneus	05 – 15

TA´S ASSOCIATED WITH PAIN SYMPTOMS IN TENNIS PLAYERS

Most *shoulder problems* were associated with overhead strokes due to TA´s of the interior rotators: Teres major, Latissimus dorsi, Pectoralis major. Rotator cuff muscles: Supraspinatus, Infraspinatus, Teres minor, Subscapularis.

The problems were previously diagnosed as: Rotator cuff syndrome, Impingement syndrome, Bursitis, Bicepital tendinitis.

Most *back problems* were associated with rotational movements and side flexion of the trunk due to TA´s involving: Multifidi, Rotatores, Longissimus, Iliocostalis, Spinalis, Quadratus lumborum. TA´s of the hip muscles: Gluteus medius, Gluteus minimus, Piriformis and TA´s of the Hamstrings.

The problems were previously diagnosed as: Intervertebral disk disease, disk protrusion, disk prolaps, Zygoapophyseal joint pain, Lumbar muscle sprain, Lumbar muscle spasm, Sciatica.

Most *knee problems* were associated with TA´s in the low back, hip, thigh and calf muscles: Multifidi, Quadratus lumborum, Quadriceps femoris, Adductors, Sartorius, Gastrocnemius.

The problems were previously diagnosed as: Meniscal injuries, Patellofemoral stress syndrome, Chondromalacia patellae, Patellar tendinitis, Iliotibial band friction syndrome.

Most *foot problems* were related to TA´s of the Triceps surae, Peroneus, Tibialis anterior and the hamstrings.

The problems were previously diagnosed as Achilles tendonitis, plantar fasciitis, heel spur, Morton's neuroma.

DIAGNOSIS AND TREATMENT

The players were screened for STA´s and RTA´s through ROM, EOM, TA – activation through contraction in a shortened position, palpation for tight muscle bands and finally specific pressure diagnostics with the TRIGGOsan Probe.

The treatments were performed with pressure application, using the TRIGGOsan Probe, either with short peak pressure, sustained pressure for 10-15 seconds or with Trigger Shockwave Therapy TST.

OUTCOME

All Tennis players were able to intensify their practice after the first treatment. The range of treatments required is listed in Table 1. The time between treatments was based on the players traveling schedule. Some had daily treatments using TST, others had one to several weeks in between treatments. All players were able to return to their normal tournament schedule after one to three treatments.

TRIGGER SHOCKWAVE THERAPY (TST)

TST is a recent development by the Author for the treatment of muscular TA´s. Instead of manual pressure application on the TA´s, radial shockwaves are directed at the TA. This approach, virtually being pain free, makes it easier on the patient as well as on the therapist. No manual force is required for the application of the probe. The technology is similar to Extracorporeal Shockwave Lithotripsy (ESWL), where externally generated shock waves are used to cause kidney stones to break up. The shockwaves cause an increase of the blood circulation in the TA´s. This leads to a reactivation of the suppressed energy cycle, allowing the muscle fiber to relax. The effects on the TA´s and the muscle are immediate, identical to the effects of manual compression.

PREVENTING INJURIES

Current concepts of injury prevention are based on flexibility, strength training, and sports biomechanics. TA´s play an all important role in all these aspects, because they reduce the ability of the muscle to perform properly. This has a negative influence on the muscles working in synergy and the ones which function as antagonists. There is no ideal flexibility for muscles infested with TA´s and increased stretching may even activate TA´s causing more shortening of the muscles. Strength training can counteract the effects of TA´s for awhile, by shifting the ratio of normal to TA- affected fibers towards the normal fibers. Over time the TA´s grow bigger because the athlete tends to train and perform above his limit, only being stopped by pain. This is a strong stimulus for TA´s to grow even further, shifting the ratio again towards the TA affected fibers. Proper sports biomechanics are based on presumed normal body function. Muscle groups compromised by TA´s will eventually counteract biomechanical efforts requiring compensatory muscle efforts, setting the stage for development of new TA´s.

CONCLUSION

Efficient injury treatment and prevention must include the diagnosis for TA´s and their early treatment. TA´s can be easily diagnosed and treated, years before any obvious symptoms occur. Modern sports medicine needs to recognize the TA´s as the causative pathology for overuse injuries and treat them directly, instead of focusing the TA- related symptoms of the muscle, tendon, bone and joint. Including TA´s in sports medicine will reduce significantly time out from practice and tournaments. It will allow the TA - muscle – tendon – bone - joint complex to heal completely, instead of settling for temporary relief of symptoms.

REFERENCES

Bauermeister, W., (1997) Schmerzfrei durch Osteopraktik, Mosaik Verlag, 16-19

Bauermeister W., (1999) Trigger – Osteopraktik, Physikalische Therapie in Theorie und Praxis, 20/8, 487-490

Brückle W. et al. (1990) Gewebe-pO_2 –Messung in der verspannten Rückenmuskulatur, Z Rheumatol 49:208-216

Merskey H., & Bogduk N. Ed., (1994) Classification of Chronic Pain, *IASP PRESS*, p 182

Simons, D.G., & Stolov, W.C., (1976) Microscopic features and transient contraction of palpable bands in canine muscle. *Am J Phys Med* 55: 65-88

Travell J.G., & Simons D.G., (1983) Myofascial Pain and Dysfunction, Williams & Wilkins , 5-164

USTA, (1998) Complete Conditioning for Tennis, *Human Kinetics*, p. 184,

Indexes

Author index

Subject index